AF796170

Jorge Angeles

Spatial Kinematic Chains

Analysis – Synthesis – Optimization

With 67 Figures

Springer-Verlag Berlin Heidelberg NewYork 1982

JORGE ANGELES
Professor of Mechanical Engineering
Universidad Nacional Autonoma de Mexico
C. Universitaria
P.O. Box 70-256
04360 Mexico, D.F., Mexico

ISBN 978-3-642-48821-4 ISBN 978-3-642-48819-1 (eBook)
DOI 10.1007/978-3-642-48819-1

This work is subject to copyright. All rights are reserved, whether the whole or part of the material is concerned, specifically those of translation, reprinting, re-use of illustrations, broadcasting, reproduction by photocopying machine or similiar means, and storage in data banks.

Under § 54 of the German Copyright Law where copies are made for other than private use, a fee is payable to 'Verwertungsgesellschaft Wort', Munich.

© Springer-Verlag Berlin, Heidelberg 1982
Softcover reprint of the hardcover 1st edition 1982

The use of registered names, trademarks, etc. in this publication does not imply, even in the absence of a specific statement, that such names are exempt from the relevant protective laws and regulations and therefore free for general use.

2061/3020 - 543210

Foreword

The author committed himself to the writing of this book soon after he started teaching a graduate course on linkage analysis and synthesis at the Universidad Nacional Autónoma de México (UNAM), in 1973. At that time he found that a great deal of knowledge on the subject, that had already been accumulated, was rather widespread and not as yet fully systematised. One exception was the work of B. Roth, of Stanford University, which already showed outstanding unity, though appearing only in the form of scientific papers in different journals. Moreover, the rate at which new results were presented either in specialised journals or at conferences all over the world, made necessary a recording of the most relevant contributions.

On the other hand, some methods of linkage synthesis, like the one of Denavit and Hartenberg (See Ch. 4), were finding a wide acceptance. It was the impression of the author, however, that the rationale behind that method was being left aside by many a researcher. Surprisingly, he found that virtually everybody was taking for granted, without giving the least explanation, that the matrix product, pertaining to a coordinate transformation from axes labelled 1 to those labelled n, should follow an order that is the inverse of the usual one. That is to say, whereas the logical representation of a coordinate transformation from axes 1 to 3, passing through those labelled 2, demands that the individual matrices $\underset{\sim}{T}_{12}$ and $\underset{\sim}{T}_{23}$ be multiplied in the order $\underset{\sim}{T}_{23}\underset{\sim}{T}_{12}$, the application of the method of Denavit and Hartenberg demands that they be placed in the inverse order, i.e. $\underset{\sim}{T}_{12}\underset{\sim}{T}_{23}$. It is explained in Chapter 4 why this is so, making use of results derived in Chapter 1. In this respect, the author departs from the common practice. In fact, while the transformations involving an affine transformation, i.e. a coordinate transformation, are usually represented by 4 x 4 matrices containing information about both the rotation and the translation, the author separates them into a matrix containing the rotation of axes and a vector containing their translation. The reason why this is done is far more than a matter of taste. As a matter of fact, it is not always necessary to carry out operations on both the rotation and the translation parts of the transformation, as is the case in dealing with spherical linkages. One more fundamental reason why the author departs from that practice is the following: in order to comprise both the rotation and the translation of axes in one single matrix, one has to define arbitrarily arrays that are not really vectors, for they contain a constant component. From the beginning, in Chapter 1, it is explained that only linear transformations are representable by matrices. Later on, in Chapter 2, it is shown that a rigid-body motion, in general, is a nonlinear transformation. This transformation is linear only if the motion is about a fixed point, which is also rigorously proven.

All through, the author has attempted to establish the rationale behind the methods of analysis, synthesis and optimisation of linkages. In this respect, Chapter 2 is crucial. In fact, it lays the foundations of the kinematics of rigid bodies in an axiomatic way, thus attempting to follow the trend of rational mechanics lead by Truesdell[1]. This Chapter in turn, is based upon Chapter 1, which outlines the facts of linear algebra, of extrema of functions and of numerical methods of solving algebraic linear and nonlinear systems, that are resorted to throughout the book. Regarding the numerical solution of equations, all possible cases are handled, i.e. algorithms are outlined that solve the said system, whether linear or nonlinear, when this is either underdetermined, determined or overdetermined. Flow diagrams illustrating the said algorithms and computer subprograms implementing them are included.

The philosophy of the book is to regard the linkages as systems capable of being modelled, analysed, synthesised, identified and optimised. Thus the methods and philosophy introduced here can be extended from linkages, i.e. closed kinematic chains, to robots and manipulators, i.e. open kinematic chains.

Back to the first paragraph, whereas early in the seventies the need to write a book on the theory and applications of the kinematics of mechanical

1. Truesdell C., "The Classical Field Theories", in Flügge S., ed., Encyclopedia of Physics, Springer-Verlag, Berlin, 1960

systems was dramatic, presently this need has been fulfilled to a great extent by the publishing of some books in the last years. Within these, one that must be mentioned in the first place is that by Bottema and Roth[2], then the one by Duffy[3] and that by Suh and Radcliffe[4], just to mention a couple of the recently published contributions to the specialised literature in the English language. The author, nevertheless, has continued with the publication of this book because it is his feeling that he has contributed with a new point of view of the subject from the very foundations of the theory to the methods for application to the analysis and synthesis of mechanisms. This contribution was given a unified treatment, thus allowing the applications to be based upon the fundamentals of the theory laid down in the first two chapters.

Although this book evolved from the work done by the author in the course of the last eight years at the Graduate Division of the Faculty of Engineering-UNAM, a substantial part of it was completed during a sabbatical leave spent by him at the Laboratory of Machine Tools of the Aachen Institute of Technology, in 1979, under a research fellowship of the Alexander von Humboldt Foundation, to whom deep thanks are due.

The book could have not been completed without the encouragement received from several colleagues, among whom special thanks go to Profs. Bernard Roth of Stanford University, Günther Dittrich of Aachen Institute of Technology, Hiram Albala of Technion-Israel Institute of Technology and Justo Nieto of Valencia (Spain) Polytechnic University. The support given by Prof. Manfred Weck of the Laboratory of Machine Tools, Aachen, during the sabbatical leave of the author is very highly acknowledged. The discussions held with Dr. Jacques M. Hervé, Head of the Laboratory of Industrial Mechanics- Central School of Arts and Manufactures of Paris, France, contributed highly to the completion of Chapter 3.

---

2 Bottema O. and Roth B., Theoretical Kinematics, North-Holland Publishing, Co., Amsterdam, 1979.

3 Duffy J., Analysis of Mechanisms and Robot Manipulators, Wiley-Interscience, Sommerset, N.J., 1980.

4 Such C. - H. and Radchiffe C.W., Kinematics and Mechanisms Design, John Wiley & Sons, Inc., N.Y., 1978.

The students of the author who, to a great extent are responsible for the writing of this book, are herewith deeply thanked. Special thanks are due to the former graduate students of the author, Messrs. Carlos López, Cándido Palacios and Angel Rojas, who are responsible for a great deal of the computer programming included here. Mrs. Carmen González Cruz and Miss Angelina Arellano typed the first versions of this work, whereas Mrs. Juana Olvera did the final draft. Their patience and very professional work is highly acknowledged.
Last, but by no means least, the support of the administration of the Faculty of Engineering-UNAM, and particularly of its Graduate Division, deserves a very special mention. Indeed, it provided the author with all the means required to complete this task.
To extend on more names of persons or institutions who somehow contributed to the completion of this book would give rise to an endless list, for which reason the author apologises for unavoidable omissions that he is forced to make.

Paris, January 1982

Jorge Angeles

Contents

1. Mathematical Preliminaries

1.0 INTRODUCTION. Some relevant mathematical results are collected in this chapter. These results find a wide application within the realm of analysis, synthesis and optimization of mechanisms. Often, rigorous proofs are not provided; however a reference list is given at the end of the chapter, where the interested reader can find the required details.

1.1. VECTOR SPACE, LINEAR DEPENDENCE AND BASIS OF A VECTOR SPACE.

A vector space, also called a linear space, over a field F (1.1)* , is a set V of objects, called vectors, having the following properties:

a) To each pair $\{\underset{\sim}{x}, \underset{\sim}{y}\}$ of vectors from the set, there corresponds one (and only one) vector, denoted $\underset{\sim}{x} + \underset{\sim}{y}$, also from V, called "the addition of $\underset{\sim}{x}$ and $\underset{\sim}{y}$" such that

   i) This addition is commutative, i.e.

$$\underset{\sim}{x} + \underset{\sim}{y} = \underset{\sim}{y} + \underset{\sim}{x}$$

   ii) It is associative, i.e., for any element $\underset{\sim}{z}$ of V,

$$\underset{\sim}{x} + (\underset{\sim}{y} + \underset{\sim}{z}) = (\underset{\sim}{x} + \underset{\sim}{y}) + \underset{\sim}{z}$$

   iii) There exists in V a unique vector $\underset{\sim}{0}$, called "the zero of V", such that, for any $\underset{\sim}{x} \in V$,

$$\underset{\sim}{x} + \underset{\sim}{0} = \underset{\sim}{x}$$

   iv) To each vector $\underset{\sim}{x} \in V$, there corresponds a unique vector $-\underset{\sim}{x}$, also in V, such that

$$\underset{\sim}{x} + (-\underset{\sim}{x}) = \underset{\sim}{0}$$

---

* Numbers in brackets designate references at the end of each chapter.

b) To each pair $\{\alpha, \underset{\sim}{x}\}$, where $\alpha \in F$ (usually called "a scalar") and $\underset{\sim}{x} \in V$, there corresponds one vector $\alpha\underset{\sim}{x} \in V$, called "the product of the scalar α times $\underset{\sim}{x}$", such that:

i) This product is associative, i.e. for any $\beta \in F$,

$$\alpha(\beta\underset{\sim}{x}) = (\alpha\beta)\underset{\sim}{x}$$

ii) For the identity 1 of F (with respect to multiplication) the following holds

$$1\underset{\sim}{x} = \underset{\sim}{x}$$

c) The product of a scalar times a vector is distributive, i.e.

i) $\alpha(\underset{\sim}{x} + \underset{\sim}{y}) = \alpha\underset{\sim}{x} + \alpha\underset{\sim}{y}$

ii) $(\alpha + \beta)\underset{\sim}{x} = \alpha\underset{\sim}{x} + \beta\underset{\sim}{x}$

Example 1.1.1. The set of triads of real numbers (x,y,z) constitute a vector space. To prove this, define two such triads, namely (x_1,y_1,z_1) and (x_2,y_2,z_2) and show that their addition is also one such triad and it is commutative as well. To prove associativity, define one third triad, (x_3,y_3,x_3), and so on.

Example 1.1.2 The set of all polynomials of a real variable, t, of degree less than or equal to n, for $0 \le t \le 1$, constitute a vector space over the field of real numbers.

Example 1.1.3 The set of tetrads of the form $(x,y,z,1)$ do not constitute a vector space (Why?)

Given the set of vectors $\{\underset{\sim}{x}_1,\underset{\sim}{x}_2,\ldots,\underset{\sim}{x}_n\} \subset V$ and the set of scalars $\{\alpha_1,\alpha_2,\ldots,\alpha_n\} \subset F$ not necessarily distinct, a linear combination of the n vectors is the vector defined as

$$\underset{\sim}{c} = \alpha_1 \underset{\sim}{x}_1 + \alpha_2 \underset{\sim}{x}_2 + \ldots + \alpha_n \underset{\sim}{x}_n$$

The said set of vectors is linearly independent (ℓ. i.) if $\tilde{c}$ equals zero implies that all α's are zero as well. Otherwise, the set is said to be linearly dependent (ℓ. d.)

Example 1.1.4 The set containing only one nonzero vector, $\{\tilde{x}\}$, is ℓ.i.

Example 1.1.5 The set containing only two vectors, one of which is the origin, $\{\tilde{x},\tilde{0}\}$, is ℓ.d.

The set of vectors $\{\tilde{x}_1,\tilde{x}_2,\ldots,\tilde{x}_n\} \subset V$ spans V if and only if every vector $\tilde{v} \in V$ can be expressed as a linear combination of the vectors of the set.

A set of vectors $B = \{x_1,x_2,\ldots,x_n\} \subset V$ is a basis for V if and only if:

i) B is linearly independent, and

ii) B spans V

All bases of a given space V contain the same number of vectors. Thus, if B is a basis for V, the number n of elements of B is the dimension of V (abreviated: n=dim V)

Example 1.1.6 In 3-dimensional Euclidean space the unit vectors $\{\tilde{i}, \tilde{j}\}$ lying parallel to the X and Y coordinate axes span the vectors in the X-Y plane, but do not span the vectors in the physical three-dimensional space.

Exercise 1.1.1 Prove that the set B given above is a basis for V if and only if each vector in V can be expressed as a unique linear combination of the elements of B.

1.2 LINEAR TRANSFORMATION AND ITS MATRIX REPRESENTATION

Henceforth, only finite-dimensional vector spaces will be dealt with and, when necessary, the dimension of the space will be indicated as an exponent of the space, i.e., V^n means dim V=n.

A transformation $\tilde{T}$, from an m-dimensional vector space U, into an n-dimensional vector space V is a rule which establishes a correspondence between an element of U and a unique element of V. It is represented as:

$$\underset{\sim}{T}: U^m \rightarrow V^n \qquad (1.2.1)$$

If $\underset{\sim}{u} \varepsilon \underset{\sim}{U}^m$ and $\underset{\sim}{v} \varepsilon V^n$ are such that $\underset{\sim}{T}: \underset{\sim}{u} \rightarrow \underset{\sim}{v}$, the said correspondence may also be denoted as

$$\underset{\sim}{v} = \underset{\sim}{T}(\underset{\sim}{u}) \qquad (1.2.3a)$$

$\underset{\sim}{T}$ is linear if and only if, for any $\underset{\sim}{u}$, $\underset{\sim}{u}_1$ and $\underset{\sim}{u}_2$ ε U, and $\alpha \varepsilon$ F,

i) $\underset{\sim}{T}(\underset{\sim}{u}_1 + \underset{\sim}{u}_2) = \underset{\sim}{T}(\underset{\sim}{u}_1) + \underset{\sim}{T}(\underset{\sim}{u}_2)$ and (1.2.3b)

ii) $\underset{\sim}{T}(\alpha\underset{\sim}{u}) = \alpha\underset{\sim}{T}(\underset{\sim}{u})$ (1.2.3c)

Space U^m over which $\underset{\sim}{T}$ is defined is called the "domain" of T, whereas the subspace of V^n containing vectors $\underset{\sim}{v}$ for which eq. (1.2.3a) holds is called the "range" of $\underset{\sim}{T}$. A subspace of a given vector space V is a subset of V and is in turn a vector space, whose dimension is less than or equal to that of V

Exercise 1.2.1 Show that the range of a given linear transformation of a vector space U into a vector space V constitutes a subspace, i.e. it satisfies properties a) to c) of Section 1.1.

For a given $\underset{\sim}{u} \varepsilon$ U, vector $\underset{\sim}{v}$, as defined by (1.2.2) is called the "image of $\underset{\sim}{u}$ under $\underset{\sim}{T}$", or, simply, the "image of $\underset{\sim}{u}$" if $\underset{\sim}{T}$ is selfunderstood.

An example of a linear transformation is an orthogonal projection onto a plane. Notice that this projection is a transformation of the three-dimensional Euclidean space onto a two-dimensional space (the plane). The domain of $\underset{\sim}{T}$ in this case is the physical 3-dimensional space, while its range is the projection plane.

If $\underset{\sim}{T}$, as defined in (1.2.1), is such that all of V contains $\underset{\sim}{v}$'s such that (1.2.2) is satisfied (for some $\underset{\sim}{u}$'s), $\underset{\sim}{T}$ is said to be "onto". If $\underset{\sim}{T}$ is such that, for all distinct $\underset{\sim}{u}_1$ and $\underset{\sim}{u}_2$, $\underset{\sim}{T}(\underset{\sim}{u}_1)$ and $\underset{\sim}{T}(\underset{\sim}{u}_2)$ are also distinct, $\underset{\sim}{T}$ is said to be one-to-one. If $\underset{\sim}{T}$ is onto and one-to-one, it is said to be <u>invertible</u>.

If $\underset{\sim}{T}$ is invertible, to each $\underset{\sim}{v} \in V$ there corresponds a unique $\underset{\sim}{u} \in U$ such that $\underset{\sim}{v} = \underset{\sim}{T}(\underset{\sim}{u})$, so one can define a mapping $\underset{\sim}{T}^{-1}: V \rightarrow U$ such that

$$\underset{\sim}{u} = \underset{\sim}{T}^{-1}(\underset{\sim}{v}) \qquad (1.2.4)$$

$\underset{\sim}{T}^{-1}$ is called the "<u>inverse</u>" of $\underset{\sim}{T}$.

<u>Exercise 1.2.2</u> Let $\underset{\sim}{P}$ be the projection of the three-dimensional Euclidean space onto a plane, say, the X-Y plane. Thus, $\underset{\sim}{v} = \underset{\sim}{P}(\underset{\sim}{u})$ is such that the vector with components (x, y, z), is mapped into the vector with components (x, y, 0).

i) Is $\underset{\sim}{P}$ a linear transformation?

ii) Is $\underset{\sim}{P}$ onto?, one-to-one?, invertible?

A very important fact concerning linear transformations of finite dimensional vector spaces is contained in the following result:

Let $\underset{\sim}{L}$ be a linear transformation from U^m into V^n. Let B_u and B_v be bases for U^m and V^n, respectively. Then clearly, for each $\underset{\sim}{u}_i \in B_u$ its image $\underset{\sim}{L}(\underset{\sim}{u}_i) \in V$ can be expressed as a linear combination of the $\underset{\sim}{v}_k$'s in B_v. Thus

$$L(\underset{\sim}{u}_i) = \alpha_{1i}\underset{\sim}{v}_1 + \alpha_{2i}\underset{\sim}{v}_2 + \ldots + \alpha_{ni}\underset{\sim}{v}_n \qquad (1.2.5)$$

Consequently, to represent the images of the m vectors of B_u, mn scalars like those appearing in (1.2.5) are required. These scalars can be arranged in the following manner:

$$[\underset{\sim}{A}] = \begin{pmatrix} \alpha_{11} & \alpha_{12} & \ldots\alpha_{1m} \\ \alpha_{21} & \alpha_{22} & \ldots\alpha_{2m} \\ \cdot & \cdot & \cdot \\ \cdot & \cdot & \cdot \\ \cdot & \cdot & \cdot \\ \alpha_{n1} & \alpha_{n2} & \ldots\alpha_{nm} \end{pmatrix} \qquad (1.2.6)$$

where the brackets enclosing $\underset{\sim}{A}$ are meant to denote a matrix, i.e. an array of numbers, rather than an abstract linear transformation.
$[\underset{\sim}{A}]$ is called "The matrix of $\underset{\sim}{L}$ referred to B_u and B_v" . This result is summarized in the following:

DEFINITION 1.2.1 The i th column of the matrix representation of $\underset{\sim}{L}$, referred to B_u and B_v, contains the scalar coefficients α_{ji} of the representation (in terms of B_v) of the image of the i th vector of B_u

Example 1.2.1 What is the representation of the reflexion $\underset{\sim}{R}$ of the 3-dimensional Euclidean space E^3 into itself, with respect to one plane, say the X-Y plane, referred to unit vectors parallel to the X,Y,Z axes?.

Solution: Let $\underset{\sim}{i}$, $\underset{\sim}{j}$, $\underset{\sim}{k}$, be unit vectors parallel to the X, Y and Z axes, respectively. Clearly,

$$\underset{\sim}{R}(\underset{\sim}{i}) = \underset{\sim}{i}$$

$$\underset{\sim}{R}(\underset{\sim}{j}) = \underset{\sim}{j}$$

$$\underset{\sim}{R}(\underset{\sim}{k}) = -\underset{\sim}{k}$$

Thus, the components of the images of $\underset{\sim}{i}$, $\underset{\sim}{j}$ and $\underset{\sim}{k}$ under $\underset{\sim}{R}$ are:

$$(\underset{\sim}{R}(\underset{\sim}{i})) = \begin{pmatrix} 1 \\ 0 \\ 0 \end{pmatrix}, \quad (\underset{\sim}{R}(\underset{\sim}{j})) = \begin{pmatrix} 0 \\ 1 \\ 0 \end{pmatrix}, \quad (\underset{\sim}{R}(\underset{\sim}{k})) = \begin{pmatrix} 0 \\ 0 \\ -1 \end{pmatrix}$$

Hence, the matrix representation of $\underset{\sim}{R}$, denoted by $[\underset{\sim}{R}]$, is

$$(\underset{\sim}{R}) = \begin{pmatrix} 1 & 0 & 0 \\ 0 & 1 & 0 \\ 0 & 0 & -1 \end{pmatrix} \tag{1.2.7}$$

Notice that, in this case, U = V and so, it is not necessary to use two different bases for U and V. Thus, $(\underset{\sim}{R})$, as given by (1.2.7), is the matrix representation of the reflection $\underset{\sim}{R}$ under consideration, referred to the basis $\{\underset{\sim}{i}, \underset{\sim}{j}, \underset{\sim}{k}\}$.

1.3 RANGE AND NULL SPACE OF A LINEAR TRANSFORMATION

As stated in Section 1.2, the set of vectors $\underset{\sim}{v} \in V$ for which there is at least one $\underset{\sim}{u} \in U$ such that $\underset{\sim}{v} = L(\underset{\sim}{u})$, as pointed out in Sect. 4.2., is called "the range of $\underset{\sim}{L}$" and is represented as $\underset{\sim}{R}(\underset{\sim}{L})$, i.e. $R(\underset{\sim}{L}) = (\underset{\sim}{v}=\underset{\sim}{L}(\underset{\sim}{u}): \underset{\sim}{u} \in U)$.

The set of vectors $\underset{\sim}{u}_0 \in U$ for which $\underset{\sim}{L}(\underset{\sim}{u}_0) = \underset{\sim}{0} \in V$ is called "the null space of L" and is represented as $N(\underset{\sim}{L})$, i.e. $N(\underset{\sim}{L}) = \{\underset{\sim}{u}_0 : \underset{\sim}{L}(\underset{\sim}{u}_0)=\underset{\sim}{0}\}$.

It is a simple matter to show that $R(\underset{\sim}{L})$ and $N(\underset{\sim}{L})$ are subspaces of V and U, respectively*.

The dimensions of $\mathrm{dom}(\underset{\sim}{L})$, $R(\underset{\sim}{L})$ and $N(\underset{\sim}{L})$ are not independent, but they are related (see [1.2]):

$$\dim \mathrm{dom}(\underset{\sim}{L})=\dim R(\underset{\sim}{L}) + \dim N(\underset{\sim}{L}) \qquad (1.3.1)$$

Example 1.3.1 In considering the projection of Exercise 1.2.1, U is E^3 and thus $R(\underset{\sim}{P})$ is the X-Y plane, $N(\underset{\sim}{P})$ is the Z axis, hence of dimension 1. The X-Y plane is two-dimensional and $\mathrm{dom}(\underset{\sim}{L})$ is three-dimensional, hence (1.3.1) holds.

Exercise 1.3.1 Describe the range and the null space of the reflection of Example 1.2.1 and verify that eq. (1.3.1) holds true.

1.4 EIGENVALUES AND EIGENVECTORS OF A LINEAR TRANSFORMATION

Let $\underset{\sim}{L}$ be a linear transformation of V into itself (such an $\underset{\sim}{L}$ is called an "endomorphism"). In general, the image $\underset{\sim}{L}(\underset{\sim}{v})$ of an element $\underset{\sim}{v}$ of V is linearly independent with $\underset{\sim}{v}$, but if it happens that a nonzero vector $\underset{\sim}{v}$ and its image under $\underset{\sim}{L}$ are linearly dependent, i.e. if

$$\underset{\sim}{L}(\underset{\sim}{v}) = \lambda\underset{\sim}{v} \qquad (1.4.1)$$

---

* The proof of this statement can be found in any of the books listed in the reference at the end of this chapter.

such a $\underset{\sim}{v}$ is said to be an eigenvector of $\underset{\sim}{L}$, corresponding to the eigenvalue λ. If $[\underset{\sim}{A}]$ is the matrix representation of $\underset{\sim}{L}$, referred to a particular basis then, dropping the brackets, eq. (1.4.1) can be rewritten as

$$\underset{\sim}{A}\underset{\sim}{v} = \lambda \underset{\sim}{v} \tag{1.4.2}$$

or else

$$(\underset{\sim}{A} - \lambda \underset{\sim}{I})\underset{\sim}{v} = \underset{\sim}{0} \tag{1.4.3}$$

where $\underset{\sim}{I}$ is the identity matrix, i.e. the matrix with the unity on its diagonal and zeros elsewhere. Equation (1.4.3) states that the eigenvectors of $\underset{\sim}{L}$(or of $\underset{\sim}{A}$, clearly) lie in the null space of $\underset{\sim}{A} - \lambda\underset{\sim}{I}$. One trivial vector $\underset{\sim}{v}$ satisfying (1.4.3) is, of course, $\underset{\sim}{0}$, but since in this context $\underset{\sim}{0}$ has been discarded, nontrivial solutions have to be sought. The condition for (1.4.3) to have nontrivial solutions is, of course, that the determinant of $\underset{\sim}{A} - \lambda\underset{\sim}{I}$ vanishes, i.e.

$$\det (\underset{\sim}{A} - \lambda \underset{\sim}{I}) = 0 \tag{1.4.4}$$

which is an <u>nth</u> order polynomial in λ, n being the order of the square matrix $\underset{\sim}{A}$ (1.3). The polynomial

$$P(\lambda) \equiv \det (\underset{\sim}{A} - \lambda \underset{\sim}{I})$$

is called "<u>the characteristic polynomial</u>" of $\underset{\sim}{A}$. Notice that its roots are the eigenvalues of $\underset{\sim}{A}$. These roots can, of course, be real or complex; in case $P(\lambda)$ has one complex root, say λ_1, then $\overline{\lambda}_1$ is also a root of $P(\lambda)$, $\overline{\lambda}_1$ being the complex conjugate of λ_1. Of course, one or several roots could be repeated. The number of times that a particular eigenvalue λ_i is repeated is called the <u>algebraic multiplicity</u> of λ_i.

In general, corresponding to each λ_i there are several linearly independent eigenvectors of $\underset{\sim}{A}$. It is not difficult to prove (Try it!) that the ℓ.i. eigenvectors associated with a particular eigenvalue span a subspace. This subspace is called the "spectral space" of λ_i, and its dimension is called "the geometric multiplicity of λ_i".

Exercise 1.4.1 Show that the geometric multiplicity of a particular eigenvalue cannot be greater than its algebraic multiplicity.

A Hermitian matrix is one which equals its transpose conjugate. If a matrix equals the negative of its transpose conjugate, it is said to be skew Hermitian. For Hermitian matrices we have the very important result:

THEOREM 1.4.1 *The eigenvalues of a Hermitian matrix are real and its eigenvectors are mutually orthogonal (i.e. the inner product, which is discussed in detail in Sec. 1.8, of two distinct eigenvectors, is zero).*

The proof of the foregoing theorem is very widely known and is not presented here. The reader can find a proof in any of the books listed at the end of the chapter.

1.5 CHANGE OF BASIS

Given a vector $\underset{\sim}{v}$, its representation $(v_1, v_2, \ldots, v_n)^T$ referred to a basis $B = \{\underset{\sim}{\beta}_1, \underset{\sim}{\beta}_2, \ldots, \underset{\sim}{\beta}_n\}$, is defined as the ordered set of scalars that produce $\underset{\sim}{v}$ as a linear combination of the vectors of B. Thus, $\underset{\sim}{v}$ can be expressed as

$$\underset{\sim}{v} = v_1\underset{\sim}{\beta}_1 + v_2\underset{\sim}{\beta}_2 + \ldots + v_n\underset{\sim}{\beta}_n \tag{1.5.1}$$

A vector $\underset{\sim}{v}$ and its representation, though isomorphic* to each other, are essentially different entities. In fact, $\underset{\sim}{v}$ is an abstract algebraic entity satisfying properties a), b) & c) of Section 1.1, whereas its representation is an array of numbers. Similarly, a linear transformation, $\underset{\sim}{L}$, and its representation, $(\underset{\sim}{L})_B$, are essentially different entities. A question that could arise naturally is: Given the representations $(\underset{\sim}{v})_B$ and $(\underset{\sim}{L})_B$ of $\underset{\sim}{v}$ and L, respectively, referred to the basis B, what are the corresponding

---

* Two sets are isomorphic to each other if similar operations can be defined on their elements.

representations referred to the basis $C = \{\underset{\sim}{\gamma}_1, \underset{\sim}{\gamma}_2, \ldots, \underset{\sim}{\gamma}_n\}$?

Let $(\underset{\sim}{A})_B$ be the matrix relating both B and C, referred to B, i.e.

$$(\underset{\sim}{A})_B = \begin{pmatrix} \alpha_{11} & \alpha_{12} \cdots & \alpha_{1n} \\ \alpha_{21} & \alpha_{22} & \alpha_{2n} \\ \vdots & \vdots & \\ \alpha_{n1} & \alpha_{n2} & \alpha_{nn} \end{pmatrix} \tag{1.5.2}$$

and

$$\underset{\sim}{\gamma}_1 = \alpha_{11}\underset{\sim}{\beta}_1 + \alpha_{21}\underset{\sim}{\beta}_2 + \ldots + \alpha_{n1}\underset{\sim}{\beta}_n$$

$$\underset{\sim}{\gamma}_2 = \alpha_{12}\underset{\sim}{\beta}_1 + \alpha_{22}\underset{\sim}{\beta}_2 + \ldots + \alpha_{n2}\underset{\sim}{\beta}_n$$

$$\vdots$$

$$\underset{\sim}{\gamma}_n = \alpha_{1n}\underset{\sim}{\beta}_1 + \alpha_{2n}\underset{\sim}{\beta}_2 + \ldots + \alpha_{nn}\underset{\sim}{\beta}_n$$

Thus, calling v_i' the <u>ith</u> component of $(\underset{\sim}{v})_C$, then

$$\underset{\sim}{v} = v_1'\underset{\sim}{\gamma}_1 + v_2'\underset{\sim}{\gamma}_2 + \ldots + v_n'\underset{\sim}{\gamma}_n \tag{1.5.4}$$

and, from (1.5.3), (1.5.4) leads to

$$\underset{\sim}{v} = \sum_j v_j' \sum_i \alpha_{ij}\underset{\sim}{\beta}_j \tag{1.5.5}$$

or, using index notation* for compactness,

$$\underset{\sim}{v} = \alpha_{ij} v_j' \underset{\sim}{\beta}_i \tag{1.5.6}$$

Comparing (1.5.1) with (1.5.6),

$$v_i = \alpha_{ij} v_j' \tag{1.5.7}$$

i.e.

$$(\underset{\sim}{v})_B = (\underset{\sim}{A})_B (\underset{\sim}{v})_C$$

---

* According to this notation, a repeated index implies that a summation over all the possible values of this index is performed.

or, equivalently,

$$\{\underset{\sim}{v}\}_C = \{\underset{\sim}{A}\}_B^{-1}\{\underset{\sim}{v}\}_B \qquad (1.5.8)$$

Now, assuming that $\underset{\sim}{w}$ is the image of $\underset{\sim}{v}$ under $\underset{\sim}{L}$,

$$\{\underset{\sim}{w}\}_B = \{\underset{\sim}{L}\}_B\, \{\underset{\sim}{v}\}_B \qquad (1.5.9)$$

or, referring eq. (1.5.9) to the basis C, instead,

$$\{\underset{\sim}{w}\}_C = \{\underset{\sim}{L}\}_C\, \{\underset{\sim}{v}\}_C \qquad (1.5.10)$$

Applying the relationship (1.5.8) to vector $\underset{\sim}{w}$ and introducing it into eq. (1.5.10),

$$\{\underset{\sim}{A}^{-1}\}_B\, \{\underset{\sim}{w}\}_B = \{\underset{\sim}{L}\}_C\, \{\underset{\sim}{A}\}_B^{-1}\{\underset{\sim}{v}\}_B$$

from which the next relationship readily follows

$$\{\underset{\sim}{w}\}_B = \{\underset{\sim}{A}\}_B\, \{\underset{\sim}{L}\}_C\, \{\underset{\sim}{A}\}_B^{-1}\, \{\underset{\sim}{v}\}_B \qquad (1.5.11)$$

Finally, comparing (1.5.9) with (1.5.11),

$$\{\underset{\sim}{L}\}_B = \{\underset{\sim}{A}\}_B\, \{\underset{\sim}{L}\}_C\, \{\underset{\sim}{A}\}_B$$

or, equivalently,

$$\{\underset{\sim}{L}\}_C = \{\underset{\sim}{A}\}_B^{-1}\, \{\underset{\sim}{L}\}_B\, \{\underset{\sim}{A}\}_B \qquad (1.5.12)$$

Relationships (1.5.8) and (1.5.12) are the answers to the question posed at the beginning of this Section. The right hand side of (1.5.12) is <u>a similarity transformation</u> of $\{\underset{\sim}{L}\}_B$

<u>Exercise 1.5.1</u> Show that, under a similarity transformation, the characteristic polynomial of a matrix remains invariant.

<u>Exercise 1.5.2</u> The trace of a matrix is defined as the sum of the elements on its diagonal. Show that the trace of a matrix remains invariant under a similarity transformation. Hint: Show first that, if $\underset{\sim}{A}$, $\underset{\sim}{B}$ and $\underset{\sim}{C}$ are nxn matrices,

$$\mathrm{Tr}(\underset{\sim}{A}\underset{\sim}{B}\underset{\sim}{C}) = \mathrm{Tr}(\underset{\sim}{B}\underset{\sim}{C}\underset{\sim}{A}).$$

1.6 DIAGONALIZATION OF MATRICES

Let $\underset{\sim}{A}$ be a symmetric nxn matrix and $\{\lambda_i\}_1^n$ its set of n eigenvalues, some of which could be repeated. Assume $\underset{\sim}{A}$ has a set of n linearly independent* eigenvectors, $\{\underset{\sim}{e}_i\}$, so that

$$\underset{\sim}{A}\underset{\sim}{e}_i = \lambda_i \underset{\sim}{e}_i \tag{1.6.1}$$

Arranging the eigenvectors of $\underset{\sim}{A}$ in the matrix

$$\underset{\sim}{Q} = \left(\underset{\sim}{e}_1, \underset{\sim}{e}_2, \ldots, \underset{\sim}{e}_n\right) \tag{1.6.2}$$

and its eigenvalues in the diagonal matrix

$$\underset{\sim}{\Lambda} = \text{diag}\,(\lambda_1, \lambda_2, \ldots, \lambda_n) \tag{1.6.3}$$

eq. (1.6.1) can be rewritten as

$$\underset{\sim}{A}\underset{\sim}{Q} = \underset{\sim}{Q}\underset{\sim}{\Lambda} \tag{1.6.4}$$

Since the set $\{\underset{\sim}{e}_i\}$ has been assumed to be ℓ.i., $\underset{\sim}{Q}$ is non-singular; hence from (1.6.4)

$$\underset{\sim}{\Lambda} = \underset{\sim}{Q}^{-1}\underset{\sim}{A}\underset{\sim}{Q} \tag{1.6.5}$$

which states that the diagonal matrix containing the eigenvalues of a matrix $\underset{\sim}{A}$ (which has as many ℓ.i. eigenvectors as its number of columns or rows) is a similarity transformation of $\underset{\sim}{A}$; furthermore, the transformation matrix is the matrix containing the components of the eigenvectors of $\underset{\sim}{A}$ as its columns. On the other hand, if $\underset{\sim}{A}$ is Hermitian, its eigenvalues are real and its eigenvectors are mutually orthogonal. If this is the case and the set $\{\underset{\sim}{e}_i\}$ is normalized, i.e., if $||\underset{\sim}{e}_i|| = 1$, for all i, then

$$\underset{\sim}{e}_i^T \underset{\sim}{e}_j = 0, \; i \neq j \tag{1.6.6a}$$

$$\underset{\sim}{e}_i^T \underset{\sim}{e}_i = 1 \tag{1.6.6b}$$

---

* Some square matrices have less than n ℓ.i eigenvectors, but these are not considered here.

where e_i^T is the transpose of e_i (e_i being a column vector, e_i^T is a row vector). The whole set of equations (1.6.6), for all i and all j can then be written as

$$Q^T Q = I \tag{1.6.7}$$

where I is the matrix with unity on its diagonal and zeros elsewhere. Eq. (1.6.7) states a very important fact about Q, namely, that it is an orthogonal matrix. Summarizing, <u>a symmetric nxn matrix A can be diagonalized via a similarity transformation, the columns of whose matrix are the eigenvectors of A</u>

The eigenvalue problem stated in (1.6.1) is solved by first finding the eigenvalues $\{\lambda_i\}_1^n$. These values are found from the following procedure: Write eq. (1.6.1) in the form

$$(A - \lambda_i I)e_i = 0 \tag{1.6.8}$$

This equation states that the set $\{e_1\}_1^n$ lies in the null space of $A - \lambda_i I$. For this matrix to have nonzero vectors in its null space, its determinant should vanish, i.e.

$$\det(A-\lambda_i I) \equiv P(\lambda_i)=0 \tag{1.6.9}$$

whose left hand side is its characteristic polynomial, which was introduced in section 1.4. This equation thus contains n roots, some of which could be repeated.

A very useful result is next summarized, though not proved.

<u>THEOREM (Cayley-Hamilton)</u>. *A square matrix satisfies its own characteristic equation, i.e. if $P(\lambda_i)$ is its characteristic polynomial, then*

$$P(A) = 0 \tag{1.6.10}$$

A proof ot this teorem can be found either in [1.3, pp. 148-150] or in [1.4, pp. 112-115]

Exercise 1.6.1 A square matrix $\underset{\sim}{A}$ is said to be strictly lower triangular (SLT) if $a_{ij}=0$, for $j\geq i$. On the other hand, this matrix is said to be nillpotent of index k if k is the lowest integer for which $\underset{\sim}{A}^k = \underset{\sim}{0}$.

i) Show that an nxn SLT matrix is nillpotent of index $k\leq n$.

ii) Show that an nxn SLT matrix $\underset{\sim}{A}$ staisfies the following indentity:

$$(\underset{\sim}{I}+\underset{\sim}{A})^{-1}=\sum_{1}^{n}(-1)^{k-1}\underset{\sim}{A}^{k-1}$$

The inverse of $\underset{\sim}{I}+\underset{\sim}{A}$ appears very often in the solution of linear algebraic systems by iterative methods.

1.7. BILINEAR FORMS AND SIGN DEFINITION OF MATRICES.

Given that the space of matrices does not constitute an ordered set (as is the case for the real, rational or integer sets), it is not possible to attribute a sign to a matrix. However, it will be shown that, if a bilinear form (in particular, a quadratic form) is associated with a matrix, then it makes sense to speak of the sign of a matrix. Before proceeding further, some definitions are needed. Let $\underset{\sim}{u}$ and $\underset{\sim}{v} \in U$, U being a vector space defined over the complex tield F. A bilinear form of $\underset{\sim}{u}$ and $\underset{\sim}{v}$, represented as $\phi(\underset{\sim}{u},\underset{\sim}{v})$ is a mapping from U into F, having the following properties:

i) It is linear in $\underset{\sim}{u}$ and conjugate linear in $\underset{\sim}{v}$:

$$\phi(\underset{\sim}{u}_1+\underset{\sim}{u}_2,\underset{\sim}{v}) = \phi(\underset{\sim}{u}_1,\underset{\sim}{v}) + \phi(\underset{\sim}{u}_2,\underset{\sim}{v}) \qquad (1.7.1a)$$

$$\phi(\underset{\sim}{u},\underset{\sim}{v}_1+\underset{\sim}{v}_2) = \phi(\underset{\sim}{u},\underset{\sim}{v}_1) + \phi(\underset{\sim}{u},\underset{\sim}{v}_2) \qquad (1.7.1b)$$

$$\phi(\alpha\underset{\sim}{u},\underset{\sim}{v})=\alpha\phi(\underset{\sim}{u},\underset{\sim}{v}) \qquad (1.7.1c)$$

$$\phi(\underset{\sim}{u},\beta\underset{\sim}{v})= \overline{\beta}\phi(\underset{\sim}{u},\underset{\sim}{v}) \qquad (1.7.1d)$$

where α and $\beta \in F$, their conjugates being $\bar{\alpha}$ and $\bar{\beta}$, respectively.

ii) $\phi(\underset{\sim}{v},\underset{\sim}{u})$ is the complex conjugate of $\phi(\underset{\sim}{u},\underset{\sim}{v})$, i.e.

$$\phi(\underset{\sim}{v},\underset{\sim}{u}) = \bar{\phi}(\underset{\sim}{u},\underset{\sim}{v}) \tag{1.7.1e}$$

The foregoing properties of conjugate bilinear forms suggest that one possible way of constructing a bilinear form is as follows:

Let
$$\phi(\underset{\sim}{u}, \underset{\sim}{v}) = \underset{\sim}{u}^*\underset{\sim}{A}\underset{\sim}{v} \tag{1.7.2}$$

provided that $\underset{\sim}{A}$ is Hermitian, i.e. $\underset{\sim}{A}=\underset{\sim}{A}^*$.

<u>Exercise 1.7.1</u> Prove that definition (1.7.2) satisfies properties (1.7.1)

If, in (1.7.2), $\underset{\sim}{v} = \underset{\sim}{u}$, the bilinear form becomes the <u>quadratic form</u>

$$\psi(\underset{\sim}{u}) = \underset{\sim}{u}^*\underset{\sim}{A}\ \underset{\sim}{u} \tag{1.7.3}$$

It will be shown that the bilinear form (1.7.2) defines a <u>scalar product</u> for a vector space under certain conditions on $\underset{\sim}{A}$.

Definition: A scalar product, $p(\underset{\sim}{u},\underset{\sim}{v})$, of two elements for a vector space U is a complex number with the following properties:

i) It is <u>Hermitian symmetric</u>:

$$p(\underset{\sim}{u},\underset{\sim}{v}) = \bar{p}(\underset{\sim}{v},\underset{\sim}{u}) \tag{1.7.4a}$$

ii) It is <u>conjugate linear in both $\underset{\sim}{u}$ and $\underset{\sim}{v}$</u>:

$$p(\underset{\sim}{u}_1+\underset{\sim}{u}_2,\underset{\sim}{v}) = p(\underset{\sim}{u}_1,\underset{\sim}{v}) + p(\underset{\sim}{u}_2,\underset{\sim}{v}) \tag{1.7.4b}$$

$$p(\underset{\sim}{u},\underset{\sim}{v}_1+\underset{\sim}{v}_2) = p(\underset{\sim}{u},\underset{\sim}{v}_1) + p(\underset{\sim}{u},\underset{\sim}{v}_2) \tag{1.7.4c}$$

$$p(\alpha\underset{\sim}{u},\underset{\sim}{v}) = \alpha p(\underset{\sim}{u},\underset{\sim}{v}) \tag{1.7.4d)*}$$

$$p(\underset{\sim}{u},\beta\underset{\sim}{v}) = \bar{\beta}p(\underset{\sim}{u},\underset{\sim}{v}) \tag{1.7.4e}$$

iii) It is real and <u>positive definite</u>:

$$p(\underset{\sim}{u},\underset{\sim}{u})>0, \text{ for } \underset{\sim}{u}, \neq \underset{\sim}{0} \tag{1.7.4f}$$

$$p(\underset{\sim}{u},\underset{\sim}{u}) = 0, \text{ if and only if } \underset{\sim}{u} = \underset{\sim}{0} \tag{1.7.4g}$$

---

* Note: conjugate linear in $\underset{\sim}{v}$

From definition (1.7.2) and properties (1.7.1), it follows that all that is needed for a bilinear form to constitute a scalar product for a vector space is that it is positive definite (and hence, real). Whether a bilinear form is positive definite or not clearly depends entirely on its matrix and not on its vectors. The following definition will be needed:

A square nxn matrix is said to be <u>positive definite</u> if (and only if), the quadratic form for any vector $\underset{\sim}{u} \neq \underset{\sim}{0}$ associated to it is real and positive and only vanishes for the zero vector. A positive definite matrix $\underset{\sim}{A}$ is symbolically designated as $\underset{\sim}{A} > 0$. If the said quadratic form vanishes for some nonzero vectors, then $\underset{\sim}{A}$ is said to be positive semidefinite, symbolically designated as $\underset{\sim}{A} \geq 0$. <u>Negative definite</u> and <u>negative semidefinite</u> matrices are similarly defined. Now:

<u>THEOREM 1.7.1</u> *Any square matrix is decomposable into the sum of a Hermitian and a skew Hermitian part (this is called the <u>Cartesian decomposition of the matrix</u>)*

Proof. Write the matrix $\underset{\sim}{A}$ in the form

$$\underset{\sim}{A} \equiv \frac{1}{2}(\underset{\sim}{A}+\underset{\sim}{A}^*) + \frac{1}{2}(\underset{\sim}{A}-\underset{\sim}{A}^*) \tag{1.7.5}$$

Clearly the first term of the right hand side is Hermitian and the second one is skew Hermitian.

<u>THEOREM 1.7.2</u> *The quadratic form associated with a matrix $\underset{\sim}{A}$ is real if and only if $\underset{\sim}{A}$ is Hermitian. It is imaginary if and only if $\underset{\sim}{A}$ is skew Hermitian.*

Proof.

("if" part) Let $\underset{\sim}{A}$ be Hermitian; then

$$\psi(\underset{\sim}{u})=\underset{\sim}{u}^*\underset{\sim}{A}^*\underset{\sim}{u}=\underset{\sim}{u}^*\underset{\sim}{A}\underset{\sim}{u}$$

and

$$\bar{\psi}(\underset{\sim}{u})=\underset{\sim}{u}^*\underset{\sim}{A}\underset{\sim}{u}$$

Since

$$\text{Im}\ \{\psi(\underset{\sim}{u})\}\equiv\frac{1}{2}\ \left(\psi(\underset{\sim}{u})-\bar{\psi}(\underset{\sim}{u})\right)$$

then

$$\text{Im}\ \{\psi(\underset{\sim}{u})\}=0$$

On the other hand, if $\underset{\sim}{A}$ is skew-Hermitian, then,

$$\psi(\underset{\sim}{u})=\underset{\sim}{u}^{*}\underset{\sim}{A}^{*}\underset{\sim}{u}=-\underset{\sim}{u}^{*}\underset{\sim}{A}\underset{\sim}{u}$$

and

$$\bar{\psi}(\underset{\sim}{u})=\underset{\sim}{u}^{*}\underset{\sim}{A}\underset{\sim}{u}$$

Since

$$\text{Re}\{\psi(\underset{\sim}{u})\}\equiv\frac{1}{2}\left(\psi(\underset{\sim}{u})+\bar{\psi}(\underset{\sim}{u})\right)$$

then

$$\text{Re}\{\psi(\underset{\sim}{u})\}=0$$

thus proving the "if" part of the theorem.

Exercise 1.7.2 Prove the "only if" part of Theorem 1.7.2

What Theorem 1.7.2 states is very important, namely that Hermitian matrices are good candidates for defining a scalar product for a vector space, since the associated quadratic form is real. What is now left to investigate is whether this form turns out to be positive definite as well. Though this is not true for any Hermitian matrix, it is (obviouly!) so for positive definite. Hermitian matrices (by definition!). Futhermore, since the quadratic form of a positive definite matrix must, in the first place, be real, and since, for the quadratic form associated with a matrix to be real, the matrix must be Hermitian (from Theorem 1.7.2), it is not necessary to refer to a positive definite (or semidefinite) matrix as being Hermitian.

Summarizing: In order for the quadratic form (1.7.3) to be a scalar product, $\underset{\sim}{A}$ must be positive definite. Next, a very important result concerning an easy characterization of positive definite (semidefinite) matrices is given.

<u>THEOREM 1.7.3</u> *A matrix is positive definite (semidefinite) if and only if its eigenvalues are all real and greater than (or equal to) zero.*

Proof. ("only if" part).

Indeed, if a matrix A is positive definite (semidefinite), it must be Hermitian. Thus, it can be diagonalized (a consequence of Theorem 1.4.1). Furthermore, once the matrix is in diagonal form, the elements on its diagonal are its eigenvalues, which are real and greater than (or equal to) zero. It takes on the form

$$\underset{\sim}{A} = \begin{pmatrix} \lambda_1 & & & & & \\ & \lambda_2 & & & \bigcirc & \\ & & \cdot & & & \\ & & & \cdot & & \\ & \bigcirc & & & \cdot & \\ & & & & & \lambda_n \end{pmatrix} \tag{1.7.10}$$

where

$$\lambda_i > (\geq)\ 0,\ i=1,2,\ldots,n$$

For any vector $\underset{\sim}{u} \neq \underset{\sim}{0}$, by definition,

$$\underset{\sim}{u}^*\underset{\sim}{A}^*\underset{\sim}{u}=\underset{\sim}{u}^*\underset{\sim}{A}\underset{\sim}{u}>(\geq)0 \tag{1.7.11}$$

where the components of $\underset{\sim}{u}$ (with respect to the basis formed with the complete set of eigenvectors of $\underset{\sim}{A}$) are

$$\underset{\sim}{u} = \begin{pmatrix} u_1 \\ u_2 \\ \cdot \\ \cdot \\ \cdot \\ u_n \end{pmatrix} \tag{1.7.12}$$

Substitution of (1.7.10) and (1.7.12) into (1.7.11) yields

$$\sum_{1}^{n}\lambda_i |u_i|^2 > (\geq) 0 \tag{1.7.13}$$

Now, assume $\underset{\sim}{u}$ is such that all but its $k^{\underline{th}}$ component vanish; in this case, (1.7.13) reduces to

$$\lambda_k |u_k|^2 > (\geq) 0$$

from which

$$\lambda_k > (\geq) 0$$

and, since λ_k can be any of the eigenvalues of $\underset{\sim}{A}$, the proof of this part is done. The proof of the "if" part is obvious and is left as an exercise for the reader.

Exercise 1.7.2 Show that, if the eigenvalues of a square matrix are all real and greater than (or equal to) zero, the matrix is positive definite (semidefinite).

A very special case of a positive definite matrix is the identity matrix, $\underset{\sim}{I}$, which yields the very well known scalar product

$$p(\underset{\sim}{u},\underset{\sim}{v}) = \underset{\sim}{u}^*\underset{\sim}{I}^*\underset{\sim}{v} = \underset{\sim}{u}^*\underset{\sim}{I}\underset{\sim}{v} = \underset{\sim}{u}^*\underset{\sim}{v} \tag{1.7.14}$$

In dealing with vector spaces over the real field, the arising inner product is real and hence, from Schwarz's inequality (1.4, p.125).

$$\frac{p(\underset{\sim}{u},\underset{\sim}{v})}{\sqrt{p(\underset{\sim}{u},\underset{\sim}{u})\,p(\underset{\sim}{v},\underset{\sim}{v})}} \leq 1$$

thus making it possible to define a "geometry" for then, the cosine of the angle between vectors $\underset{\sim}{u}$ and $\underset{\sim}{v}$ can be defined as

$$\cos(\underset{\sim}{u},\underset{\sim}{v}) = \frac{p(\underset{\sim}{u},\underset{\sim}{v})}{\sqrt{p(\underset{\sim}{u},\underset{\sim}{u})\,p(\underset{\sim}{v},\underset{\sim}{v})}}$$

For vector spaces over the complex field, such an angle cannot be defined, for then the inner product is a complex number.

1.8 NORMS, ISOMETRIES, ORTHOGONAL AND UNITARY MATRICES.

Given a vector space V, a norm for $\underset{\sim}{v} \varepsilon V$ is defined as a real-valued mapping from $\underset{\sim}{v}$ into a real number, represented by $||\underset{\sim}{v}||$, such that this norm

i) is positive definite, i.e.

$$||\underset{\sim}{v}|| > 0, \text{ for any } \underset{\sim}{v} \neq \underset{\sim}{0}$$

$$||\underset{\sim}{v}|| = 0 \text{ if and only if } \underset{\sim}{v} = \underset{\sim}{0}$$

ii) is linear homogeneous, i.e., for some $\alpha \varepsilon F$ (the field over which V is defined),

$$||\alpha\underset{\sim}{v}|| = |\alpha|\,||\underset{\sim}{v}||$$

$|\alpha|$ being the modulus (or the absolute value, in case α is real) of α.

iii) satisfies the triangle inequality, i.e. for $\underset{\sim}{u}$ and $\underset{\sim}{v} \varepsilon V$,

$$||\underset{\sim}{u}+\underset{\sim}{v}|| < ||\underset{\sim}{u}|| + ||\underset{\sim}{v}||$$

Example 1.8.1 Let v_i be the ith component of a vector $\underset{\sim}{v}$ of a space over the complex field. The following are well defined norms for $\underset{\sim}{v}$:

$$||\underset{\sim}{v}|| = \max_{1<i<n} |v_i| \tag{1.8.1}$$

$$||\underset{\sim}{v}|| = \left(\sum_1^n |x_i|^p\right)^{1/p} \tag{1.8.2}$$

where p is a positive integer. For p = 2 in (1.8.2) the corresponding norm is the Euclidean norm, or the "magnitude" of $\underset{\sim}{v}$.

Norm (1.8.1) is easy and fast to compute, and hence it is widely used in numerical computations. However, it is not suitable for physical or geometrical problems since it is not invariant*, i.e. it depends on the coordinate axes being used. The Euclidean norm has the advantage that it is invariant.

---

* Besides, there is no inner product associated with it and hence obviously no "geometry"

However, computing it requires n (the dimensio the space to which the vector under consideration belongs) multiplications (i.e. n square raisings), n-1 additions and one square root computation. In order to proceed further, some more definitions are needed.

An invertible linear transformation is called an "isometry" if it preserves the following scalar product

$$p(\underset{\sim}{x},\underset{\sim}{y}) = \underset{\sim}{x}^*\underset{\sim}{y} \tag{1.8.3}$$

It is a very simple matter to show that, in order for a transformation $\underset{\sim}{P}$ to be an isometry, it is required that its transpose conjugate, $\underset{\sim}{P}^*$, equals its inverse, i.e.,

$$\underset{\sim}{P}^*=\underset{\sim}{P}^{-1} \tag{1.8.4}$$

If $\underset{\sim}{P}$ is defined over the complex field and meets condition (1.8.4), then it is said to be unitary. If $\underset{\sim}{P}$ is defined over the real field, then $\underset{\sim}{P}^*=\underset{\sim}{P}^T$, the transpose of $\underset{\sim}{P}$ and, if it satisfies (1.8.4), it is said to be orthogonal.

Exercise 1.8.1 Show that in order for $\underset{\sim}{P}$ to be an isometry, it is necessary and sufficient that $\underset{\sim}{P}$ satisfies (1.8.4), i.e., show that under the similarity transformation

$$\underset{\sim}{\zeta}=\underset{\sim}{P}\underset{\sim}{x},\ \underset{\sim}{\eta} = \underset{\sim}{P}\underset{\sim}{y},\ \underset{\sim}{B}=\underset{\sim}{P}\underset{\sim}{A}\underset{\sim}{P}^{-1}$$

the scalar product (1.8.3) is preserved if $\underset{\sim}{P}$ meets condition (1.8.4).

1.9 PROPERTIES OF UNITARY AND ORTHOGONAL MATRICES.

Some important facts about unitary and orthogonal matrices are discussed in this section. Notice that all results concerning unitary matrices apply to orthogonal matrices, for the latter are a special case of the former.

THEOREM 1.9.1 *The set of eigenvalues of a unitary matrix lies on the unit circle $|z|^2 = 1$, centered at the origin of the complex plane.*

Proof: Let $\underset{\sim}{U}$ be an nxn unitary matrix. Let λ be one of its eigenvalues and $\underset{\sim}{e}$ a corresponding eigenvector, so that

$$\underset{\sim}{U}\underset{\sim}{e} = \lambda \underset{\sim}{e} \qquad (1.9.1)$$

Taking the transpose conjugate of both sides of (1.9.1),

$$\underset{\sim}{e}^*\underset{\sim}{U}^* = \bar{\lambda}\underset{\sim}{e}^* \qquad (1.9.2)$$

Performing the corresponding products on both sides of eqs. (1.9.1) and (1.9.2),

$$\underset{\sim}{e}^*\underset{\sim}{U}^*\underset{\sim}{U}\underset{\sim}{e} = \lambda\bar{\lambda}\underset{\sim}{e}^*\underset{\sim}{e} \qquad (1.9.3)$$

But, since $\underset{\sim}{U}$ is unitary, (1.9.3) leads to

$$\underset{\sim}{e}^*\underset{\sim}{e} = |\lambda|^2 \underset{\sim}{e}^*\underset{\sim}{e}$$

from which

$$|\lambda|^2 = 1, \text{q.e.d.}$$

Corollary 1.9.1 If an nxn unitary matrix is of odd order (i.e. n is odd), then it has at least one real eigenvalue, which is either + 1 or -1,

Exercise 1.9.1 Prove Corollary 1.9.1

1.10 STATIONARY POINTS OF SCALAR FUNCTIONS OF A VECTOR ARGUMENT.

Let $\phi = \phi(\underset{\sim}{x})$ be a (scalar) real function of a vector argument, $\underset{\sim}{x}$, assumed to be continuous and differentiable up to second derivatives within a certain neighborhood around some $\underset{\sim}{x}_0$. The stationary points of this function are defined as those values $\underset{\sim}{x}_0$ of $\underset{\sim}{x}$ where the gradient of ϕ, $\phi'(\underset{\sim}{x})$ vanishes. Each stationary point can be an extremum or a saddle point. An extremum, in turn, can be either a local maximum or minimum. The function ϕ attains a local maximum at $\underset{\sim}{x}_0$ if and only if

$$f(\underset{\sim}{x}_0) \geq f(\underset{\sim}{x})$$

for any $\underset{\sim}{x}$ in the neighborhood of $\underset{\sim}{x}_0$, i.e., for any $\underset{\sim}{x}$ such that

$$||\underset{\sim}{x}-\underset{\sim}{x}_0|| \leq \varepsilon$$

ε being an arbitrarily small positive number. A local minimum is correspondingly defined. If an extremum is neither a local maximum nor a local minimum, it is said to be a saddle point. Criteria to decide whether an extremum is a maximum, a minimum or a saddle point are next derived.

An expansion of ϕ around $\underset{\sim}{x}_0$ in a Taylor series illustrates the kind of stationary point at hand. In fact, the Taylor expansion of ϕ is

$$\phi(\underset{\sim}{x}) = \phi(\underset{\sim}{x}_0) + \phi'(\underset{\sim}{x}_0)^T(\underset{\sim}{x}-\underset{\sim}{x}_0) + \frac{1}{2}(\underset{\sim}{x}-\underset{\sim}{x}_0)^T \phi''(\underset{\sim}{x}_0)(\underset{\sim}{x}-\underset{\sim}{x}_0) + R \qquad (1.10.1)$$

where R is the residual, which contains terms of third and higher orders.

Then the increment of ϕ at x_0, for a given increment $\Delta x = x-x_0$, is given by

$$\Delta\phi=\phi'(\underset{\sim}{x}_0)^T\Delta\underset{\sim}{x}+\frac{1}{2}\Delta\underset{\sim}{x}^T\phi''(\underset{\sim}{x}_0)\Delta\underset{\sim}{x} \qquad (1.10.2)$$

if terms of third and higher orders are neglected.

From eq. (1.10.2) it can be concluded that the linear pant of $\Delta\phi$ vanishes at a stationary point, which makes clear why such points are called stationary. Whether $\underset{\sim}{x}_0$ constitutes an extremum or not, depends on the sign of $\Delta\phi$. It is a maximum if $\Delta\phi$ is nonpositive for arbitrary $\Delta\underset{\sim}{x}$. It is a minimum if the said increment is nonnegative for arbitrary $\Delta\underset{\sim}{x}$. If the sign of the increment depends on $\Delta\underset{\sim}{x}$, then $\underset{\sim}{x}_0$ is a saddle point for reasons which are brought up in the following. Eq. (1.10.2) shows that the sign of $\Delta\phi$ depends entirely on the quadratic term, at a stationary point. Whether this term is nonpositive or nonnegative, it is sufficient that the Hessian matrix $\phi''(x)$ be sign semidefinite at x_0. Notice, however, that this condition on the Hessian matrix is only sufficient, but not necessary, for it is based on Eq. (1.10.2), which is truncated after third-order terms. In fact, a function whose Hessian at a stationary point is sign-semidefinite can constitute either a maximum, a minimum, or a saddle point as shown next.

From the foregoing discussion, the following theorem is concluded.

THEOREM 1.10.1 *Extrema and saddle points of a differentiable function occur at stationary points. For a stationary point to constitute a local maximum (minimum) it is sufficient, although not necessary, that the corresponding Hessian matrix be negative (positive) semidefinite. For the said point to constitute a saddle point, it is sufficient that the corresponding Hessian matrix sign-indefinite at this stationary point.*

A hypersurface in an n-dimensional space resembles a hyperbolic paraboloid at a saddle point, the resemblance lying in the fact that, at its stationary point, the sign of the curvature of the surface is different for each direction. To illustrate this, consider the hyperbolic paraboloid of Fig 1.10.1 for which, when seen from the X-axis, its stationary point (the origin) appears as a minimum (positive curvature), whereas, if seen from the Y-axis, it appears as a maximum (negative curvature). In fact, it is none of these.

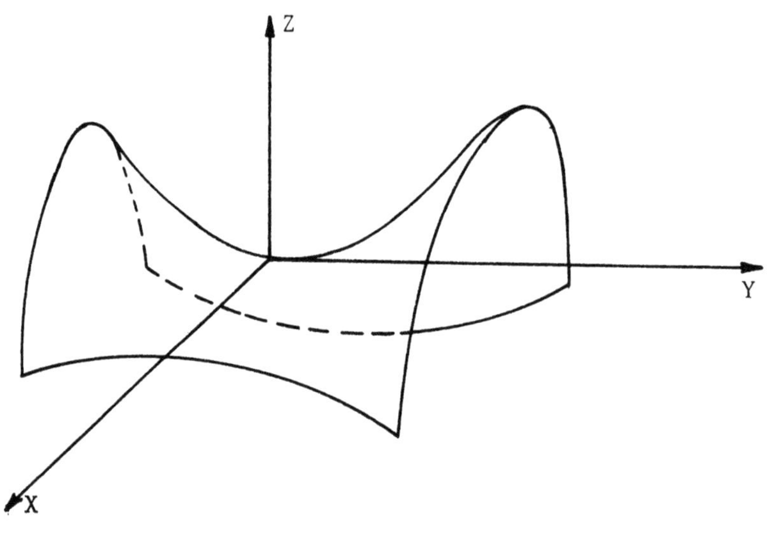

Fig. 1. 10.1 Saddle point of a 3-dimensional surface

<u>*Corollary 1.10.7*</u> *The quadratic form*

$$\psi(\underset{\sim}{x}) = \underset{\sim}{x}^T \underset{\sim}{A} \underset{\sim}{x} + \underset{\sim}{b}^T \underset{\sim}{x} + c$$

has a unique extremum at $x_0 = -\frac{1}{2} \underset{\sim}{A}^{-1} \underset{\sim}{b}$, *if* $\underset{\sim}{A}^{-1}$ *exists. This is a maximum (minimum) if* $\underset{\sim}{A}$ *is negative (positive) semidefinite*

Exercise 1.10.1 Prove Corollary 1.10.7

Example 1.10.1 The function $\phi = x_1^4 + x_2^4 + \ldots + x_n^4$ has a local minimum at $x_1 = x_2 = \ldots = x_n = 0$. The Hessian matrix of this function, however, vanishes at this minimum.

Example 1.10.2 The function $= x_1^4 - x_2^4$ has a stationary point at the origin, which is a saddle point. Its Hessian matrix, however, vanishes at this point.

Example 1.10.3 The function $x_1^2 + x_2^4$ has a minimum at (0,0). At this point its Hessian matrix is positive semidefinite.

1.11 LINEAR ALGEBRAIC SYSTEMS

Let $\underset{\sim}{A}$ be an mxn matrix and $\underset{\sim}{x}$ and $\underset{\sim}{b}$ be n-and m-dimensional vectors where, in general, $m \neq n$. Equation

$$\underset{\sim}{A}\underset{\sim}{x}=\underset{\sim}{b} \qquad (1.11.1)$$

is a linear algebraic system. It is linear because, if $\underset{\sim}{x}_1$ and $\underset{\sim}{x}_2$ are solutions to it for $\underset{\sim}{b}=\underset{\sim}{b}_1$ and $\underset{\sim}{b}=\underset{\sim}{b}_2$, and α and β are scalars, then $\alpha\underset{\sim}{x}_1+\beta\underset{\sim}{x}_2$ is a solution for $\underset{\sim}{b}=\alpha\underset{\sim}{b}_1+\beta\underset{\sim}{b}_2$. It is <u>algebraic</u> as opposed to differential or dynamic because it does not involve derivatives. There are three different cases regarding the solution of eq. (1.11.1), depending upon whether m is equal to, greater than or less than n. These are discussed next:

i) m=n. This is the best-known case and an extensive discussion of it can be found in any elementary linear algebra textbook. The most important result in this case states that if $\underset{\sim}{A}$ is of full rank, i.e. if det $\underset{\sim}{A} \neq 0$, then the system has a unique solution, which is given by

$$\underset{\sim}{x}=\underset{\sim}{A}^{-1}\underset{\sim}{b}$$

ii) m>n. In this case the number of equations is greater than that of unknowns. The system is overdetermined and there is no guarantee of the existence of a certain $\underset{\sim}{x}_0$ such that $\underset{\sim}{A}\underset{\sim}{x}_0=\underset{\sim}{b}$.

A very simple example of such a system is the following:

$$x_1=5 \tag{1.11.1a}$$

$$x_1=3 \tag{1.11.1b}$$

where m=2 and n=1. If $x_1=5$, the first equation is satisfied but the second one is not. If, on the other hand, $x_1=3$, the second equation is satisfied, but the first one is not. However, a system with m>n could have a solution, which could even be unique if, out of the m equations involved, only n are linearly independent, the remaining m-n being linearly dependent on the n ℓ.i. equations. As an example, consider the following system

$$x_1+x_2=5 \tag{1.11.2a}$$

$$x_1-x_2=3 \tag{1.11.2b}$$

$$3x_1+x_2=13 \tag{1.11.2c}$$

whose (unique) solution is

$$x_1=4, x_2=1 \tag{1.11.3}$$

Here equation (1.11.2c) is linearly dependent on (1.11.2a) and (1.11.2b). In general, however, for m>n it is not possible to satisfy all the equations of a system with more equations than unknowns; but it is possible to "satisfy" them with the minimum possible error. Assume that $\underset{\sim}{x}_0$ does not satisfy all the equations of a mxn system, with m>n, but satisfies the system with the least possible error. Ley $\underset{\sim}{e}$ be the said error, i.e.

$$\underset{\sim}{e}=\underset{\sim}{A}\underset{\sim}{x}_0-\underset{\sim}{b} \tag{1.11.4}$$

The Euclidean norm of $\underset{\sim}{e}$ is

$$||\underset{\sim}{e}||^2=(\underset{\sim}{A}\underset{\sim}{x}_0-\underset{\sim}{b})^T(\underset{\sim}{A}\underset{\sim}{x}_0-\underset{\sim}{b}) \tag{1.11.5}$$

Expanding $||\underset{\sim}{e}||^2$, it is noticed that it is a quadratic form of $\underset{\sim}{x}_0$, i.e.

$$\phi(\underset{\sim}{x}_0)=||\underset{\sim}{e}||^2=\underset{\sim}{x}_0^T\underset{\sim}{A}^T\underset{\sim}{A}\underset{\sim}{x}_0-2\underset{\sim}{b}^T\underset{\sim}{A}\underset{\sim}{x}_0+\underset{\sim}{b}^T\underset{\sim}{b} \tag{1.11.6}$$

The latter quadratic form has an extremum where $\phi'(\underset{\sim}{x}_0)$ vanishes. The corresponding value of $\underset{\sim}{x}$, $\underset{\sim}{x}_0$, is found by setting $\phi'(\underset{\sim}{x}_0)$ equal to zero, i.e.

$$\phi'(\underset{\sim}{x}_0)=2\underset{\sim}{A}^T\underset{\sim}{A}\underset{\sim}{x}_0-2\underset{\sim}{A}^T\underset{\sim}{b}=\underset{\sim}{0} \qquad (1.11.7)$$

If $\underset{\sim}{A}$ is of full rank, i.e., if rank $(\underset{\sim}{A})=n$, then $\underset{\sim}{A}^T\underset{\sim}{A}$, a nxn matrix, is also of rank n (1.4), i.e. $\underset{\sim}{A}^T\underset{\sim}{A}$ is invertible and so, from eq. (1.11.5),

$$\underset{\sim}{x}_0=(\underset{\sim}{A}^T\underset{\sim}{A})^{-1}\underset{\sim}{A}^T\underset{\sim}{b}=\underset{\sim}{A}^I\underset{\sim}{b} \qquad (1.11.8)$$

where $\underset{\sim}{A}^I$ is a "pseudo-inverse" of $\underset{\sim}{A}$, called the "Moore-Penrose generalised inverse" of $\underset{\sim}{A}$. A method to determine $\underset{\sim}{x}_0$ that does not require the computation of $\underset{\sim}{A}^I$ is given in (1.5) and (1.6). In (1.7), an iterative method to compute $\underset{\sim}{A}^I$ is proposed. The numerical solution of this problem is presented in section 1.12. This problem arises in such fields as control theory, curve-fitting (regressions) and mechanism synthesis.

iii) $m<n$. In this case the number of equations is less than that of unknowns. Hence, if the system is consistent*, it has an infinity of solutions. For instance, the system

$$x+y=3 \qquad (1.11.9)$$

in which $m=1$ and $n=2$, admits infinitely many solutions, namely all points lying on the line

$$y=x+3 \qquad (1.11.10)$$

Now consider the system

$$x+y+z=1 \qquad (1.11.11a)$$

$$x+y-z=1 \qquad (1.11.11b)$$

with $m=2$ and $n=3$. This system admits an infinity of solutions, provided $z=0$. Otherwise, none.

In case a system with $m<n$ admits a solution, it in fact admits infinitely many, which is not difficult to prove. Indeed, partition matrix $\underset{\sim}{A}$ and vector $\underset{\sim}{x}$ in the form

---

* i.e. if $\underset{\sim}{b} \in R(\underset{\sim}{A})$

$$A = \left(A_1 \,\vdots\, A_2 \right) \updownarrow m, \quad x = \begin{pmatrix} x_1 \\ - - \\ x_2 \end{pmatrix} \begin{matrix} \updownarrow m \\ \updownarrow n-m \end{matrix} \qquad \text{(1.11.12)}$$

$$\overleftrightarrow{m}\;\overleftrightarrow{n-m}$$

Thus, eq. (1.11.1) is equivalent to

$$A_1x_1+A_2x_2=b \qquad (1.11.13)$$

In the latter equation, in rank $(A_1) = m$, A_1^{-1} exists and a solution to (1.11.13) is

$$x_1=A_1^{-1}b, \qquad A_2x_2=0 \qquad (1.11.14)$$

where x_1 is unique, as was stated for the case m=n, and x_2 is a vector lying in the null space of A_2. Clearly, there are as many linearly independent solutions (1.11.12) as there are linearly independent vectors in the null space of A_2.

From the foregoing discussion, in m<n, system (1.11.1) admits an infinity of solutions. However, among those infinitely many solutions, there is exactly one whose Euclidean norm is a minimum. That "optimal" solution is found next, via a <u>quadratic programming problem</u>, namely,

$$\text{Min}\,\phi(x)=x^Tx \qquad (1.11.15)$$

subject to

$$Ax=b \qquad (1.11.16)$$

Applying the Lagrange multiplier technique (1.8), let λ be an m-dimensional vector whose components are called Lagrange multipliers. Define, then, the new quadratic form

$$\psi(x)=x^Tx+\lambda^T(Ax-b) \qquad (1.11.17)$$

which reduces to the original one (1.11.15), when (1.11.16) is satisfied. $\psi(x)$ has an extremum where its gradient $\psi'(x)$ vanishes. The condition is

$$\psi'(x)=2x+A^T\lambda=0 \qquad (1.11.18)$$

from which

$$\underset{\sim}{x} = -\frac{1}{2}\underset{\sim}{A}^T\underset{\sim}{\lambda} \qquad (1.11.19)$$

However, $\underset{\sim}{\lambda}$ is yet unknown. Substituting the value of $\underset{\sim}{x}$ given in (1.11.19), in (1.11.16), one obtains

$$-\frac{1}{2}\underset{\sim}{A}\underset{\sim}{A}^T\underset{\sim}{\lambda}=\underset{\sim}{b} \qquad (1.11.20)$$

From which, if $\underset{\sim}{A}\underset{\sim}{A}^T$ is of full rank,

$$\underset{\sim}{\lambda} = -2(\underset{\sim}{A}\underset{\sim}{A}^T)^{-1}\underset{\sim}{b} \qquad (1.11.21)$$

Finally, substituting the latter value of $\underset{\sim}{\lambda}$ into eq. (1.11.19),

$$\underset{\sim}{x}=\underset{\sim}{A}^T(\underset{\sim}{A}\underset{\sim}{A}^T)^{-1}\underset{\sim}{b}=\underset{\sim}{A}^+\underset{\sim}{b} \qquad (1.11.22)$$

where

$$\underset{\sim}{A}^+\equiv\underset{\sim}{A}^T(\underset{\sim}{A}\underset{\sim}{A}^T)^{-1}$$

is another pseudo-inverse of $\underset{\sim}{A}$.

Exercise 1.11.1 Can both pseudo-inverses of $\underset{\sim}{A}$, the one given in (1.11.8) and that of (1.11.23) exist for a given matrix $\underset{\sim}{A}$? Explain.

The foregoing solution (1.11.22) has many interpretations: in control theory it yields the control taking a system from a known initial state to a desired final one while spending the minimum amount of energy. In Kinematics it finds two interpretations which will be given in Ch. 2, together with applications to hypoid gear design.

Exercise 1.11.2 Show that the image of the error (1.11.4) is perpendicular to $\underset{\sim}{x}_0$ as given by (1.11.8). This result is known as the "Projection Theorem" and finds extensive applications in optimisation theory (1.9).

1.12 NUMERICAL SOLUTION OF LINEAR ALGEBRAIC SYSTEMS

Consider the system (1.11.1) for all three cases discussed in section 1.11.

i) m=n. There are many methods to solve a linear algebraic system for as many equations as unknowns, but all of them fall into one of two categories, namely, a) direct methods and b) iterative methods. Because the first ones are more suitable to be applied in nonlinear algebraic systems, which will be discussed in section 1.13, only direct methods will be treated here. There is an extensive literature dealing with interative methods, of which the treatise by Varga (1.10) discusses the topic very extensively.

As to direct methods, Gauss'algorithm is the one which has received most attention (1.11), (1.12). In (1.11) the LU decomposition algorithm is presented and, with further refinements, in (1.12). The solution is obtained in two steps:

In the first step the matrix of the system, $\underset{\sim}{A}$, is factored into the product of a lower triangular matrix, $\underset{\sim}{L}$, times an upper triangular one, $\underset{\sim}{U}$, in the form

$$\underset{\sim}{A} = \underset{\sim}{L}\underset{\sim}{U} \tag{1.12.1}$$

where the diagonal of $\underset{\sim}{L}$ contains ones in all its entries. Matrix $\underset{\sim}{U}$ contains the <u>singular</u> values of $\underset{\sim}{A}$ on its diagonal, and all its elements below the main diagonal are zero. The singular values of a matrix $\underset{\sim}{A}$ are the nonnegative square roots of the eigenvalues of $\underset{\sim}{A}^T\underset{\sim}{A}$. These are real and nonnegative, which is not difficult to prove.

<u>Exercise 1.12.1</u> Show that if $\underset{\sim}{A}$ is a nonsingular nxn matrix, $\underset{\sim}{A}^T\underset{\sim}{A}$ is positive definite, and if it is singular, then $\underset{\sim}{A}^T\underset{\sim}{A}$ is positive semi-definite. (Hint: Compute the norm of $\underset{\sim}{A}\underset{\sim}{x}$, for arbitraty $\underset{\sim}{x}$).

The $\underset{\sim}{L}\underset{\sim}{U}$ decomposition of $\underset{\sim}{A}$ is performed via the DECOMP subprogram appearing in (1.12). If $\underset{\sim}{A}$ happens to be singular, DECOMP detects this by computing det $\underset{\sim}{A}$, which is done performing the product of the singular values of $\underset{\sim}{A}$, and if this product turns out to be zero, sends a message to the user thereby warning him that he cannot proceed any further.

If $\underset{\sim}{A}$ is not singular, the user calls the SOLVE subprogram, which computes the solution to the system by back substitution, i.e. from (1.12.1) in the following manner: The equation

$$\underset{\sim}{L}\underset{\sim}{U}\underset{\sim}{x}=\underset{\sim}{b} \tag{1.12.2}$$

can be written as

$$\underset{\sim}{L}\underset{\sim}{y}=\underset{\sim}{b}$$

by setting $\underset{\sim}{U}\underset{\sim}{x}=\underset{\sim}{y}$. Thus

$$\underset{\sim}{y}=\underset{\sim}{L}^{-1}\underset{\sim}{b}=\underset{\sim}{c} \tag{1.12.3}$$

where $\underset{\sim}{L}^{-1}$ exists since det $\underset{\sim}{L}$ (the product of the elements on the diagonal of $\underset{\sim}{L}$) is equal to one (1.11). Substituting (1.12.3) into $\underset{\sim}{U}\underset{\sim}{x}=\underset{\sim}{y}$, one obtains the final solution:

$$\underset{\sim}{x}=\underset{\sim}{U}^{-1}\underset{\sim}{c}$$

where $\underset{\sim}{U}^{-1}$ exists because $\underset{\sim}{A}$ has been detected to be nonsingular*.

The flow diagram of the whole program appears in Fig 1.12.1 and the listings of DECOMP and SOLVE in Figs. 1.12.2 and 1.12.3

ii) m>n. Next, the numerical solution of the overdetermined linear system $\underset{\sim}{A}\underset{\sim}{x}=\underset{\sim}{b}$ is discussed. In this case the number of equations is greater than that of unknowns and hence the sought "solution" is that $\underset{\sim}{x}_0$ which minimizes the Euclidean norm of the error $\underset{\sim}{A}\underset{\sim}{x}_0-\underset{\sim}{b}$. This is done by application of Householder reflections (1.5) to both $\underset{\sim}{A}$ and $\underset{\sim}{b}$. A Householder reflection is an orthogonal transformation $\underset{\sim}{H}$ which has the property that

$$\underset{\sim}{H}^{-1}=\underset{\sim}{H}^{T}=\underset{\sim}{H} \tag{1.12.4}$$

Given an m-vector $\underset{\sim}{a}$ with components $a_1, a_2, \ldots, a_m$, the Householder reflection $\underset{\sim}{H}$ (a function of $\underset{\sim}{a}$) defined as

* In fact, there is no need to explicitely compute $\underset{\sim}{L}^{-1}$ and $\underset{\sim}{U}^{-1}$, for the triangular structure of $\underset{\sim}{L}$ and $\underset{\sim}{U}$ permits a recursive solution.

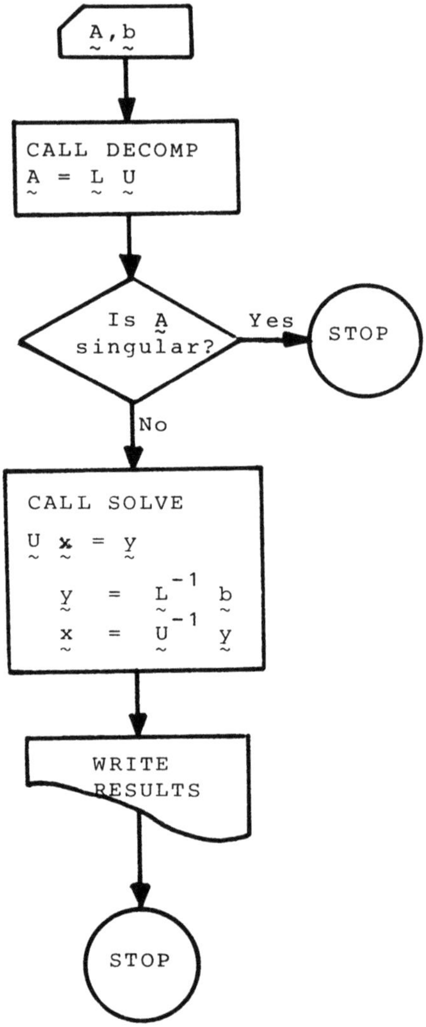

Fig. 1.12.1 Flow diagram for the direct solution of a linear algebraic system with equal number of equations as unknowns

```
      SUBROUTINE DECOMP(N,NDIM,A,IP)
      REAL A(NDIM,NDIM),T
      INTEGER IP(NDIM)
C
C      MATRIX TRIANGULARIZATION BY GAUSSIAN ELIMINATION
C
C     INPUT :
C     N      = ORDER OF MATRIX
C     NDIM   = DECLARED DIMENSION OF ARRAY A. IN THE MAIN PROGRAM
C     A      = MATRIX TO BE TRIANGULARIZED
C
C     OUTPUT :
C     A(I,J), I.LE.J   = UPPER TRIANGULAR FACTOR, U
C     A(I,J), I.GT.J   =MULTIPLIERS = LOWER TRIANGULAR FACTOR, I-L
C     IP(K), K.LT.N    =INDEX OF K-TH PIVOT ROW
C     IP(N)            = (-1)**(NUMBER OF INTERCHANGES) OR 0.
C     USE 'SOLVE' TO OBTAIN SOLUTION OF LINEAR SISTEM
C     DETERM(A)        = IP(N)*A(1,1)*A(2,2)*...*A(N,N)
C     IF IP(N)=0, A IS SINGULAR, 'SOLVE' WILL DIVIDE BY ZERO
C     INTERCHANGES FINISHED IN U, ONLY PARTLY IN L
C
      IP(N)=1
      DO 60 K=1,N
        IF(K.EQ.N) GO TO 50
        KP1=K+1
        M=K
        DO 10 I=KP1,N
        IF(ABS(A(I,K)).GT.ABS(A(M,K))) M=I
 10     CONTINUE
        IP(K)=M
        IF(M.NE.K) IP(N)=-IP(N)
        T=A(M,K)
        A(M,K)=A(K,K)
        A(K,K)=T
        IF(T.EQ.0) GO TO 50
        DO 20 I=KP1,N
 20       A(I,K)=-A(I,K)/T
        DO 40 J=KP1,N
          T=A(M,J)
          A(M,J)=A(K,J)
          A(K,J)=T
          IF(T.EQ.0.) GO TO 40
          DO 30 I=KP1,N
 30         A(I,J)=A(I,J)+A(I,K)*T
 40     CONTINUE
 50     IF(A(K,K).EQ.0.) IP(N)=0
 60   CONTINUE
      RETURN
      END
*
```

Fig. 1.12.2 Listing of SUBROUTINE DECOMP

Copyright 1972, Association for Computing Machinery, Inc., reprinted by permission from [1.12]

```
      SUBROUTINE SOLVE(N,NDIM,A,B,IP)
      REAL A(NDIM,NDIM),B(NDIM),T
      INTEGER IP(NDIM)
C
C     SOLUTION OF LINEAR SYSTEM, A*X = B
C
C     INPUT :
C     N    = ORDER OF MATRIX.
C     NDIM = DECLARED DIMENSION OF ARRAY A. IN THE MAIN PROGRAM
C     A    = TRIANGULARIZED MATRIX OBTAINED FROM "DECOMP"
C     B    = RIGHT HAND SIDE VECTOR
C     IP   = PIVOT VECTOR OBTAINED FROM "DECOMP"
C     DO NOT USE "SOLVE" IF "DECOMP" HAS SET IP(N)=0
C
C     OUTPUT :
C     B    = SOLUTION VECTOR, X
C
      IF(N.EQ.1) GO TO 90
      NM1=N-1
      DO 70 K=1,NM1
         KP1=K+1
         M=IP(K)
         T=B(M)
         B(M)=B(K)
         B(K)=T
         DO 70 I=KP1,N
   70 B(I)=B(I)+A(I,K)*T
      DO 80 KB=1,NM1
         KM1=N-KB
         K=KM1+1
         B(K)=B(K)/A(K,K)
         T=-B(K)
         DO 80 I=1,KM1
   80 B(I)=B(I)+A(I,K)*T
   90 B(1)=B(1)/A(1,1)
      RETURN
      END
#
```

Fig. 1.12.3 Listing of SUBROUTINE SOLVE

Copyright 1972, Association for Computing Machinery, Inc., reprinted by permission from [1.12]

$$\alpha = \text{sgn}\ (a_1)\,||\underset{\sim}{a}|| \qquad (1.12.5a)$$

$$\underset{\sim}{u} = \underset{\sim}{a}+\alpha\underset{\sim}{e}_1 \qquad (1.12.5b)$$

$$\beta = \alpha u_1 \qquad (1.12.5c)$$

$$\underset{\sim}{H} = \underset{\sim}{I}- \frac{1}{\beta}\,\underset{\sim}{u}\underset{\sim}{u}^T \qquad (1.12.5b)$$

transforms $\underset{\sim}{a}$ into $-\alpha\underset{\sim}{e}_1$, and reflects any other vector $\underset{\sim}{b}$ about a hyperplane perpendicular to $\underset{\sim}{u}$.

On the other hand, if $\underset{\sim}{H}_k$ is defined as

$$\alpha_k = \text{sgn}(a_k)\,(a_k^2 + a_{k+1}^2+\ldots+a_m^2)^{\frac{1}{2}} \qquad (1.12.6a)$$

$$\underset{\sim}{u}_k = \left(0,\ldots,0,a_k+\alpha_k,\ a_{k+1},\ldots,a_m\right)^T \qquad (1.12.6b)$$

$$\beta_k = \alpha_k(\underset{\sim}{u}_k)_k \qquad (1.12.6c)$$

$$\underset{\sim}{H}_k = \underset{\sim}{I} - \frac{1}{\beta_k}\underset{\sim}{u}_k\underset{\sim}{u}_k^T \qquad (1.12.6d)$$

then $\underset{\sim}{H}_k\underset{\sim}{a}$ is a vector whose first k-1 components are identical to those of $\underset{\sim}{a}$, its $k^{\underline{th}}$ component is $-\alpha_k$ and its remaining m-k components are all zero. Furthermore, if v is any other vector, then

$$\underset{\sim}{H}_k\underset{\sim}{v} = \underset{\sim}{v} - \gamma\underset{\sim}{u}$$

where

$$\gamma = \frac{\underset{\sim}{v}^T\underset{\sim}{v}}{\beta_k}$$

and if, in particular, $v_k = v_{k+1} = \ldots = v_m = 0$, then

$$\underset{\sim}{H}_k\underset{\sim}{v} = \underset{\sim}{v}$$

Let now $\underset{\sim}{H}_i$ be the Householder reflection wich cancels the last m-i components of the $i^{\underline{th}}$ column of $\underset{\sim}{H}_{i-1}\underset{\sim}{A}$, while leaving its i-1 components unchanged and setting its $i^{\underline{th}}$ component equal to $-\alpha_i$, for i=1,...,n. By application of the n Householder reflections thus defined, on $\underset{\sim}{A}$ and $\underset{\sim}{b}$ in the form

$$\underset{\sim}{H}_n\underset{\sim}{H}_{n-1}\cdots\underset{\sim}{H}_2\underset{\sim}{H}_1\underset{\sim}{A}\underset{\sim}{x}_0=\underset{\sim}{H}_n\underset{\sim}{H}_{n-1}\cdots\underset{\sim}{H}_2\underset{\sim}{H}_1\underset{\sim}{b} \qquad (1.12.7)$$

the original system is transformed into the following two systems

$$\underset{\sim}{A}_1'\underset{\sim}{x}_0 = \underset{\sim}{b}_1'$$

$$\underset{\sim}{A}_2'\underset{\sim}{x}_0 = \underset{\sim}{b}_2'$$

where $\underset{\sim}{A}_1'$ is nxn and upper triangular, whereas $\underset{\sim}{A}_2'$ is the (m-n)xn zero matrix and $\underset{\sim}{b}_2'$ is of dimension m-n and different from zero. Once the system is in upper triangular form, it is a simple matter to find the values of the components of $\underset{\sim}{x}_0$ by back substitution. Let a^*_{ij} and b^*_k be the values of the (i, j) element of $\underset{\sim}{A}_1'$ and the kth component of $\underset{\sim}{b}_1'$ respectively. Then, starting from the nth equation of system (1.12.7),

$$a^*_{nn}x_n = b^*_n$$

x_n is obtained as

$$x_n = \frac{b^*_n}{a^*_{nn}}$$

Substituting this value into the (n-1) st equation,

$$a^*_{n-1,n-1}x_{n-1} + a^*_{n-1,n}\frac{b^*_n}{a^*_{nn}} = b^*_{n-1}$$

from which

$$x_{n-1} = \frac{b^*_{n-1} - \dfrac{b^*_n}{a^*_{nn}}}{a^*_{n-1,n-1}}$$

Proceeding similarly with the (n-2)nd,...,2nd and 1st equations, the n components of $\underset{\sim}{x}_0$ are found. Clearly, then, $\underset{\sim}{b}_2'$ is the error in the approximation and $||\underset{\sim}{b}_2'|| = ||\underset{\sim}{A}\,\underset{\sim}{x} - \underset{\sim}{b}||$.

The foregoing Householder reflection method can be readily implemented in a digital computer via the HECOMP and HOLVE subroutines appearing in (1.14), whose listings are reproduced in Figs 1.12.4 and 1.12.5.

Exercise 1.12.2 Show that, for any n-vector $\underset{\sim}{x}$

$$\det(\underset{\sim}{I} + \underset{\sim}{x}\underset{\sim}{x}^T) = 1 + \underset{\sim}{x}^T\underset{\sim}{x}$$

```
#
      SUBROUTINE HECOMP(MDIM,M,N,A,U)
      INTEGER MDIM,M,N
      REAL A(MDIM,N),U(M)
      REAL ALPHA,BETA,GAMMA,SQRT
C
C   HOUSEHOLDER REDUCTION OF RECTANGULAR MATRIX TO UPPER
C   TRIANGULAR FORM.  USE WITH HOLVE FOR LEAST-SQUARE
C   SOLUTIONS OF OVERDETERMINED SYSTEMS.
C
C   MDIM= DECLARED ROW DIMENSION OF A
C   M   = NUMBER OF ROWS OF A
C   N   = NUMBER OF COLUMNS OF A
C   A   = M-BY-N MATRIX WITH M.>.N
C         INPUT :
C                 MATRIX TO BE REDUCED
C         OUTPUT:
C                 REDUCED MATRIX AND INFORMATION ABOUT REDUCTION
C   U   = M-VECTOR
C         INPUT :
C                 IGNORED
C         OUTPUT:
C                 INFORMATION ABOUT REDUCTION
C
C         FIND REFLECTION WHICH ZEROES A(I,K), I= K+1,.........,M
C
      DO 6 K= 1,N
         ALPHA= 0.0
         DO 1 I= K,M
            U(I)= A(I,K)
            ALPHA= ALPHA+U(I)*U(I)
 1       CONTINUE
         ALPHA= SQRT(ALPHA)
         IF(U(K).LT.0.0) ALPHA= -ALPHA
         U(K)= U(K)+ALPHA
         BETA= ALPHA*U(K)
         A(K,K)= -ALPHA
         IF(BETA.EQ.0.0.OR.K.EQ.N) GO TO 6
C
C   APPLY REFLECTION TO REMAINING COLUMNS OF A
         KP1= K+1
         DO 4 J= KP1,N
            GAMMA= 0.0
            DO 2 I= K,M
               GAMMA= GAMMA+U(I)*A(I,J)
 2          CONTINUE
            GAMMA= GAMMA/BETA
            DO 3 I= K,M
               A(I,J)= A(I,J)-GAMMA*U(I)
 3          CONTINUE
 4       CONTINUE
 6    CONTINUE
      RETURN
C
C   TRIANGULAR RESULT STORED IN A(I,J), I.LE.J
C   VECTORS DEFINING REFLECTIONS STORED IN U AND REST OF A
      END
#
```

Fig. 1.12.4 Listing of SUBROUTINE HECOMP (Reproduced from [1.14])

```
      SUBROUTINE HOLVE(MDIM,M,N,A,U,B)
      INTEGER MDIM,M,N
      REAL A(MDIM,N),U(M),B(M)
      REAL BETA,GAMMA,T
C
C
C   LEAST-SQUARE SOLUTION OF OVERDETERMINED SYSTEMS
C   FIND X THAT MINIMIZES NORM(A*X- B)
C
C   MDIM,M,N,A,U.  RESULTS FROM HECOMP
C   B= M-VECTOR
C      INPUT :
C              RIGHT HAND SIDE
C      OUTPUT:
C              FIRST N COMPONENTS = THE SOLUTION, X
C              LAST M-N COMPONENTS= TRANSFORMED RESIDUAL
C   DIVISION BY ZERO IMPLIES A NOT OF FULL RANK
C
C   APPLY REFLECTIONS TO B
C
C
      DO 3 K= 1,N
         T= A(K,K)
         BETA= -U(K)*A(K,K)
         A(K,K)= U(K)
         GAMMA= 0.0
         DO 1 I= K,M
            GAMMA= GAMMA+A(I,K)*B(I)
 1       CONTINUE
         GAMMA= GAMMA/BETA
         DO 2 I= K,M
            B(I)= B(I)-GAMMA*A(I,K)
 2       CONTINUE
         A(K,K)= T
 3    CONTINUE
C
C   BACK SUBSTITUTION
C
      DO 5 KB= 1,N
         K= N+1-KB
         B(K)= B(K)/A(K,K)
         IF(K.EQ.1) GO TO 5
         KM1= K-1
         DO 4 I= 1,KM1
            B(I)= B(I)-A(I,K)*B(K)
 4       CONTINUE
 5    CONTINUE
      RETURN
      END
#
```

Fig. 1.12.5 Listing of SUBROUTINE HOLVE (Reproduced from 1.14)

Exercise 1.12.3* Show that $\underset{\sim}{H}$, as defined in eqs. (1.12.5) is in fact a reflection, i.e. show that $\underset{\sim}{H}$ is orthogonal and the value of its determinant is -1. (Hint: Use the result of Exercise 1.12.2).

iii) m<n: Now, the linear system of equations $\underset{\sim}{A}\underset{\sim}{x}=\underset{\sim}{b}$ is studied when the number of unknowns is greater that the number of equations.

In this case, the system is underdetermined and has an infinity of solutions. However, as was discussed in Section 1.11, among those solutions, there is one, say $\underset{\sim}{x}_0$ whose Euclidean norm is a minimum. This is given by eq. (1.11.22), repeated here for ready reference.

$$\underset{\sim}{x}_0=\underset{\sim}{A}^T(\underset{\sim}{A}\underset{\sim}{A}^T)^{-1}\underset{\sim}{b} \tag{1.12.8}$$

One possible way of computing $\underset{\sim}{x}_0$ is given next:

a) Write eq. (1.11.20) in the form

$$\underset{\sim}{A}\underset{\sim}{A}^T\underset{\sim}{\lambda}=\underset{\sim}{b} \tag{1.12.9}$$

b) Using the LU decomposition method, find $\underset{\sim}{\lambda}$ from (1.12.9)

c) With $\underset{\sim}{\lambda}$ known from step ii), compute $\underset{\sim}{x}_0$ by matrix multiplication, as appearing in (1.11.19), i.e.

$$\underset{\sim}{x}_0=\underset{\sim}{A}^T\underset{\sim}{\lambda} \tag{1.12.10}$$

1.13 NUMERICAL SOLUTION OF NONLINEAR ALGEBRAIC SYSTEMS.

For several reasons, nonlinear systems are more difficult to deal with than linear systems. Considering the simplest case of equal number of equations and unknowns, there is no guarantee that the nonlinear system has a unique solution; in fact, there is no guarantee that the system has a solution at all.

---

* See Section 2.3 for more details on reflections.

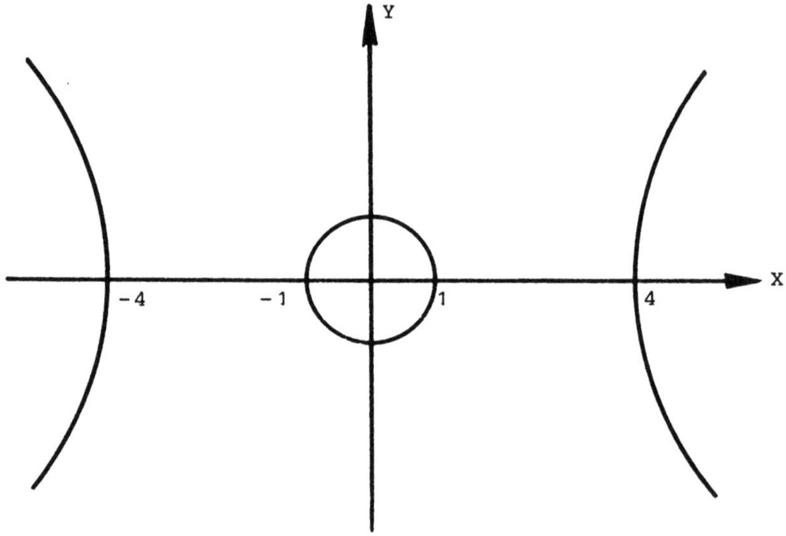

Fig 1.13.1 Non-intersecting hyperbola and circle

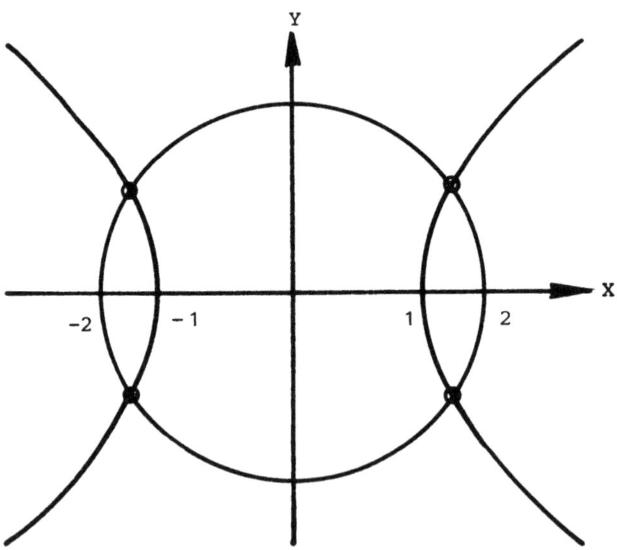

Fig 1.13.2 Intersections of a hyperbola and a circle

Example 1.13.1 The 2nd order nonlinear algebraic system

$$x^2 - y^2 = 16 \quad \text{(a)}$$

$$x^2 + y^2 = 1 \quad \text{(b)}$$

has no solution, for the hyperbola (a) does not intersect the circle (b), as is shown in Fig. 1.13.1

Example 1.13.2 The 2nd order nonlinear algebraic system

$$x^2 - y^2 = 1 \quad \text{(c)}$$

$$x^2 + y^2 = 4 \quad \text{(d)}$$

has four solutions, namely

$$x_1 = \sqrt{\frac{5}{2}}, \; y_1 = \sqrt{\frac{3}{2}}$$

$$x_2 = \sqrt{\frac{5}{2}}, \; y_2 = -\sqrt{\frac{3}{2}}$$

$$x_3 = -\sqrt{\frac{5}{2}}, \; y_3 = -\sqrt{\frac{3}{2}}$$

$$x_4 = -\sqrt{\frac{5}{2}}, \; y_4 = \sqrt{\frac{3}{2}}$$

which are the four points where the hyperbola (c) intersects the circle (d). These intersections appear in Fig. 1.13.2

The most popular method of solving a nonlinear algebraic system is the so-called Newton-Raphson method. First, the system of equations has to be written in the form

$$\underset{\sim}{f}(\underset{\sim}{x}) = \underset{\sim}{0} \quad (1.13.1)$$

where $\underset{\sim}{f}$ and $\underset{\sim}{x}$ are m- and n- dimensional vectors. For example, system (a), (b) of Example 1.13.1 can be written in the form

$$f_1(x_1,x_2) = x_1^2 - x_2^2 - 16 = 0 \quad \text{(a')}$$

$$f_2(x_1,x_2) = x_1^2 + x_2^2 - 1 = 0 \quad \text{(b')}$$

Here f_1 and f_2 are the components of the 2-dimensional vectors $\underset{\sim}{f}$ and x_1 and

x_2 (clearly, x and y have been replaced by x_1 and x_2, respectively) are the components of the 2-dimensional vector $\underset{\sim}{x}$. Next, the three cases, m=n, m>n and m<n, are discussed

First case: m=n

Let $\underset{\sim}{x}_0$ be known to be a "good" approximation to the solutions $\underset{\sim}{x}_r$ or a "guess". The expansion of $\underset{\sim}{f}(\underset{\sim}{x})$ about $\underset{\sim}{x}_0$ in a Taylor series yields

$$\underset{\sim}{f}(\underset{\sim}{x}_0+\Delta\underset{\sim}{x}) = \underset{\sim}{f}(\underset{\sim}{x}_0) + \underset{\sim}{f}'(\underset{\sim}{x}_0)\Delta\underset{\sim}{x} + \underset{\sim}{R} \tag{1.13.1}$$

If $\underset{\sim}{x}_0 + \Delta\underset{\sim}{x}$ is an even better approximation to $\underset{\sim}{x}_r$, then $\Delta\underset{\sim}{x}$ must be small and so, only linear terms could be retained in (1.13.2) and, of course, $\underset{\sim}{f}(\underset{\sim}{x}_0+\Delta\underset{\sim}{x})$ must be closer to $\underset{\sim}{0}$ than is $\underset{\sim}{f}(\underset{\sim}{x}_0)$. Under these assumptions, $\underset{\sim}{f}(\underset{\sim}{x}_0+\Delta\underset{\sim}{x})$ can be assumed to be zero and (1.13.2) leads to

$$\underset{\sim}{f}(\underset{\sim}{x}_0) + \underset{\sim}{f}'(\underset{\sim}{x}_0)\Delta\underset{\sim}{x} = \underset{\sim}{0} \tag{1.13.3}$$

In the above equation $\underset{\sim}{f}'(\underset{\sim}{x}_0)$ is the value of the gradient of $\underset{\sim}{f}(\underset{\sim}{x})$, $\underset{\sim}{f}'(\underset{\sim}{x})$, at $\underset{\sim}{x} = \underset{\sim}{x}_0$. This gradient is an nxn matrix, $\underset{\sim}{J}$, whose (k,ℓ) element is

$$J_{k\ell} = \frac{\partial f_k}{\partial x_\ell} \tag{1.13.4}$$

If the Jacobian matrix $\underset{\sim}{J}$ is nonsingular, it can be inverted to yield

$$\Delta\underset{\sim}{x} = -\underset{\sim}{J}^{-1}(\underset{\sim}{x}_0)\underset{\sim}{f}(\underset{\sim}{x}_0) \tag{1.13.5}$$

Of course, $\underset{\sim}{J}$ need not actually be inverted, for $\Delta\underset{\sim}{x}$ can be obtained via the LU decomposition method from eq. (1.13.3) written in the form

$$\underset{\sim}{J}(\underset{\sim}{x}_0)\Delta\underset{\sim}{x}=-\underset{\sim}{f}(\underset{\sim}{x}_0) \tag{1.13.6}$$

With the value of $\Delta\underset{\sim}{x}$ thus obtained, the improved value of $\underset{\sim}{x}$, is computed as

$$\underset{\sim}{x}_1 = \underset{\sim}{x}_0 + \Delta\underset{\sim}{x}$$

In general, at the kth iteration, the new value $\underset{\sim}{x}_{k+1}$ is computed from the formula

$$\underset{\sim}{x}_{k+1} = \underset{\sim}{x}_k - \underset{\sim}{J}(\underset{\sim}{x}_k)^{-1}\underset{\sim}{f}(\underset{\sim}{x}_k) \tag{1.13.7}$$

which is the Newton-Raphson iterative scheme. The procedure is stopped when a convergence criterion is met. One possible criterion is that the norm of $\underset{\sim}{f}(\underset{\sim}{x}_k)$ reaches a value below certain prescribed tolerance, i.e.

$$||\underset{\sim}{f}(\underset{\sim}{x}_k)|| \leq \varepsilon \qquad (1.13.8)$$

where ε is the said tolerance. On the other hand, it can also happen that at iteration k, the norm of the increment becomes smaller than the tolerance. In this case, even if the convergence criterion (1.13.8) is not met, it is useless to perform more interations. Thus, it is more reasonable to verify first that the norm of the correction does not become too small before proceeding further, and stop the procedure if both $||\underset{\sim}{f}(\underset{\sim}{x}_k)||$ and $||\Delta\underset{\sim}{x}_k||$ are small enough, in which case, convergence is reached.

If only $||\Delta\underset{\sim}{x}_k||$ goes below the imposed tolerance, do not accept the corresponding $\underset{\sim}{x}_k$ as the solution. The conditions under which the procedure converges are discussed in (1.15). These conditions, however, cannot be verified easily, in general. What is advised to do is to try different initial guesses $\underset{\sim}{x}_0$ till convergence is reached and to stop the procedure if either

i) too many iterations have been performed

or

ii) $||\Delta\underset{\sim}{x}_k|| \leq \varepsilon$ but $||\underset{\sim}{f}(\underset{\sim}{x}_k)|| > \varepsilon$

If the method of Newton-Raphson converges for a given problem, it does so <u>quadratically</u>, i.e. two digits are gained per iteration during the aproximation to the solution. It can happen, however, that the procedure does not converge monotonically, in which case,

$$||\underset{\sim}{f}(\underset{\sim}{x}_{k-1})|| > ||\underset{\sim}{f}(\underset{\sim}{x}_k)||$$

thus giving rise to strong oscillations and, possibly, divergence. One way to cope with this situation is to introduce <u>damping</u>, i.e. instead of using the whole computed increment $\Delta \underset{\sim}{x}_k$, use a fraction of it, i.e. at the kth iteration, for i=0,1,..., max, instead of using formula (1.13.7) to compute the next value $\underset{\sim}{x}_{k+1}$, use

$$\underset{\sim}{x}_{k+1} = \underset{\sim}{x}_k - \alpha^i \underset{\sim}{J}(\underset{\sim}{x}_k)^{-1} \underset{\sim}{f}(\underset{\sim}{x}_k) \qquad (1.13.9)$$

where α is a real number between 0 and 1. For a given k, eq. (1.13.9) represents the damping part of the procedure, which is stopped when

$$||\underset{\sim}{f}(\underset{\sim}{x}_{k+1})_i|| < ||\underset{\sim}{f}(\underset{\sim}{x}_k)||$$

The algorithm is summarized in the flow chart of Fig 1.13.3 and implemented in the subroutine NRDAMP appearing in Fig 1.13.4

Second case: m>n

In this case the system is overdetermined and it is not possible, in general, to satisfy all the equations. What can be done, however, is to find that $\underset{\sim}{x}_0$ which minimizes $||\underset{\sim}{f}(\underset{\sim}{x})||$.

This problem arises, for example, when one tries to design a planar four-bar linkage to guide a rigid body through more than five configurations.

To find the minimizing $\underset{\sim}{x}_0$, define first which norm of $\underset{\sim}{f}(\underset{\sim}{x})$ is desired to minimize. One norm which has several advantages is the Euclidean norm, already discussed in case i of Section 1.11, where the linear least-square problem was discussed. In the context of nonlinear systems of equations, minimizing the quadratic norm of $\underset{\sim}{f}(\underset{\sim}{x})$ leads to the nonlinear least-square problem. The problem is then to find the minimum of the scalar function

$$\phi(\underset{\sim}{x}) = ||\underset{\sim}{f}(\underset{\sim}{x})||^2 = \underset{\sim}{f}^T(\underset{\sim}{x})\underset{\sim}{f}(\underset{\sim}{x}) \qquad (1.13.10)$$

As already discussed in Section 1.10, for this function to reach a minimum, it must first reach a stationary point, i.e. its gradient must vanish. Thus,

$$\phi'(\underset{\sim}{x}) = 2\,\underset{\sim}{J}^T(\underset{\sim}{x})\underset{\sim}{f}(\underset{\sim}{x}) \qquad (1.13.11)$$

where $\underset{\sim}{J}(\underset{\sim}{x})$ is the Jacobian matrix of $\underset{\sim}{f}$ with respect to $\underset{\sim}{x}$, i.e. an mxn matrix

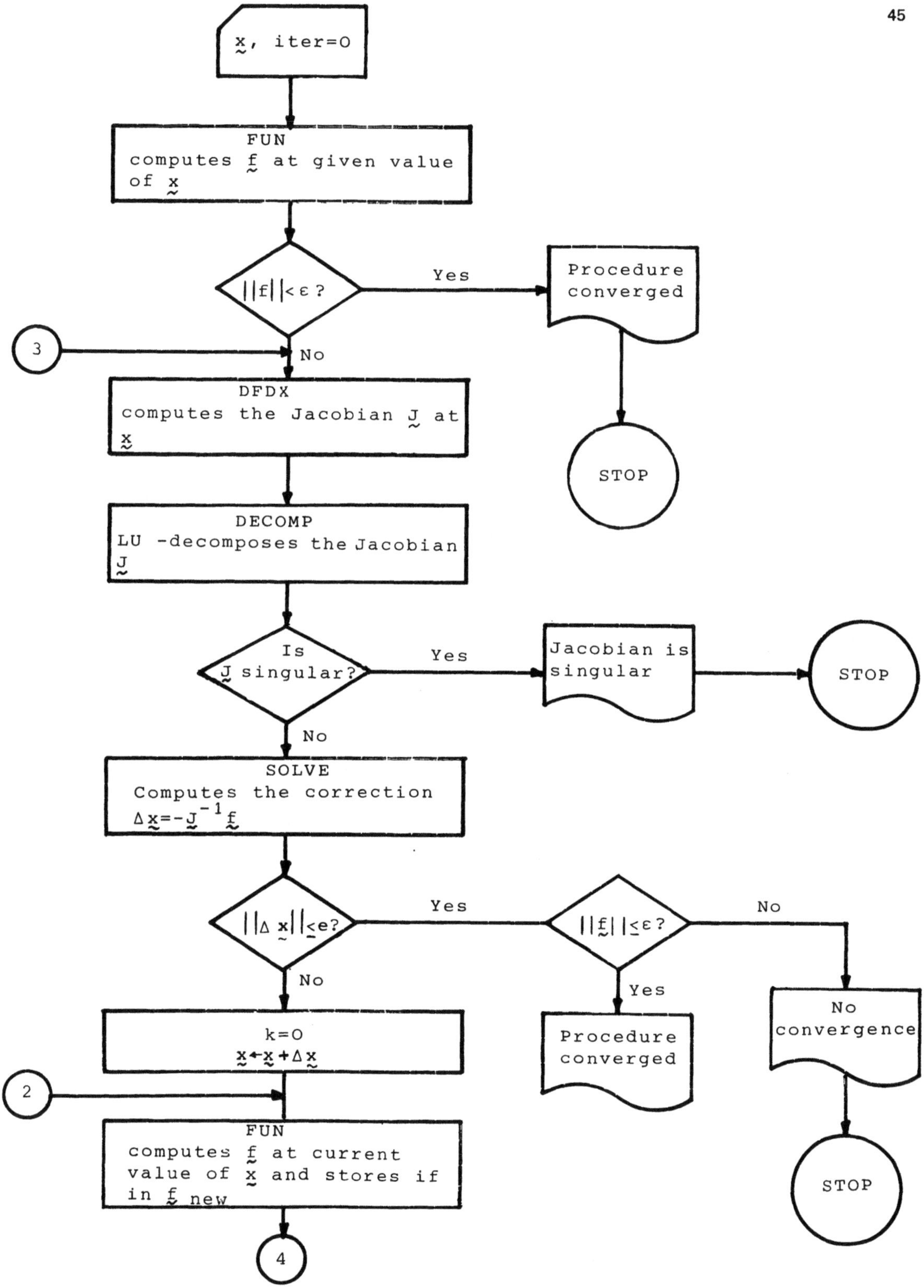

Fig. 1.13.3 Flow diagram to solve a nonlinear algebraic system with as many equations as unknowns, via the method of Newton-Raphson with damping (first part)

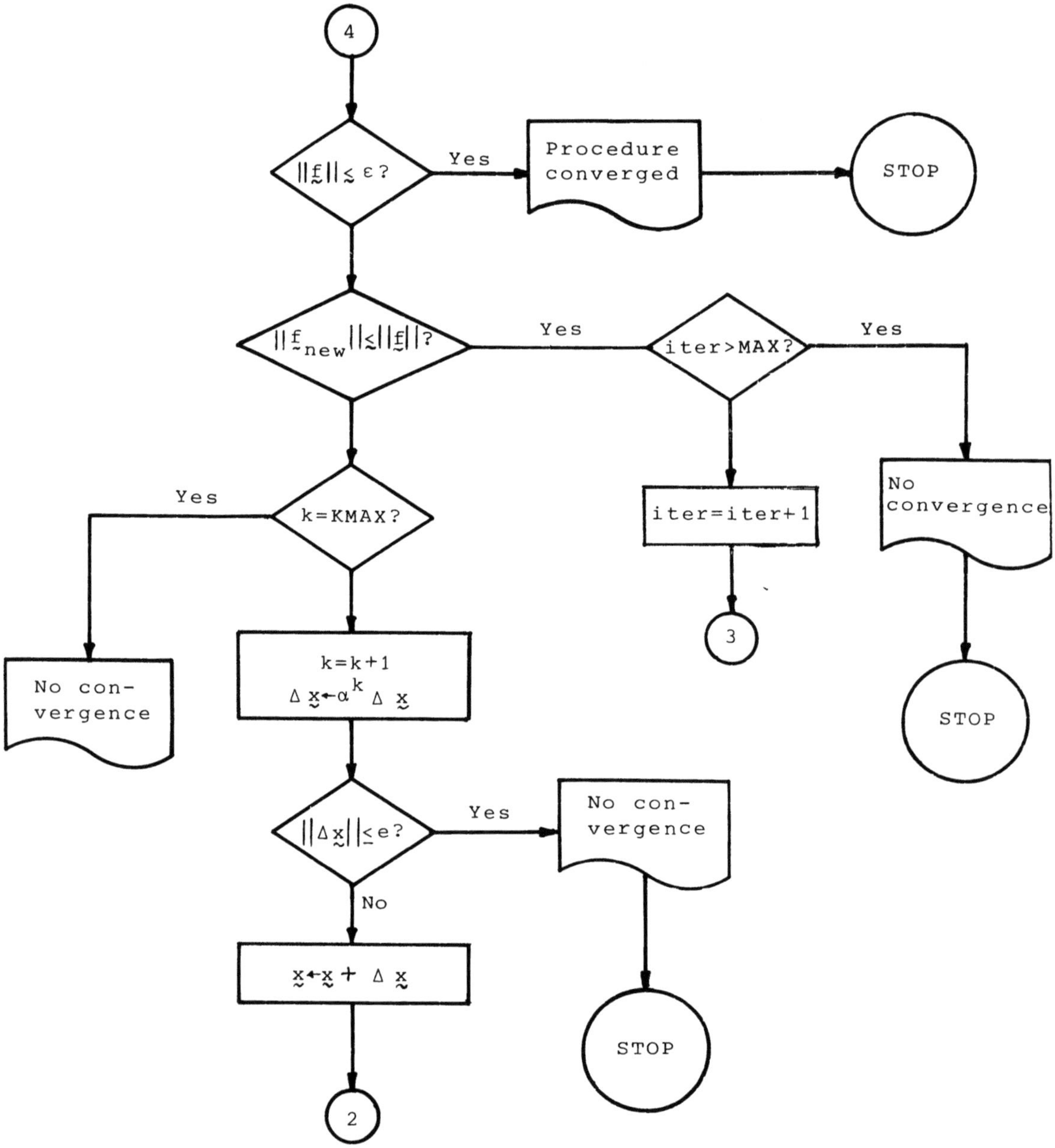

Note:

ε = tolerance imposed on $\underset{\sim}{f}$

e = tolerance imposed on $\underset{\sim}{x}$

Fig. 1.13.3 Flow diagram to solve a nonlinear algebraic system with as many equations as unknowns, via the method of Newton-Raphson with damping (second part)

```
      SUBROUTINE NRDAMP(X,FUN,DFDX,P,TOLX,TOLF,DAMP,N,ITER,MAX,KMAX)
      REAL X(1),P(1),DF(12,12),DELTA(12),F(12)
      INTEGER IP(12)
C  THIS SUBROUTINE FINDS THE ROOTS OF A NONLINEAR ALGEBRAIC SYSTEM OF
C  ORDER N, VIA NEWTON-RAPHSON METHOD(ISAACSON E. AND KELLER H. B.
C  ANALYSIS OF NUMERICAL METHODS, JOHN WILEY AND SONS, INC., NEW YORK
C  1966,PP. 85-123)WITH DAMPING. SUBROUTINE PARAMETERS
C  X          = N-VECTOR OF UNKNOWS.
C  FUN        = EXTERNAL SUBROUTINE WHICH COMPUTES VECTOR F, CONTAINING
C               THE FUNTIONS WHOSE ROOTS ARE OBTAINED.
C  DFDX       = EXTERNAL SUBROUTINE WHICH COMPUTES THE JACOBIAN MATRIX
C               OF VECTOR F WHIT RESPECT TO X.
C  P          = AN AUXILIAR VECTOR OF SUITABLE DIMENSION. IT CONTAINS
C               THE PARAMETERS THAT EACH PROBLEM MAY REQUIERE.
C  TOLX       = POSITIVE SCALAR, THE TOLERANCE IMPOSED ON THE APPROXIMA-
C               TION TO X.
C  TOLF       = POSITIVE SCALAR, THE TOLERANCE IMPOSED ON THE APPROXIMA-
C               TION TO F.
C  DAMP        =THE DAMPING VALUE, PROVIDED BY THE USER SUCH THAT
C               0.LT.DAMP.LT.1 .
C  ITER        =NUMBER OF ITERATION BEING EXECUTED.
C  MAX         =MAXIMUM NUMBER OF ALLOWED ITERATIONS.
C  KMAX        =MAXIMUM NUMBER OF ALLOWED DAMPINGS PER ITERATION. IT IS
C               PROVIDED BY THE USER.
C  FUN AND DFDX ARE SUPPLIED BY THE USER.
C  SUBROUTINES 'DECOMP' AND 'SOLVE' SOLVE THE NTH, ORDER LINEAR
C  ALGEBRAIC SYSTEM DF(X)*DELTA=F(X), DELTA BEING THE CORRECTION TO
C  THE K-TH ITERATION. THE METHOD USED IS THE LU DECOMPOSITION (MOLER
C  C.B. MATRIX COMPUTATIONS WITH FORTRAN AND PAGING. COMMUNICATIONS OF
C  THE A.C.M., VOLUME 15, NUMBER 4, APRIL 1972).
C
      KONT=1
      ITER=0
      CALL FUN(X,F,P,N)
      FNOR1=FNORM(F,N)
      IF(FNOR1.LE.TOLF) GO TO 4
    1 CALL DFDX(X,DF,P,N)
      CALL DECOMP(N,N,DF,IP)
      K=0
C
C  IF THE JACOBIAN MATRIX IS SINGULAR, THE SUBROUTINE RETURNS TO THE
C  MAIN PROGRAM,. OTHERWISE, IT PROCEEDS FURTHER.
C
      IF(IP(N).EQ.0) GO TO 14
      CALL SOLVE (N,N,DF,F,IP)
      DO 2 I=1,N
    2 DELTA(I)=F(I)
      DELNOR=FNORM(DELTA,N)
      IF(DELNOR.LT.TOLX) GO TO 4
      DO 3 I=1,N
    3 X(I)=X(I)-DELTA(I)
      GO TO 5
#
```

Fig 1.13.4 Listing of SUBROUTINE NRDAMP

```
    4 FNOR2=FNOR1
      GO TO 6
    5 CALL FUN(X,F,P,N)
      KONT=KONT+1
      FNOR2=FNORM(F,N)
    6 IF(FNOR2.LE.TOLF) GO TO 11
C
C   TESTING THE NORM OF THE FUNCTION. IF THIS DOES NOT DICREASE
C   THEN DAMPING IS INTRODUCED.
C
      IF(FNOR2.LT.FNOR1) GO TO 10
      IF(K.EQ.KMAX) GO TO 16
      K=K+1
      DO 8 I=1,N
            IF(K.GE.2) GO TO 7
            DELTA(I)=(DAMP-1.)*DELTA(I)
            GO TO 8
    7       DELTA(I)=DAMP*DELTA(I)
    8 CONTINUE
      DELNOR=FNORM(DELTA,N)
      IF(DELNOR.LE.TOLX) GO TO 16
      DO 9 I=1,N
    9 X(I)=X(I)-DELTA(I)
      GO TO 5
   10 IF(ITER.GT.MAX) GO TO 16
      ITER=ITER+1
      FNOR1=FNOR2
      GO TO 1
   11 WRITE(6,110) ITER,FNOR2,KONT
   12 DO 13 I=1,N
   13 WRITE(6,120) I,X(I)
      RETURN
   14 WRITE(6,130) ITER,KONT
      GO TO 12
   16 WRITE(6,140) ITER,FNOR2,KONT
      GO TO 12
  110 FORMAT(5X,'AT ITERATION NUMBER ',I3,' THE NORM OF THE FUNCTION IS'
     -,E20.6/5X,'THE FUNCTION WAS EVALUATED ',I3,' TIMES'/
     -5X,'PROCEDURE CONVERGED, THE SOLUTION BEING :'/)
  120 FORMAT(5X,'X(',I3,')=',E20.6)
  130 FORMAT(5X,'AT ITERATION NUMBER ',J3,'THE JACOBIAN MATRIX'
     -' IS SINGULAR.'/5X,'THE FUNCTION WAS EVALUATED ',I3,' TIMES'/
     -5X,'THE CURRENT VALUE OF X IS :',/)
  140 FORMAT(10X,'PROCEDURE DIVERGES AT ITERATION NUMBER ',I3/10X,
     -'THE NORM OF THE FUNCTION IS ',E20.6/10X,
     -'THE FUNCTION WAS EVALUATED ',I3,' TIMES'/10X,
     -'THE CURRENT VALUE OF X IS :'/)
      END
#
```

Fig. 1.13.4 Listing of SUBROUTINE NRDAMP (Continued).

Exercise 1.13.1. Derive the expression (1.13.11)

In order to compute the value of $\underset{\sim}{x}$ that zeroes the gradient (1.13.11) proceed iteratively, as next outlined. Expand $\underset{\sim}{f}(\underset{\sim}{x})$ around $\underset{\sim}{x}_0$:

$$\underset{\sim}{f}(\underset{\sim}{x}_0+\Delta\underset{\sim}{x}) = \underset{\sim}{f}(\underset{\sim}{x}_0) + \underset{\sim}{f}'(\underset{\sim}{x}_0)\Delta\underset{\sim}{x} + \underset{\sim}{R} \tag{1.13.12}$$

If $\underset{\sim}{x}_0+\Delta\underset{\sim}{x}$ is a better approximation to the value that minimizes the Euclidean norm of $\underset{\sim}{f}(\underset{\sim}{x})$, and if in addition $||\Delta\underset{\sim}{x}||$ is small enough, $\underset{\sim}{R}$ can be neglected in eq. (1.13.12) and as trying to set the whole expression equal to zero, the following equation is obtained

$$\underset{\sim}{f}'(\underset{\sim}{x}_0)\Delta\underset{\sim}{x} = -\underset{\sim}{f}(\underset{\sim}{x}_0)$$

or, denoting by $\underset{\sim}{J}$ the Jacobian matrix $\underset{\sim}{f}'(\underset{\sim}{x})$,

$$\underset{\sim}{J}(\underset{\sim}{x}_0)\Delta\underset{\sim}{x} = -\underset{\sim}{f}(\underset{\sim}{x}_0)$$

which is an overdetermined linear system. As discussed in Section 1.11, such a system has in general no solution, but a value of $\Delta\underset{\sim}{x}$ can be computed which minimizes the quadratic norm of the error $J(\underset{\sim}{x}_0)\Delta\underset{\sim}{x} + \underset{\sim}{f}(\underset{\sim}{x}_0)$. This value is given by the expression (1.11.8) as

$$\Delta\underset{\sim}{x} = -(\underset{\sim}{J}^T\underset{\sim}{J})^{-1}\underset{\sim}{J}^T\underset{\sim}{f}$$

In general, at the kth iteration, compute $\Delta\underset{\sim}{x}_k$ as

$$\Delta\underset{\sim}{x}_k = -\left(\underset{\sim}{J}^T(\underset{\sim}{x}_k)\underset{\sim}{J}(\underset{\sim}{x}_k)\right)^{-1}\underset{\sim}{J}^T(\underset{\sim}{x}_k)\underset{\sim}{f}(\underset{\sim}{x}_k) \tag{1.13.13}$$

and stop the procedure when $||\Delta\underset{\sim}{x}_k||$ becomes smaller than a prescribed tolerance, thus indicating that the procedure converged. In fact, if $\Delta\underset{\sim}{x}_k$ vanishes, unless $(\underset{\sim}{J}^T\underset{\sim}{J})^{-1}$ becomes infinity, this means that $\underset{\sim}{J}^T\underset{\sim}{f}$ vanishes. But if this product vanishes, then from eq. (1.13.11), the gradient $\phi'(\underset{\sim}{x})$ also vanishes, thus obtaining a stationary point of the quadratic norm of $\underset{\sim}{f}(\underset{\sim}{x})$.

In order to accelerate the convergence of the procedure, damping can also be introduced. This way, instead of computing $\Delta\underset{\sim}{x}_k$ from eq. (1.13.13), compute it from

$$\Delta x_k = -\alpha^i \left(J^T(x_k) J(x_k)\right)^{-1} J^T(x_k) f(x_k) \tag{1.13.14}$$

for i = 0, 1,..., max and stop the damping when

$$||\phi(x_{k+1,i})|| < ||\phi(x_{k,i})||$$

The algorithm is illustrated with the flow diagram of Fig 1.13.5 and implemented with the subroutine NRDAMC, appearing in Fig 1.13.6

Third case: m<n

The system, in this case, is underdetermined and infinitely many solutions can be expected to exist. Out of these solutions, however, one can choose that with a minimum norm, thus converting the problem into a nonlinear quadratic programming problem, stated as

$$\text{Minimize } x^T x \tag{1.13.15a}$$

$$\text{subject to } f(x) = 0 \tag{1.13.15b}$$

One way to find the minimizing x_0, of problem (1.13.15) is via the method of Lagrange multipliers. Thus, define a new objective function

$$\psi(x) = x^T x + \lambda^T f(x) \tag{1.13.16}$$

which is stationary at x_0 where its gradient vanishes. Thus,

$$\psi(x_0) = 2x_0 + f'^T(x_0)\lambda = 0 \tag{1.13.17}$$

The systems of equations (1.13.15b) and (1.13.17) now represent a larger system of m+n equations (m in (1.13.10b) and n in (1.13.12)) in m+n unknowns (m components of λ and n components of x). Hence, the problem now reduces to the first case and so can be solved by application of the subroutine NRDAMP.

Exercise 1.13.2 Let

$$f(x) = \frac{1}{2}x_1^2 + (x_1-x_2)^2 + (x_2-x_3)^2 + (x_3-x_4)^2 + \\ +\beta\left(\frac{7}{2}\cos x_1 + \frac{5}{2}\cos x_2 + \frac{3}{2}\cos x_3 + \frac{1}{2}\cos x_4\right)$$

be a scalar function of a vector argument $x = (x_1, x_2, x_3, x_4)^T$. Find its

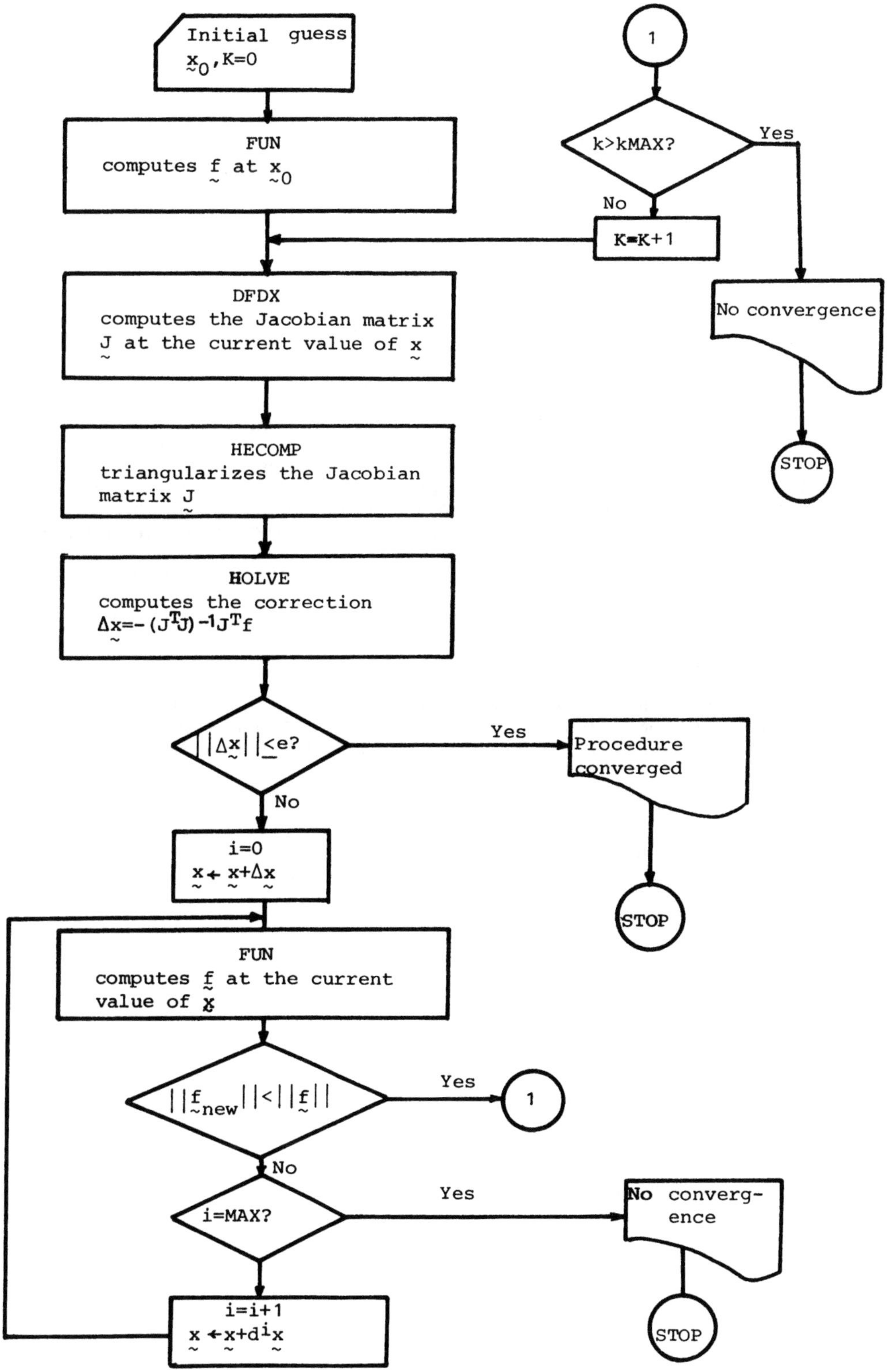

Fig 1.13.5 Flow diagram to compute the least-square solution to an overdetermined nonlinear algebraic system.

```
      SUBROUTINE NRDAMC(X,FUN,DFDX,P,TOL,DAMP,N,M,ITER,MAX,KMAX)
      REAL X(2),F(3),DF(3,2),P,U(3),DELTA(3),FNORM1,FNORM2,
     -     DELNOR
C
C   THIS PROGRAM OBTAINS THE LEAST SQUARE SOLUTION TO THE NONLINEAR
C   SYSTEM F(X)= 0, WHERE F AND X ARE M-AND N-DIMENSIONAL VECTORS, M
C   BEING GREATER THAN N. THE PROCEDURE IS ITERATIVE, AND AT EACH
C   ITERATION FINDS THE LEAST SQUARE SOLUTION TO THE LINEAR SYSTEM
C   DF*DELTA= -F, WHERE DF IS THE JACOBIAN M X N MATRIX OF THE ORIGINAL
C   SYSTEM, COMPUTED AT THE CURRENT VALUE OF X. THE LINEAR LEAST SQUARE
C   SOLUTION AT EACH ITERATION IS FOUND VIA HOUSEHOLDER REFLECTIONS (
C   BJOERCK A. AND G. DAHLQUIST, NUMERICAL METHODS, PRENTICE HALL,
C   ENGLEWOOD CLIFFS, N.J., 1974, PP. 201-206, 443-444).
C   PARAMETERS :
C       X     ..= N-DIMENSIONAL VECTOR OF UNKNOWNS.
C       F     ..= M-DIMENSIONAL VECTOR OF FUNCTIONS WHOSE EUCLIDEAN NORM
C                 IS TO BE MINIMIZED.
C       DF    ..= M X N JACOBIAN MATRIX OF F WITH RESPECT TO X.
C       P     ..= VECTOR OF PARAMETERS APPEARING IN FUN AND/OR DFDX. ITS
C                 DIMENSION VARIES FROM PROBLEM TO PROBLEM.
C       TOL   ..= A REAL POSITIVE "SMALL" VARIABLE DENOTING THE IMPOSED
C                 TOLERANCE. IT IS SUPPLIED BY THE USER.
C       DAMP  ..= A REAL POSITIVE VARIABLE, 0.LT. DAMP .LT.1. IT DENOTES
C                 THE DAMPING FACTOR AND IS SUPPLIED BY THE USER.
C       ITER  ..= AN INTEGER VARIABLE DENOTING THE NUMBER OF THE CURRENT
C                 ITERATION.
C       MAX   ..= AN INTEGER VARIABLE DENOTING THE MAXIMUM NUMBER OF
C                 ALLOWED ITERATIONS.
C       KMAX  ..= AN INTEGER VARIABLE DENOTING THE MAXIMUM NUMBER OF
C                 ALLOWED DAMPINGS.
C   SUBSIDIARY SUBROUTINES :
C
C   HECOMP  = TRIANGULARIZES A RECTANGULAR MATRIX BY HOUSEHOLDER
C             REFLECTIONS (MOLER C. B., MATRIX EIGENVALUE AND LEAST-
C             SQUARE COMPUTATIONS, COMPUTER SCIENCE DEPARTAMENT,
C             STANFORD UNIVERSITY. MARCH, 1973.)
C   HOLVE   = SOLVES TRIANGULARIZED SYSTEM BY BACK-SUBSTITUTION (MOLER
C             C. B., OP. CIT.)
C   FUN     = COMPUTES F.
C   DFDX    = COMPUTES DF.
C   FNORM   = COMPUTES THE MAXIMUM NORM OF A VECTOR.
C
C
      ITER=0
      CALL FUN(X,F,P,M,N)
 1    ITER=ITER+1
      IF(ITER.GT.MAX) GO TO 10
C
C   FORMS LINEAR LEAST SQUARE PROBLEM
      FNORM1=FNORM(F,M)
      CALL DFDX(X,DF,P,M,N)
      CALL HECOMP(M,M,N,DF,U)
      CALL HOLVE(M,M,N,DF,U,F)
```

Fig 1.13.6 Listing of SUBROUTINE NRDAMC

```
C
C    COMPUTES CORRECTION BETWEEN TWO SUCCESSIVE ITERATIONS
        DO 2 I=1,M
             DELTA(I)=F(I)
   2    CONTINUE
        DELNOR=FNORM(DELTA,N)
        IF(DELNOR.LT.TOL) GO TO 8
        K=1
C
C    IF DELNOR IS STILL LARGE, PERFORMS CORRECTION TO VECTOR X
   3    DO 4 I=1,N
             X(I)=X(I)-DELTA(I)
   4    CONTINUE
        CALL FUN(X,F,P,M,N)
        FNORM2=FNORM(F,M)
C
C    TESTING THE NORM OF THE FUNCTION F AT CURRENT VALUE OF X. IF THIS
C    DOES NOT DECREASE, THEN DAMPING IS INTRODUCED.
        IF(FNORM2.LT.TOL) GO TO 8
        IF(FNORM2.LT.FNORM1) GO TO 1
        IF(K.GT.KMAX) GO TO 7
        DO 6 I=1,N
             IF(K.GE.2) GO TO 5
             DELTA(I)=(DAMP-1.)*DELTA(I)
             GO TO 6
   5         DELTA(I)=DAMP*DELTA(I)
   6    CONTINUE
        K=K+1
        GO TO 3
   7    WRITE(6,101)DAMP
C
C    AT THIS ITERATION THE NORM OF THE FUNCTION CANNOT BE DECREASED
C    AFTER KMAX DAMPINGS, DAMP IS SET EQUAL TO -1 AND THE SUBROUTINE
C    RETURNS TO THE MAIN PROGRAM.
        DAMP=-1.
        RETURN
   8    WRITE(6,102)FNORM2,ITER,K
        DO 9 I=1,N
             WRITE(6,103) I,X(I)
   9    CONTINUE
        RETURN
  10    WRITE(6,104)ITER
        RETURN
 101    FORMAT(5X,'DAMP =',F10.5,5X,'NO CONVERGENCE WITH THIS DAMPING',
     -          ' VALUE'/)
 102    FORMAT(/5X,'CONVERGENCE REACHED. NORM OF THE FUNCTION :',
     -          F15.6//5X,'NUMBER OF ITERATIONS :',I3,5X,'NUMBER OF ',
     -          'DAMPINGS AT THE LEAST ITERATION :',I3//5X,'THE SOLUTION'
     -          ,' IS :'/)
 103    FORMAT(5X,2HX(I2,3H)= F15.5/)
 104    FORMAT(10X,'NO CONVERGENCE WITH',I3,' ITERATIONS'/)
        END
#
```

Fig 1.13.6 Listing of SUBROUTINE NRDAMC (Continued)

stationary points and decide whether each is either a maximum, a minimum or a saddle point, for $\beta = 1, 10, 50$.

Note: $f(\tilde{x})$ could represent the potential energy of a mechanical system. In this case the stationary points correspond to the following equilibrium states: minima yield a stable equilibrium state, whereas maxima and saddle points yield unstable states.

Example 1.13.3 Find the point closest to all three curves of Fig 1.13.7. These curves are the parabola(P), the circle(C) and the hyperbola(H) with the following equations:

$$y = \frac{1}{2.4} x^2 \qquad \text{(P)}$$

$$x^2 + y^2 = 4 \qquad \text{(C)}$$

$$x^2 - y^2 = 1 \qquad \text{(H)}$$

From Fig 1.13.7 it is clear that no single pair (x,y) satisfies all three equations simultaneously. There exist points of coordinates x_0, y_0, however, that minimize the quadratic norm of the error of the said equations. These can be found with the aid of SUBROUTINE NRDAMC. A program was written that calls NRDAMC, HECOMP and HOLVE to find the least-square solution to eqs. (P), (C) and (H). The found solutions were:

First solution: x=-1.61537, y=1.17844

Second solution: x= 1.61537, y=1.17844

which are shown in Fig 1.13.7. These points have symmetrical locations, as expected, and lie almost on the circle at abount equal distances from A_i and C_i and B_i and D_i (i=1,2)

The maximum error of the foregoing approximation was computed as 0.22070

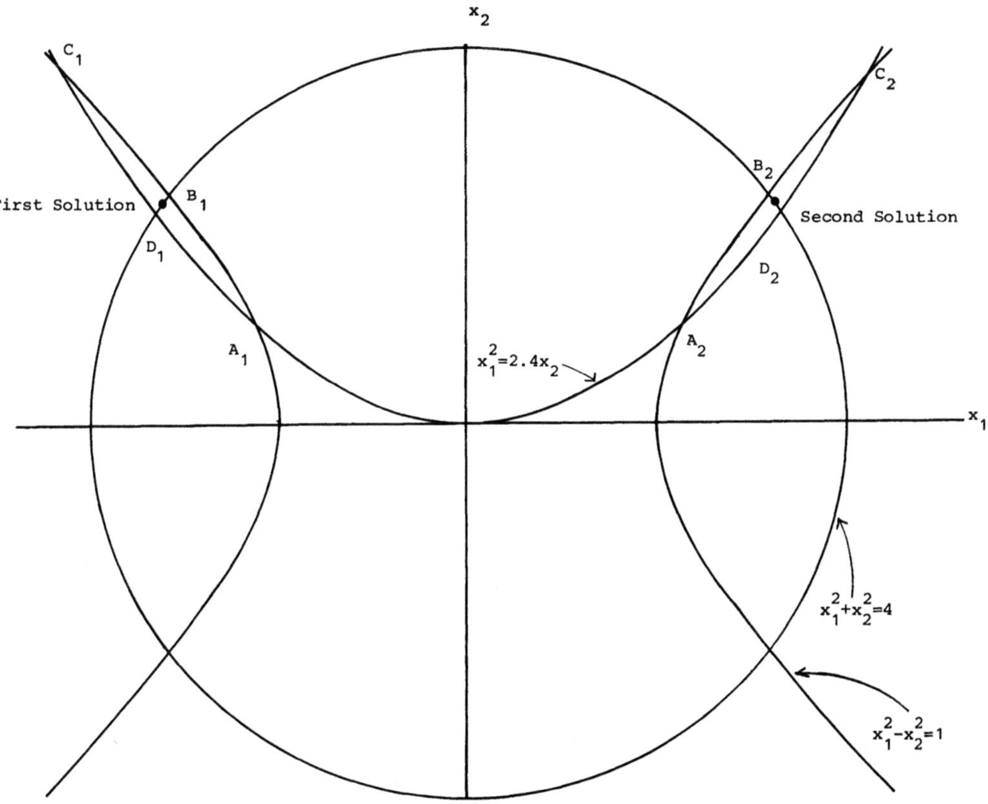

Fig 1.13.7 Location of the point closest to a parabola, a circle and a hyperbola.

REFERENCES

1.1 Lang S., Linear Algebra, Addison-Wesley Publishing Co., Menlo Park, 1970, pp. 39 and 40.

1.2 Lang S., op. cit., pp. 99 and 100

1.3 Finkbeiner, D.F., Matrices and Linear Transformations, W.H. Freeman and Company, San Francisco, 1960, pp. 139-142

1.4 Halmos, P.R., Finite-Dimensional Vector Spaces, Springer-Verlag, N. York, 1974.

1.5 Businger P. and G.H. Golub, "Linear Least Squares Solutions by Householder Transformations", in Wilkinson J.H. and C. Reinsch, eds., Handbook for Automatic Computation, Vol. II, Springer-Verlag, N. York, 1971, pp. 111-118

1.6 Stewart, G.W., Introduction to Matrix Computations, Academic Press, N.York, 1973, pp. 208-249.

1.7 Soderstrom T. and G.W. Stewart, "On the numerical properties of an iterative method for computing the Moore-Penrose generalized inverse", SIAM J. on Numerical Analysis, Vol. II, No. 1, March 1974.

1.8 Brand L., Advanced Calculus, John Wiley and Sons, Inc., N. York, 1955, pp. 147-197.

1.9 Luenberger, D.G., Optimization by Vector Space Methods, John Wiley and Sons, Inc., N. York, 1969, pp. 8, 49-52

1.10 Varga, R.S., Matrix Iterative Analysis, Prentice Hall, Inc., Englewood Cliffs, 1962, pp. 56-160

1.11 Forsythe, G.E. and C.B. Moler, Computer Solution of Linear Algebraic Systems, Prentice Hall, Inc., Englewood Cliffs, 1967, pp. 27-33

1.12 Moler C.B., "Algorithm 423. Linear Equation Solver (F 4)" Communications of the ACM, Vol. 15, Number 4, April 1973, p. 274.

1.13 Björck Å. and G. Dahlquist, Numerical Methods, Prentice-Hall, Inc., Englewood Cliffs, 1974, pp. 201-206.

1.14 Moler C.B., Matrix Eigenvalue and Least Square Computations, Computer Science Departament, Stanford University, Stanford, California, 1973 pp. 4.1-4.15

1.15 Isaacson, E. and H. B. Keller, Analysis of Numerical Methods, John Wiley and Sons, Inc., N. York, 1966, pp. 85-123

1.16 Angeles, J., "Optimal synthesis of linkages using Householder reflections", Proceedings of the Fifth World Congress on the Theory of Machines and Mechanisms, vol. I, Montreal, Canada, July 8-13, 1979, pp. 111-114.

2. Fundamentals of Rigid-Body Three-Dimensional Kinematics

2.1 INTRODUCTION. The rigid body is defined as a continuum for which, under any physically possible motion, the distance between any pair of its points remains unchanged. The rigid body is a mathematical abstraction which models very accurately the behaviour of a wide variety of natural and man-made mechanical systems under certain conditions. However, as such it does not exist in nature, as neither do the elastic body nor the perfect fluid. The theorems related to rigid-body motions are rigorously proved and the foundations for the analysis of the motion of systems of coupled rigid bodies (linkages) are laid down. The main results in this chapter are the theorems of Euler, Chasles, the one on the existence of an instantaneous screw, the Theorem of Aronhold-Kennedy and that of Coriolis.

2.2 MOTION OF A RIGID BODY.

Consider a subset D of the Euclidean three-dimensional physical space occupied by a rigid body, and let $\underset{\sim}{x}$ be the position vector of a point of that body. A rigid-body motion is a mapping $\underset{\sim}{M}$ which maps every point $\underset{\sim}{x}$ of D into a unique point $\underset{\sim}{y}$ of a set D', called "the image" of D under $\underset{\sim}{M}$,

$$\underset{\sim}{M} : \underset{\sim}{x} \rightarrow \underset{\sim}{y} \tag{2.2.1}$$

such that, for any pair $\underset{\sim}{x}_1$ and $\underset{\sim}{x}_2$, mapped by $\underset{\sim}{M}$ into $\underset{\sim}{y}_1$ and $\underset{\sim}{y}_2$, respectively, one has

$$||\underset{\sim}{x}_2 - \underset{\sim}{x}_1|| = ||\underset{\sim}{y}_2 - \underset{\sim}{y}_1|| \tag{2.2.2}$$

The symbol $||.||$ denotes the Euclidean norm* of the space under consideration.

It is next shown that, under the above definition, a rigid-body motion preserves the angle between any two lines of a body. Indeed, let $\underset{\sim}{x}_1$, $\underset{\sim}{x}_2$

---

* See Section 1.8

and $\underset{\sim}{x}_3$ be three noncollinear points of a rigid body. Let $\underset{\sim}{M}$ map these points into $\underset{\sim}{y}_1$, $\underset{\sim}{y}_2$ and $\underset{\sim}{y}_3$, respectively. Clearly,

$$||\underset{\sim}{x}_3-\underset{\sim}{x}_2||^2 = (\underset{\sim}{x}_3-\underset{\sim}{x}_2,\underset{\sim}{x}_3-\underset{\sim}{x}_2) = ((\underset{\sim}{x}_3-\underset{\sim}{x}_1) - (\underset{\sim}{x}_2-\underset{\sim}{x}_1),\ (\underset{\sim}{x}_3-\underset{\sim}{x}_1)-(\underset{\sim}{x}_2-\underset{\sim}{x}_1)) =$$
$$=||\underset{\sim}{x}_3-\underset{\sim}{x}_1||^2 -2(\underset{\sim}{x}_3-\underset{\sim}{x}_1,\underset{\sim}{x}_2-\underset{\sim}{x}_1)+||\underset{\sim}{x}_2-\underset{\sim}{x}_1||^2$$

Similarly,

$$||\underset{\sim}{y}_3-\underset{\sim}{y}_2||^2 =(\underset{\sim}{y}_3-\underset{\sim}{y}_2,\underset{\sim}{y}_3-\underset{\sim}{y}_2) = ||\underset{\sim}{y}_3-\underset{\sim}{y}_1||^2-2(\underset{\sim}{y}_3-\underset{\sim}{y}_1,\underset{\sim}{y}_2,-\underset{\sim}{y}_1) + ||\underset{\sim}{y}_2-\underset{\sim}{y}_1||^2$$

From the definition of a rigid-body motion, however,

$$||\underset{\sim}{y}_3-\underset{\sim}{y}_2||^2 = ||\underset{\sim}{x}_3-\underset{\sim}{x}_2||^2$$

Thus,

$$||\underset{\sim}{x}_3-\underset{\sim}{x}_1||^2-2(\underset{\sim}{x}_3-\underset{\sim}{x}_1,\underset{\sim}{x}_2-\underset{\sim}{x}_1) +||\underset{\sim}{x}_2-\underset{\sim}{x}_1||^2=||\underset{\sim}{y}_3-\underset{\sim}{y}_1||^2 -2(\underset{\sim}{y}_3-\underset{\sim}{y}_1,\underset{\sim}{y}_2-\underset{\sim}{y}_1)$$
$$+ ||\underset{\sim}{y}_2-\underset{\sim}{y}_1||^2 \qquad (2.2.3)$$

Again, from the aforementioned definition,

$$||\underset{\sim}{x}_3-\underset{\sim}{x}_1||^2 = ||\underset{\sim}{y}_3-\underset{\sim}{y}_1||^2 \qquad (2.2.4)$$

and

$$||\underset{\sim}{x}_2-\underset{\sim}{x}_1||^2 = ||\underset{\sim}{y}_3-\underset{\sim}{y}_1||^2 \qquad (2.2.5)$$

Thus clearly, from (2.2.3), (2.2.4) and (2.2.5),

$$(\underset{\sim}{x}_3-\underset{\sim}{x}_1,\underset{\sim}{x}_2-\underset{\sim}{x}_1) = (\underset{\sim}{y}_3-\underset{\sim}{y}_1,\underset{\sim}{y}_2-\underset{\sim}{y}_1) \qquad (2.2.6)$$

which states that the angle (See Section 1.7) between vectors x_3-x_1 and x_2-x_1 remains unchanged.

The foregoing mapping $\underset{\sim}{M}$ is, in general, nonlinear, but there exists a class $\underset{\sim}{Q}$ of mappings $\underset{\sim}{M}$, leaving one point of a body fixed, that are linear.

In fact, let O be a point of a rigid body which remains fixed under $\underset{\sim}{Q}$, its position vector being the zero vector $\underset{\sim}{0}$ of the space under study (this can always be rearranged since one has the freedom to place the origin of coordinates in any suitable position). Let $\underset{\sim}{x}_1$ and $\underset{\sim}{x}_2$ be any two points of this rigid body.

From the previous results,

$$||\mathbf{x}_i|| = ||\mathbf{Q}(\mathbf{x}_i)||, \; i = 1, 2 \tag{2.2.7}$$

$$(\mathbf{x}_i, \mathbf{x}_j) = (\mathbf{Q}(\mathbf{x}_i), \; \mathbf{Q}(\mathbf{x}_j)), \; i,j=1,2 \tag{2.2.8}$$

Assume for a moment that $\mathbf{Q}$ is not linear. Thus, let

$$\mathbf{e} \equiv \mathbf{Q}(\mathbf{x}_1+\mathbf{x}_2) - \left(\mathbf{Q}(\mathbf{x}_1)+\mathbf{Q}(\mathbf{x}_2)\right)$$

Then

$$\begin{aligned}||\mathbf{e}||^2 &= ||\mathbf{Q}(\mathbf{x}_1+\mathbf{x}_2)||^2 + ||\mathbf{Q}(\mathbf{x}_1)+\mathbf{Q}(\mathbf{x}_2)||^2 - 2(\mathbf{Q}(\mathbf{x}_1+\mathbf{x}_2), \mathbf{Q}(\mathbf{x}_1)+\mathbf{Q}(\mathbf{x}_2)) = \\ &= ||\mathbf{x}_1+\mathbf{x}_2||^2 + ||\mathbf{Q}(\mathbf{x}_1)||^2 + ||\mathbf{Q}(\mathbf{x}_2)||^2 + 2(\mathbf{Q}(\mathbf{x}_1), \mathbf{Q}(\mathbf{x}_2)) \\ &\quad -2(\mathbf{Q}(\mathbf{x}_1+\mathbf{x}_2), \mathbf{Q}(\mathbf{x}_1)) - 2(\mathbf{Q}(\mathbf{x}_1+\mathbf{x}_2), \mathbf{Q}(\mathbf{x}_2))\end{aligned}$$

where the rigidity condition has been applied, i.e. the condition that states that, under a rigid-body motion, any two points of the body remain equidistant. Applying this condition again, together with the condition of constancy of the angle between any two lines of the rigid body (eq. (2.2.6)),

$$\begin{aligned}||\mathbf{e}||^2 &= ||\mathbf{x}_1||^2 + ||\mathbf{x}_2||^2 + 2(\mathbf{x}_1, \mathbf{x}_2) + ||\mathbf{x}_1||^2 + ||\mathbf{x}_2||^2 + 2(\mathbf{x}_1, \mathbf{x}_2) \\ &\quad -2(\mathbf{x}_1+\mathbf{x}_2, \mathbf{x}_1) - 2(\mathbf{x}_1+\mathbf{x}_2, \mathbf{x}_2) = \\ &= 2||\mathbf{x}_1||^2 + 2||\mathbf{x}_2||^2 + 4(\mathbf{x}_1, \mathbf{x}_2) - \left(2||\mathbf{x}_1||^2 + 2||\mathbf{x}_2||^2 + 4(\mathbf{x}_1, \mathbf{x}_2)\right) \\ &= 0\end{aligned}$$

From the positive-definiteness of the norm, then

$$\mathbf{e} = \mathbf{0}$$

thereby showing that

$$\mathbf{Q}(\mathbf{x}_1+\mathbf{x}_2) = \mathbf{Q}(\mathbf{x}_1) + \mathbf{Q}(\mathbf{x}_2)$$

i.e. $\underset{\sim}{Q}$ is an <u>additive</u> operator*

On the other hand, since $\underset{\sim}{Q}$ preserves the angle between any pair of lines of a rigid body, for any given real number $\alpha>0$, $\underset{\sim}{Q}(\underset{\sim}{x})$ and $\underset{\sim}{Q}(\alpha\underset{\sim}{x})$ are parallel, i.e. linearly dependent (for $\underset{\sim}{x}$ and $\alpha\underset{\sim}{x}$ are parallel as well). Hence,

$$\underset{\sim}{Q}(\alpha\underset{\sim}{x}) = \beta\underset{\sim}{Q}(\underset{\sim}{x}), \quad \beta>0 \tag{2.2.9}$$

Since $\underset{\sim}{Q}$ preserves the Euclidean norm,

$$||\underset{\sim}{Q}(\alpha\underset{\sim}{x})||=||\alpha\underset{\sim}{x}||=|\alpha|.||\underset{\sim}{x}|| \tag{2.2.10}$$

On the other hand, from eq. (2.2.9),

$$||\underset{\sim}{Q}(\alpha\underset{\sim}{x})||=||\beta\underset{\sim}{Q}(\underset{\sim}{x})||=|\beta|.||\underset{\sim}{Q}(\underset{\sim}{x})||=|\beta|.||\underset{\sim}{x}|| \tag{2.2.11}$$

Hence, equating (2.2.10) and (2.2.11), and dropping the absolute-value brackets, for $\alpha,\beta>0$,

$$\alpha = \beta$$

and

$$\underset{\sim}{Q}(\alpha\underset{\sim}{x}) = \alpha\underset{\sim}{Q}(\underset{\sim}{x}) \tag{2.2.12}$$

and hence, $\underset{\sim}{Q}$ is a <u>homogeneous</u> operator. Being homogeneous and additive, $\underset{\sim}{Q}$ is <u>linear</u>. The following has thus been proved.

<u>THEOREM 2.2.1</u> *If $\underset{\sim}{Q}$ is a rigid-body motion that leaves a point fixed, then $\underset{\sim}{Q}$ is a linear transformation.*

From the foregoing discussion, $\underset{\sim}{Q}$ is representable by means of a 3x3 matrix referred to a certain basis (Theorem 1.2.1)

If $B=\{e_1,e_2,e_3\}$ is an orthonormal** basis for the 3-dimensional Euclidean

---

* This proof is due to Prof. G.S. Sidhu, Institute for Applied Mathematics and Systems Research, U. of Mexico (IIMAS-UNAM)

** $(\underset{\sim}{e}_i,\underset{\sim}{e}_j) = \delta_{ij}$ (The Kronecker delta)

space, the ith column of the matrix $\underset{\sim}{Q}$ is formed from the coefficients of $\underset{\sim}{Q}(\underset{\sim}{e}_i)$ expressed in terms of B according to Definition 1.2.1. In fact, the resulting matrix is orthogonal. Since $\underset{\sim}{Q}$ is linear, $\underset{\sim}{Q}(\underset{\sim}{x})$ can be expressed simply as $\underset{\sim}{Q}\underset{\sim}{x}$. Now if

$$\underset{\sim}{y} = \underset{\sim}{Q}\underset{\sim}{x}$$

then

$$||\underset{\sim}{y}||=||\underset{\sim}{x}||$$

Hence

$$\underset{\sim}{y}^T\underset{\sim}{y}=\underset{\sim}{x}^T\underset{\sim}{Q}^T\underset{\sim}{Q}\underset{\sim}{x} = \underset{\sim}{x}^T\underset{\sim}{x}, \text{ for any } \underset{\sim}{x}$$

Hence, clearly

$$\underset{\sim}{Q}^T\underset{\sim}{Q} = \underset{\sim}{I}$$

the identity matrix. This result can then be stated as

THEOREM 2.2.2 *A rigid-body motion leaving one point fixed is represented with respect to an orthonormal basis by an orthogonal matrix.*

2.3 THE THEOREM OF EULER AND THE REVOLUTE MATRIX.

In the previous sections it was shown that the motion of a rigid body which keeps one of its points fixed can be represented by an orthogonal 3 x 3 matrix. In view of Sect. 1.9 there are two classes of orthogonal matrices, depending on whether their determinant is plus or minus unity. Orthogonal matrices whose determinant is +1 are called proper orthogonal and those whose determinant is-1 are called improper orthogonal.

Proper orthogonal matrices represent rigid-body rotations, whereas improper orthogonal matrices represent reflections. Indeed, consider the rotation of axes X_1,Y_1,Z_1 into X_2,Y_2,Z_2 as shown in Fig 2.3.1

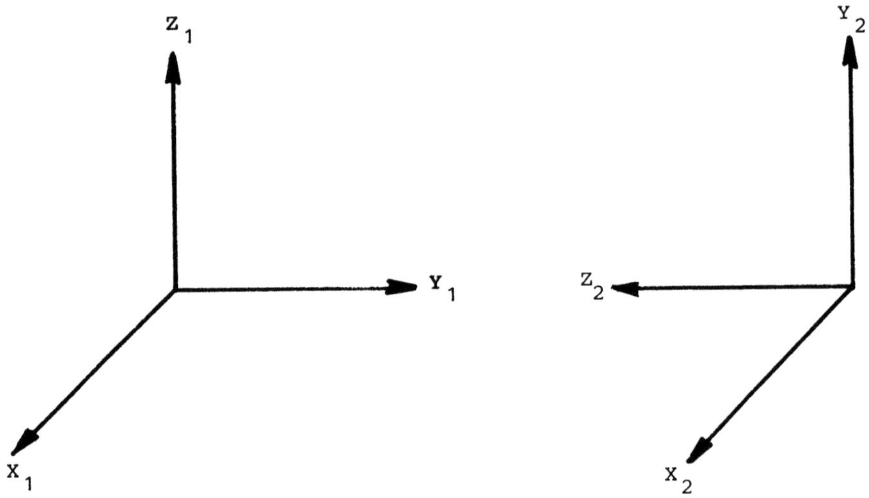

Fig 2.3.1 Rotation of axes

The matrix representation of the above rotation is obtained from the relationship

$$\underset{\sim}{x}_2=\underset{\sim}{x}_1 \tag{2.3.1}$$

$$\underset{\sim}{y}_2=\underset{\sim}{z}_1$$

$$\underset{\sim}{z}_2=-\underset{\sim}{y}_1$$

where $\underset{\sim}{x}_1$, $\underset{\sim}{x}_2$, etc. represent unit vectors along the X_1, X_2, etc. axes, respectively. From eqs. (2.3.1),

$$(\underset{\sim}{Q})_1 = \begin{pmatrix} 1 & 0 & 0 \\ 0 & 0 & -1 \\ 0 & 1 & 0 \end{pmatrix} \tag{2.3.2}$$

$(\underset{\sim}{Q})_1$ means the rotation expressed in terms of the basis $\{\underset{\sim}{x}_1,\underset{\sim}{y}_1,\underset{\sim}{z}_1\}$. Clearly,

$$\det \underset{\sim}{Q}=+1$$

and thus it is a proper orthogonal matrix.

On the other hand, consider the reflection of axes $\underset{\sim}{x}_1,\underset{\sim}{y}_1,\underset{\sim}{z}_1$ into

$\{X_2, Y_2, Z_2\}$, as shown in Fig 2.3.2

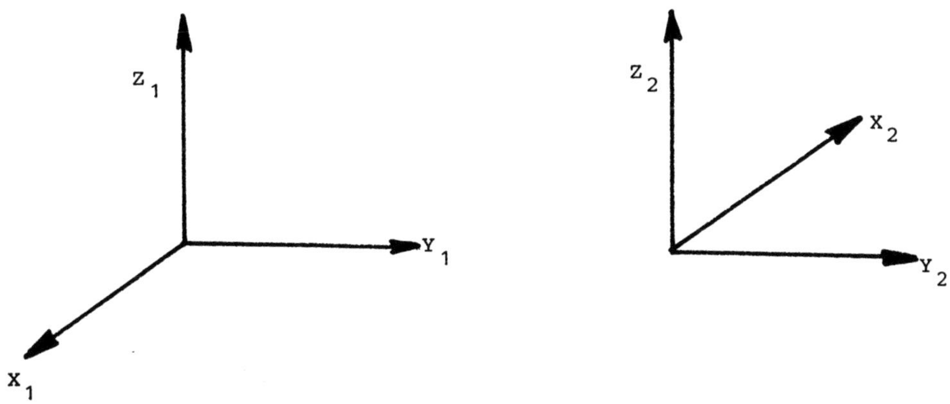

Fig. 2.3.2 Reflection of axes

Now,

$$x_2 = -x_1 \qquad (2.3.3)$$
$$y_2 = y_1$$
$$z_2 = z_1$$

Hence,

$$(\underset{\sim}{Q})_1 = \begin{pmatrix} -1 & 0 & 0 \\ 0 & 1 & 0 \\ 0 & 0 & 1 \end{pmatrix}$$

and so,

$$\det \underset{\sim}{Q} = -1$$

i.e. $\underset{\sim}{Q}$, as obtained from (2.3.3) is a reflection. Applications of reflections were studied in Sect. 1.12.

From Corollary 1.9.1 it can be seen that a 3 x 3 proper orthogonal matrix has exactly one eigenvalue equal to +1. Now if $\underset{\sim}{e}$ is the eigenvector of

$\underset{\sim}{Q}$ corresponding to the eigenvalue +1, it follows that

$$\underset{\sim}{Q}\underset{\sim}{e} = \underset{\sim}{e}$$

and, furthermore, for any scalar α,

$$\underset{\sim}{Q}\alpha\underset{\sim}{e} = \alpha\underset{\sim}{e}$$

Hence all points of the rigid body located along a line parallel to $\underset{\sim}{e}$ passing through the fixed point O, remain fixed under the rotation $\underset{\sim}{Q}$. Hence, the following result, due to Euler (2.1) :

THEOREM 2.3.1 (Euler). *If a rigid body undergoes a displacement leaving one of its points fixed, then there exists a line passing through the fixed point, such that all of the points on that line remain fixed during the displacement. This line is called "the axis of rotation" and the angle of rotation is measured on a plane perpendicular to the axis.*

The matrix representing a rotation is sometimes referred to as "the revolute". Clearly, the revolute is completely determined by a scalar parameter, the angle of rotation and a vector, the direction of the axis of rotation*. From the foregoing discussion it is clear that the direction vector of the revolute is obtained as the (unique linearly independent) eigenvector of the revolute associated with its + 1 eigenvalue. The angle of rotation is obtained as follows:

From Euler's Theorem, it is always possible to obtain an orthonormal basis $B = \{\underset{\sim}{b}_1, \underset{\sim}{b}_2, \underset{\sim}{b}_3\}$ such that, say $\underset{\sim}{b}_3$, is parallel to the axis of rotation; $\underset{\sim}{b}_1$ and $\underset{\sim}{b}_2$ thus lie in a plane perpendicular to this axis. The rotation would then rotate the vectors through an angle θ. Let $\underset{\sim}{b}'_1$ and $\underset{\sim}{b}'_2$ be the corresponding images of $\underset{\sim}{b}_1$ and $\underset{\sim}{b}_2$ after the rotation under consideration, represented graphically in Fig 2.3.3

---

* These parameters are also called "the invariants" of the revolute, for they remain unchanged under different choices of coordinate axes.

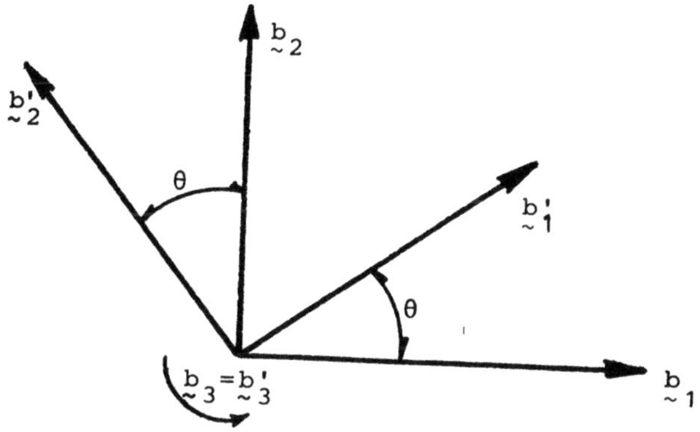

Fig 2.3.3 Rotation through an angle θ about axis b_3.

Then

$$b'_1=\cos\theta b_1+\sin\theta b_2$$

$$b'_2=-\sin\theta b_1+\cos\theta b_2 \tag{2.3.4}$$

$$b'_3=b_3$$

and it follows that

$$(Q)_B = \begin{pmatrix} \cos\theta & -\sin\theta & 0 \\ \sin\theta & \cos\theta & 0 \\ 0 & 0 & 1 \end{pmatrix} \tag{2.3.5}$$

Due to its simple and illuminating form, it seems justified to call matrix (2.3.5) a "canonical form" of the rotation matrix.

Exercise 2.3.1 Devise an algorithm to carry any orthogonal matrix into its canonical form (2.3.5).

Let a revolute matrix Q be given refered to an arbitrary orthonormal basis $A= \{a_1, a_2, a_3\}$, different from B as defined above. Furthermore, let

$$(P)_A = (b_1 \mid b_2 \mid b_3) \tag{2.3.6}$$

where

$$\underset{\sim}{b}_j = (b_{1j}, b_{2j}, b_{3j})^T, \quad j = 1,2,3$$

b_{ij} being the ith component of $\underset{\sim}{b}_j$ referred to the basis A, i.e.

$$\underset{\sim}{b}_j = b_{ij}\underset{\sim}{a}_1 + b_{2j}\underset{\sim}{a}_2 + b_{3j}\underset{\sim}{a}_3$$

Since both A and B are orthonormal, $(\underset{\sim}{P})_A$ is an orthogonal matrix. Thus, the canonical form can be obtained from the following similarity transformation

$$(\underset{\sim}{Q})_B = (\underset{\sim}{P}^T)_A (\underset{\sim}{Q})_A (\underset{\sim}{P})_A \tag{2.3.7}$$

From the canonical form given above, it is apparent that

$$\mathrm{Tr}(\underset{\sim}{Q})_B = 1 + 2\cos\theta$$

from which

$$\theta = \cos^{-1}\{\tfrac{1}{2}(\mathrm{Tr}(\underset{\sim}{Q})_B - 1)\} \tag{2.3.8}$$

is readily obtained. It should be pointed out that, since the trace is invariant under similarity transformations, i.e. since

$$\mathrm{Tr}(\underset{\sim}{Q})_B = \mathrm{Tr}\ (\underset{\sim}{Q})_A$$

one can compute the rotation angle without transforming the revolute matrix into its canonical form.

Eq. (2.3.8), however, yields the angle of rotation through the cos function, which is even, i.e. $\cos(-x) = \cos(x)$; hence, the said formula does not provide the sign of the angle. This is next determined by application of Theorem 2.3.2. The proof of this theorem needs some background, which is now laid down.

In what follows, dyadic notation will be used*. Let L be the axis of a

---

* For readers unfamiliar with this notation, a short account of algebra of dyadics is provided in Appendix 1.

rotation about point O, whose existence is guarenteed by Euler's Theorem. Moreover, let θ be the corresponding angle of rotation, as indicated in Fig 2.3.4, and $\underset{\sim}{e}$ a unit vector parallel to L.

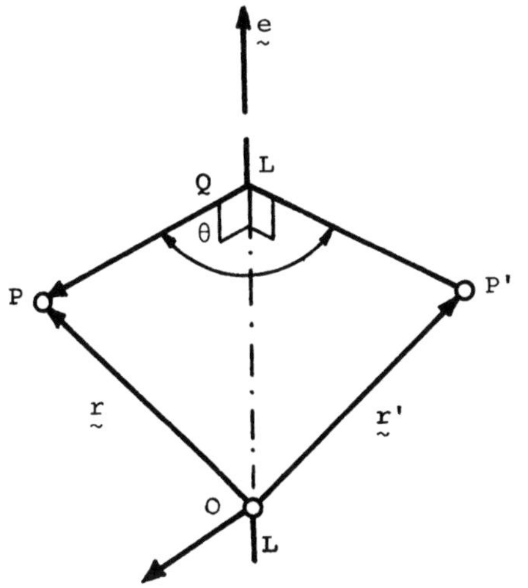

Fig 2.3.4 Rotation about a point.

In Fig 2.3.4 P' is the rotated position of point P. If PQ is perpendicular to L, so is P'Q, because rotations preserve angles of rigid bodies. Thus points P, P' and Q determine a plane perpendicular to L, on which the angle of rotation, θ, is measured. From that figure,

$$\underset{\sim}{r}' = \overrightarrow{OQ} + \overrightarrow{QP}'$$

and

$$\overrightarrow{OQ} = \underset{\sim}{r} - \overrightarrow{QP}$$

Hence

$$\underset{\sim}{r}' = \underset{\sim}{r} - \overrightarrow{QP} + \overrightarrow{QP}' \tag{2.3.9}$$

Let QP" be a line contained in plane PP'Q, at right angles with line PQ and of length equal to that of QP. Thus, vector $\overrightarrow{QP}'$ can be expressed as a linear combination of vectors $\overrightarrow{QP}$ and $\overrightarrow{QP}''$. But

$$\overrightarrow{QP}'' = \underset{\sim}{e} \times \underset{\sim}{r} \tag{2.3.10}$$

whereas

$$\vec{QP} = - \underset{\sim}{e} \times \vec{QP}'' = - \underset{\sim}{e} \times (\underset{\sim}{e} \times \underset{\sim}{r}) \qquad (2.3.11)$$

which can readily be proved. Besides, $\vec{QP}'$ can be expressed as

$$\vec{QP}' = \vec{QP} \cos\theta + \vec{QP}'' \sin\theta$$

which, in view of eqs. (2.3.10) and (2.3.11), yields

$$\vec{QP}' = - \cos\theta \underset{\sim}{e} \times (\underset{\sim}{e} \times \underset{\sim}{r}) + \sin\theta \underset{\sim}{e} \times \underset{\sim}{r} \qquad (2.3.12)$$

Substituting eqs. (2.3.11) and (2.3.12) into eq. (2.3.9) leads to

$$\underset{\sim}{r}' = \underset{\sim}{r} + \underset{\sim}{e} \times (\underset{\sim}{e} \times \underset{\sim}{r}) - \cos\theta \underset{\sim}{e} \times (\underset{\sim}{e} \times \underset{\sim}{r}) + \sin\theta \underset{\sim}{e} \times \underset{\sim}{r} \qquad (2.3.13)$$

But

$$\underset{\sim}{e} \times (\underset{\sim}{e} \times \underset{\sim}{r}) = (\underset{\sim}{e} \cdot \underset{\sim}{r})\underset{\sim}{e} - (\underset{\sim}{e} \cdot \underset{\sim}{e})\underset{\sim}{r} = (\underset{\sim}{e}\underset{\sim}{e} - \underset{\sim}{1}) \cdot \underset{\sim}{r} \qquad (2.3.14)$$

where $\underset{\sim}{1}$ is the identity dyadic, i.e. a dyadic that is isomorphic to the identity matrix. Furthermore

$$\underset{\sim}{e} \times \underset{\sim}{r} = \underset{\sim}{1} \cdot \underset{\sim}{e} \times \underset{\sim}{r} = \underset{\sim}{1} \times \underset{\sim}{e} \cdot \underset{\sim}{r} \qquad (2.3.15)$$

where the dot and the point have been exchanged, what is possible to do by virtue of the algebra of cartesian vectors. Substituting eqs. (2.3.14) and (2.3.15) into eq. (2.3.13) one obtains

$$\begin{aligned} \underset{\sim}{r}' &= \underset{\sim}{r} + (1-\cos\theta)(\underset{\sim}{e}\underset{\sim}{e} - \underset{\sim}{1}) \cdot \underset{\sim}{r} + \sin\theta \underset{\sim}{1} \times \underset{\sim}{e} \cdot \underset{\sim}{r} = \\ &= \left((1-\cos\theta)\underset{\sim}{e}\underset{\sim}{e} + \cos\theta \underset{\sim}{1} + \sin\theta \underset{\sim}{1} \times \underset{\sim}{e}\right) \cdot \underset{\sim}{r} = \\ &= \underset{\sim}{Q} \cdot \underset{\sim}{r} \end{aligned} \qquad (2.3.16)$$

i.e. $\underset{\sim}{r}'$ has been expressed as a linear transformation of vector $\underset{\sim}{r}$. The dyadic $\underset{\sim}{Q}$ is, then, isomorphic to the rotation matrix defined in Section 2.2. That is

$$\underset{\sim}{Q} = \underset{\sim}{e}\underset{\sim}{e} + (\underset{\sim}{1} - \underset{\sim}{e}\underset{\sim}{e})\cos\theta + \sin\theta \underset{\sim}{1} \times \underset{\sim}{e} \qquad (2.3.17)$$

One can now prove the following

<u>THEOREM 2.3.2</u> *Let a rigid body undergo a pure rotation about a fixed point O and let $\underset{\sim}{r}$ and $\underset{\sim}{r}'$ be the initial and the final position vectors of a point of the body (measured from O) not lying on the axis of rotation*

Furthermore let θ and $\underset{\sim}{e}$ be the angle of rotation and the unit vector pointing in the direction of the rotation. Then

$$sgn(\underset{\sim}{r}x\underset{\sim}{r}'.\underset{\sim}{e})=sgn(\theta)$$

Proof.

Application of eq. (2.3.16) leads to

$$\underset{\sim}{r}x\underset{\sim}{r}'=(1-\cos\theta)(\underset{\sim}{e}.\underset{\sim}{r})\underset{\sim}{r}x\underset{\sim}{e}+\sin\theta\underset{\sim}{r}x(\underset{\sim}{e}x\underset{\sim}{r})=$$

$$=(1-\cos\theta)(\underset{\sim}{e}.\underset{\sim}{r})\underset{\sim}{r}x\underset{\sim}{e}+\sin\theta\left(r^2\underset{\sim}{e}-(\underset{\sim}{r}.\underset{\sim}{e})\underset{\sim}{r}\right)$$

where

$$r^2\equiv||\underset{\sim}{r}||^2$$

Thus,

$$\underset{\sim}{r}x\underset{\sim}{r}'.\underset{\sim}{e}=\sin\theta\left(r^2-(\underset{\sim}{r}.\underset{\sim}{e})^2\right)$$

which can be reduced to

$$\underset{\sim}{r}x\underset{\sim}{r}'.\underset{\sim}{e} = r^2\sin\theta\sin^2(\underset{\sim}{r},\underset{\sim}{e})$$

where $(\underset{\sim}{r},\underset{\sim}{e})$ is the angle between vectors $\underset{\sim}{r}$ and $\underset{\sim}{e}$. Hence,

$$sgn(\underset{\sim}{r}x\underset{\sim}{r}'.\underset{\sim}{e})=sgn(\sin\theta)$$

But

$$sgn(\sin\theta)=sgn(\theta)$$

for sin( ) is an odd function, i.e. sin (-x)=-sin(x).

Finally, then

$$sgn(\underset{\sim}{r}x\underset{\sim}{r}'.\underset{\sim}{e})=sgn(\theta),q.e.d \qquad (2.3.18)$$

In conclusion, Theorem 2.3.2 allows to distinguish whether a rotation in the specified direction $\underset{\sim}{e}$ is either through an angle θ or through an angle $-\theta$.

Exercise 2.3.2 Let $\underset{\sim}{p}$ and $\underset{\sim}{p}'$ be the initial and the final position vectors of a point P of a rigid body undergoing a rotation whose matrix is $\underset{\sim}{Q}$. Show that the displacement $\underset{\sim}{p}'-\underset{\sim}{Q}\underset{\sim}{p}$ lies in the null space of $\underset{\sim}{Q}-\underset{\sim}{I}$.

Exercise 2.3.3 Show that the trace of a matrix is invariant under similarity transformations.

Exercise 2.3.4 Show that a revolute matrix $\underset{\sim}{Q}$ has two complex conjugate eigenvalues, λ and $\bar{\lambda}$ ($\bar{\lambda}$ = complex conjugate of λ).

Furthermore, show that

$$\mathrm{Re}\{\lambda\} = \frac{1}{2}(\mathrm{Tr}\underset{\sim}{Q}-1)$$

What is the relationship between the complex eigenvalues of the revolute matrix and its angle of rotation?.

In the foregoing paragraphs the revolute matrix was analysed, i.e. it was shown how to obtain its invariants when the matrix is known.

The inverse problem is discussed next: Given the axis and the angle of rotation, obtain the revolute matrix referred to a specified set of coordinate axes.

It is apparent that the most convenient basis (or coordinate axes) for representing the revolute matrix is the one for which this takes on its canonical form. Let $B = \{\underset{\sim}{b}_1, \underset{\sim}{b}_2, \underset{\sim}{b}_3\}$ be this basis, where $\underset{\sim}{b}_3$ coincides with the given revolute axis, and $\underset{\sim}{b}_1$ and $\underset{\sim}{b}_2$ are any pair of orthonormal vectors lying in the plane perpendicular to $\underset{\sim}{b}_3$.

Hence, $(\underset{\sim}{Q})_B$ appears as in eq. (2.3.5), with θ given. Let $A = \{a_1, a_2, a_3\}$ be an orthonormal basis with respect to which $\underset{\sim}{Q}$ is to be represented, and let

$$(\mathbf{P})_A = (\mathbf{b}_1 \;\vdots\; \mathbf{b}_2 \;\vdots\; \mathbf{b}_3)_A$$

be a matrix formed with the vectors of B. Then, it is clear that

$$(\mathbf{Q})_A = (\mathbf{P})_A (\mathbf{Q})_B (\mathbf{P}^T)_A$$

<u>Example 2.3.1</u> Let

$$\mathbf{Q} = \frac{1}{3}\begin{pmatrix} 2 & 1 & 2 \\ -2 & 2 & 1 \\ -1 & -2 & 2 \end{pmatrix}$$

Verify whether it is orthogonal. If it is, does it represent a rotation?.

If so, describe the rotation

<u>Solution:</u>

$$\mathbf{Q}\mathbf{Q}^T = \frac{1}{9}\begin{pmatrix} 2 & 1 & 1 \\ -2 & 2 & 1 \\ -1 & -2 & 2 \end{pmatrix}\begin{pmatrix} 2 & -2 & -1 \\ 1 & 2 & -2 \\ 2 & 1 & 2 \end{pmatrix}$$

$$= \frac{1}{9}\begin{pmatrix} 9 & 0 & 0 \\ 0 & 9 & 0 \\ 0 & 0 & 9 \end{pmatrix} = \mathbf{I}$$

Hence $\mathbf{Q}$ is in fact orthogonal. Next,

$$\det \mathbf{Q} = \frac{2}{3}\begin{vmatrix} \frac{2}{3} & \frac{1}{3} \\ -\frac{2}{3} & \frac{2}{3} \end{vmatrix} + \frac{2}{3}\begin{vmatrix} \frac{1}{3} & \frac{2}{3} \\ -\frac{2}{3} & \frac{2}{3} \end{vmatrix} - \frac{1}{3}\begin{vmatrix} \frac{1}{3} & \frac{2}{3} \\ \frac{2}{3} & \frac{1}{3} \end{vmatrix}$$

$$= \frac{2}{3}\left(\frac{4}{9} + \frac{2}{9}\right) + \frac{2}{3}\left(\frac{2}{9} + \frac{4}{9}\right) - \frac{1}{3}\left(\frac{1}{9} - \frac{4}{9}\right) = +1$$

Thus $\mathbf{Q}$ is a proper orthogonal matrix and, consequently, represents a rotation. To find the axis of the rotation it is necessary to find a unit vector $\mathbf{e} = (e_1, e_2, e_3)^T$ such that

$$Q e = e,$$

i.e.

$$\frac{1}{3}\begin{pmatrix} 2 & 1 & 2 \\ -2 & 2 & 1 \\ -1 & -2 & 2 \end{pmatrix} \begin{pmatrix} e_1 \\ e_2 \\ e_3 \end{pmatrix} = \begin{pmatrix} e_1 \\ e_2 \\ e_3 \end{pmatrix}.$$

Hence

$$\begin{aligned} -e_1 + e_2 + 2e_3 &= 0 \\ -2e_1 - e_2 + e_3 &= 0 \\ -e_1 - 2e_2 - e_3 &= 0 \end{aligned}$$

from which

$$e_1 = e_3$$

$$e_2 = -e_3$$

and so

$$e = \begin{pmatrix} 1 \\ -1 \\ 1 \end{pmatrix} e_3.$$

Setting $||e||=1$, it follows that $e_3 = \frac{\sqrt{3}}{3}$, and

$$e = \frac{\sqrt{3}}{3}\begin{pmatrix} 1 \\ -1 \\ 1 \end{pmatrix}$$

Thus, the axis of rotation is parallel to the vector e given above.

To find the angle of rotation is an even simpler matter:

$$\text{Tr } Q = \frac{1}{3}(2+2+2) = 1 + 2\cos\theta$$

Thus $\theta = \cos^{-1}(\frac{1}{2}) = -60°$

where use was made of Theorem 2.3.2 to find the sign of θ.

Example 2.3.2. Determine the revolute matrix representing a rotation of 90° about an axis having three equal direction cosines with respect to the X,Y,Z axes. The matrix should be expressed with respect to these axes.

Solution:

Let $\underset{\sim}{B} = \{\underset{\sim}{b}_1, \underset{\sim}{b}_2, \underset{\sim}{b}_3\}$ be an orthonormal basis with respect to which the revolute is represented in its canonical form. Let $\underset{\sim}{b}_3$ be coincident with the axis of rotation. Clearly

$$\underset{\sim}{b}_3 = \frac{\sqrt{3}}{3}\begin{pmatrix}1\\1\\1\end{pmatrix}$$

It remains only to determine $\underset{\sim}{b}_1$ and $\underset{\sim}{b}_2$. Clearly, these must satisfy

$$\underset{\sim}{b}_1 \cdot \underset{\sim}{b}_2 = \underset{\sim}{b}_1 \cdot \underset{\sim}{b}_3 = \underset{\sim}{b}_2 \cdot \underset{\sim}{b}_3 = 0.$$

Let

$$(\underset{\sim}{b}_1) = \begin{pmatrix}\alpha\\\beta\\\gamma\end{pmatrix}, \quad (\underset{\sim}{b}_2) = \begin{pmatrix}\lambda\\\mu\\\nu\end{pmatrix}.$$

Thus, the components of $\underset{\sim}{b}_1$ must satisfy

$$\alpha+\beta+\gamma=0,$$
$$\alpha^2+\beta^2+\gamma^2=1.$$

It is apparent that one component can be freely chosen. Let, for example,

$$\alpha = 0$$

Hence,

$$\beta + \gamma = 0$$
$$\beta^2 + \gamma^2 = 1$$

from which

$$2\,\beta^2=1. \quad \text{Thus } \beta = \frac{\pm\sqrt{2}}{2}, \gamma = \mp\frac{\sqrt{2}}{2}$$

Thus, choosing the + sign for β,

$$\underset{\sim}{b}_1=\frac{1}{2}\begin{pmatrix}0\\ \sqrt{2}\\ -\sqrt{2}\end{pmatrix}$$

$\underset{\sim}{b}_2$ can be obtained now very easily from the fact that $\underset{\sim}{b}_1, \underset{\sim}{b}_2$ and $\underset{\sim}{b}_3$ constitute an orthonormal right-hand triad, i.e.

$$\underset{\sim}{b}_2=\underset{\sim}{b}_3 \times \underset{\sim}{b}_1=\frac{\sqrt{6}}{6}(-2,1,1)^T$$

With respect to this basis, then, from eq. (2.3.5) the rotation matrix has the form

$$(\underset{\sim}{Q})_B=\begin{pmatrix}0 & -1 & 0\\ 1 & 0 & 0\\ 0 & 0 & 1\end{pmatrix}$$

Thus, letting A be the basis defined by the given X, Y and Z axes,

$$(P)_A=\begin{pmatrix}0 & -\sqrt{6}/3 & \sqrt{3}/3\\ \sqrt{2}/2 & \sqrt{6}/6 & \sqrt{3}/3\\ -\sqrt{2}/2 & \sqrt{6}/6 & \sqrt{3}/3\end{pmatrix}$$

and, from eq. (1.5.12), defining the following similarity transformation,

$$(\underset{\sim}{Q})_A=(\underset{\sim}{P})_A(\underset{\sim}{Q})_B(\underset{\sim}{P}^T)_A$$

With $(\underset{\sim}{Q})_B$ in its canonical form, the revolute matrix $\underset{\sim}{Q}$, expressed with respect to the X,Y,Z axes, is found to be

$$(\underset{\sim}{Q})_A=\frac{1}{3}\begin{pmatrix}1 & 1-\sqrt{3} & 1+\sqrt{3}\\ 1+\sqrt{3} & 1 & 1-\sqrt{3}\\ 1-\sqrt{3} & 1+\sqrt{3} & 1\end{pmatrix}$$

<u>Exercise 2.3.5</u> If the plane

$$x + y + z + 1 = 0$$

is rotated through 60° about an axis passing through the point (-1, -1, -1) and with direction cosines $\frac{1}{\sqrt{3}}$, $-\frac{1}{\sqrt{3}}$, $\frac{1}{\sqrt{3}}$, what is the equation of the plane in its new position?.

Exercise 2.3.6. The four vertices of an equilateral tetrahedron are labelled A, B, C, and D. If the tetrahedron is rotated in such a way that A, B, C, and D are mapped into C, B, D, and A, respectively, find the axis and the angle of the rotation.

What are the other rotations similar to the previous one, i.e., which map every vertex of the tetrahedron into another vertex?.

All these rotations, together with the identity rotation (the one leaving the vertices of the tetrahedron unchanged), constitute the symmetry group* of the tetrahedron.

Exercise 2.3.7 Given an axis A whose direction cosines are $(\frac{\sqrt{2}}{2}, \frac{1}{2}, \frac{1}{2})$, with respect to a set of coordinate axes XYZ, what is the matrix representation, with respect to these coordinate axes, of a rotation about A through an angle $2\pi/n$?

Exercise 2.3.8 A square matrix $\underset{\sim}{A}$ is said to be idempotent of index k whenever k is the smallest integer for which the kth power of $\underset{\sim}{A}$ becomes the identity matrix. Explain why the matrix obtained in Exercise 2.3.8 should be idempotent of index n.

Exercise 2.3.9 Show that any rotation matrix $\underset{\sim}{Q}$ can be expressed as

$$\underset{\sim}{Q}=e^{\underset{\sim}{A}\theta}$$

where $\underset{\sim}{A}$ is a nilpotent matrix and θ is the rotation angle. What is the relationship between matrix $\underset{\sim}{A}$ and the axis of rotation of $\underset{\sim}{Q}$?

---

*See Sect. 2.4 for the definition of this term.

Exercise 2.3.10 The equation of a three-axes ellipsoid is given as

$$\frac{x^2}{a^2} + \frac{y^2}{b^2} + \frac{z^2}{c^2} = 1$$

what is its equation after rotating it through an angle θ about an axis of direction numbers (a,b,c)?

2.4 GROUPS OF ROTATIONS.

A group is a set $\mathcal{g}$ with a binary operation $\mathcal{O}$ such that

i) if a and b ε $\mathcal{g}$, then $a \mathcal{O} b \varepsilon \mathcal{g}$

ii) if a,b,c ε $\mathcal{g}$, then $a\mathcal{O}(b\mathcal{O}c)=(a\mathcal{O}b)\mathcal{O}c$

iii) $\mathcal{g}$ contains an element i, called the identity of $\mathcal{g}$ under $\mathcal{O}$, such that, for every a ε $\mathcal{g}$

$$a\mathcal{O}i = i\mathcal{O}a = a$$

iv) for every a ε $\mathcal{g}$, there exists an element denoted $a^{-1} \varepsilon \mathcal{g}$, called the inverse of a under $\mathcal{O}$ such that

$$a\mathcal{O}a^{-1} = a^{-1}\mathcal{O}a = i$$

Notice that in the above definition it is not required that the group be commutative, i.e. that $a\mathcal{O}b=b\mathcal{O}a$ for all $a,b\varepsilon\mathcal{g}$. Commutative groups are a special class of groups, called abelian groups.

Some examples of groups are:

a) The natural numbers 1,2,..., 12 on the face of a (mechanical, not quartz or similar) clock and the operation $k\mathcal{O}m$ corresponding to "shift the clock hand from location k to location k + m", where k and m are natural numbers between 1 and 12. Of course, if k + m>12, the resulting operation is meant to be (k + m)(mod 12).

b) The set of rational numbers with the usual multiplication operation.

c) The set of integers with the usual algebraic addition operation.

The set of integers with the multiplication operation do not constitute a group (Why?)

Exercise 2.4.1 Show that the set of all those rotations referred to in Exercise 2.3.5 actually constitute a group.

Exercise 2.4.2 What is the symmetry group* of

i) an icosahedron?

ii) a regular pentagonal prism?

iii) a circular cylinder?

iv) a sphere?

It is clear, from the above discussion, that the set of all orthogonal matrices constitutes a group under matrix multiplication. In particular, the set of proper orthogonal matrices constitutes a group under matrix multiplication, but the improper set does not (Why?).

As an application of the group property of rotations or, equivalently, of proper orthogonal matrices, arbitrary rotations can be formed by the composition of successive simple rotations (See Example 2.4.1).

Another application is found in the compositon of rotations using Euler's angles (2.2)

Example 2.4.1 Referring to Fig 2.4.1, find the matrix representation, with respect to the X_1, Y_1, Z_1 axes, of the rotation that carries vertices A and B of the cube into A' and B', respectively, while leaving vertex O fixed. A' and B' lie in the Y_1Z_1 plane and points A', O and D, are collinear, as are B', F and E.

---

* See Exercise 2.3.6 for a definition of a symmetry group.

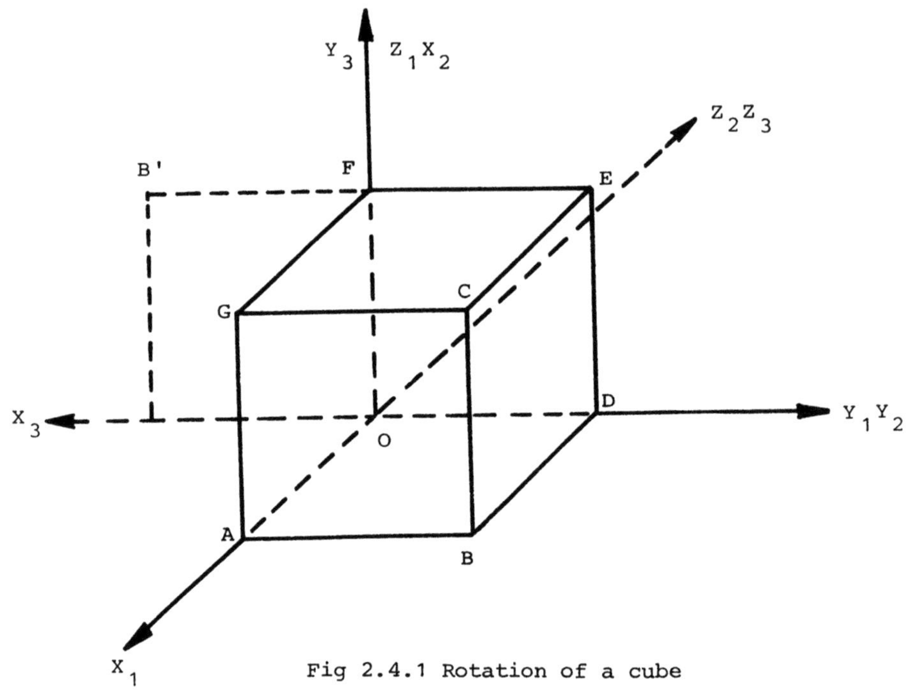

Fig 2.4.1 Rotation of a cube

Solution:

Let $(Q_{12})_1$ be the matrix representing the rotation of axes labelled 1 into those labelled 2 (referred to axes 1). Then, letting $\mathbf{x}_i$, $\mathbf{y}_i$ and $\mathbf{z}_i$ be unit vectors directed along the X_i-, Y_i- and Z_i axes, respectively,

$$\mathbf{Q}_{12}\mathbf{x}_1 = \mathbf{x}_2 = \mathbf{z}_1$$

$$\mathbf{Q}_{12}\mathbf{y}_1 = \mathbf{y}_2 = \mathbf{y}_1$$

$$\mathbf{Q}_{12}\mathbf{z}_1 = \mathbf{z}_2 = -\mathbf{x}_1$$

from which

$$(Q_{12})_1 = \begin{pmatrix} 0 & 0 & -1 \\ 0 & 1 & 0 \\ 1 & 0 & 0 \end{pmatrix}$$

Next, rotate axes labelled 2 into axes labelled 3. Call this rotation $\mathbf{Q}_{23}$. This rotation would leave axis X_1 fixed whereas it would carry axis Y_1 into Z_1 and axis Z_1 into $-Y_1$. Hence,

$$Q_{23} x_1 = x_1$$

$$Q_{23} y_1 = z_1$$

$$Q_{23} z_1 = y_1$$

and so,

$$(Q_{23})_1 = \begin{pmatrix} 1 & 0 & 0 \\ 0 & 0 & -1 \\ 0 & 1 & 0 \end{pmatrix}$$

Let Q_{13} be the rotation meant to be obtained. Its matrix can be computed then as

$$(Q_{13})_1 = (Q_{23})_1 (Q_{12})_1 = \begin{pmatrix} 0 & 0 & -1 \\ -1 & 0 & 0 \\ 0 & 1 & 0 \end{pmatrix}$$

which could also have been obtained by noticing that

$$Q_{13} x_1 = x_3 = -y_1$$

$$Q_{13} y_1 = y_3 = z_1$$

$$Q_{13} z_1 = z_3 = -x_1$$

Matrix $(Q_{13})_1$ represents a rotation through an angle $\theta = 120°$about an axis with direction cosines $-\alpha, \alpha, \alpha$. Although in this example the rotation could be obtained by an alternate method, in many cases, such as the one in Exercise 2.4.3, the use of rotation composition seems to be the simplest method.

Exercise 2.4.3 Determine the axis and the angle of the rotation carrying axes X,Y,Z into axes ξ, η, ζ , as shown in Fig. 2.4.2

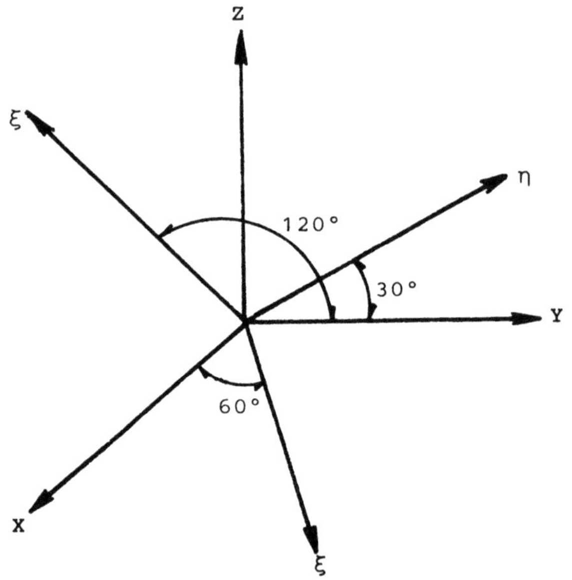

Fig 2.4.2 Rotation of axes

Exercise 2.4.4 The cube appearing in Fig 2.4.1 is rotated through 45° about diagonal OC. Find the matrix representation, with respect to X_1, Y_1, Z_1, of this rotation and the distance that vertex B is displaced through.

2.5 RODRIGUES' FORMULA AND CARTESIAN DECOMPOSITION OF THE ROTATION MATRIX.

The image $\underset{\sim}{r}_2$ of a Cartesian vector $\underset{\sim}{r}_1$ under a rotation through an angle θ about an axis parallel to the unit vector $\underset{\sim}{e}$ passing through the origin of coordinates was shown to be (See Section 2.3)

$$\underset{\sim}{r}_2 = \left((1-\cos\theta)\underset{\sim}{e}\underset{\sim}{e}+\cos\theta\underset{\sim}{1}+\sin\theta\underset{\sim}{1}\times\underset{\sim}{e}\right).\underset{\sim}{r}_1 \tag{2.5.1}$$

Multiplying both sides of eq. (2.5.1) times $\underset{\sim}{e}\times$ yields

$$\underset{\sim}{r}_2-\underset{\sim}{r}_1=\tan\frac{\theta}{2}\underset{\sim}{e}\times(\underset{\sim}{r}_1+\underset{\sim}{r}_2) \tag{2.5.2}$$

which is called Rodrigues' formula (2.3,2.4)

Form (2.5.1) of the rotation dyadic is advantageous since it shows explicitly the invariants $\underset{\sim}{e}$ and θ of the rotation.

Other useful expression of the rotation matrix is now derived. Letting

$$\underset{\sim}{e} = (u,v,w)^T \tag{2.5.3}$$

the rotation matrix can be written as (2.5)

$$\underset{\sim}{Q} = \underset{\sim}{R} + \underset{\sim}{T}\cos\theta + \underset{\sim}{P}\sin\theta \tag{2.5.4}$$

where

$$\underset{\sim}{R} = \begin{pmatrix} u^2 & uv & uw \\ uv & v^2 & vw \\ uw & vw & w^2 \end{pmatrix}, \quad \underset{\sim}{T} = \begin{pmatrix} v^2+w^2 & -uv & -uw \\ -uv & u^2+w^2 & -vw \\ -uw & -vw & u^2+v^2 \end{pmatrix} \tag{2.5.5a}$$

and

$$\underset{\sim}{P} = \begin{pmatrix} 0 & -w & v \\ w & 0 & -u \\ -v & u & 0 \end{pmatrix} \tag{2.5.5b}$$

In fact, computing the dyadics involved in expression (2.3.17),

$$\begin{aligned} \underset{\sim}{e}\underset{\sim}{e} &= (u\underset{\sim}{i} + v\underset{\sim}{j} + w\underset{\sim}{k})(u\underset{\sim}{i} + v\underset{\sim}{j} + w\underset{\sim}{k}) = \\ &= u^2\underset{\sim}{i}\underset{\sim}{i} + uv\underset{\sim}{i}\underset{\sim}{j} + uw\underset{\sim}{i}\underset{\sim}{k} + \\ &+ uv\underset{\sim}{j}\underset{\sim}{i} + v^2\underset{\sim}{j}\underset{\sim}{j} + vw\underset{\sim}{j}\underset{\sim}{k} + \\ &+ uw\underset{\sim}{k}\underset{\sim}{i} + vw\underset{\sim}{k}\underset{\sim}{j} + w^2\underset{\sim}{k}\underset{\sim}{k} \end{aligned} \tag{2.5.6}$$

$$\underset{\sim}{1} = \underset{\sim}{i}\underset{\sim}{i} + \underset{\sim}{j}\underset{\sim}{j} + \underset{\sim}{k}\underset{\sim}{k} \tag{2.5.7}$$

Hence

$$\begin{aligned} \underset{\sim}{1} \times \underset{\sim}{e} &= (\underset{\sim}{i}\underset{\sim}{i} + \underset{\sim}{j}\underset{\sim}{j} + \underset{\sim}{k}\underset{\sim}{k}) \times (u\underset{\sim}{i}+v\underset{\sim}{j}+w\underset{\sim}{k}) = \\ &= u\underset{\sim}{i}\underset{\sim}{i}\times\underset{\sim}{i} + v\underset{\sim}{i}\underset{\sim}{i}\times\underset{\sim}{j} + w\underset{\sim}{i}\underset{\sim}{i}\times\underset{\sim}{k} + \\ &+ u\underset{\sim}{j}\underset{\sim}{j}\times\underset{\sim}{i} + v\underset{\sim}{j}\underset{\sim}{j}\times\underset{\sim}{j} + w\underset{\sim}{j}\underset{\sim}{j}\times\underset{\sim}{k} + \\ &+ u\underset{\sim}{k}\underset{\sim}{k}\times\underset{\sim}{i} + v\underset{\sim}{k}\underset{\sim}{k}\times\underset{\sim}{j} + w\underset{\sim}{k}\underset{\sim}{k}\times\underset{\sim}{k} \end{aligned} \tag{2.5.8}$$

But

$$\underset{\sim}{i} \times \underset{\sim}{i} = \underset{\sim}{j}\times\underset{\sim}{j} = \underset{\sim}{k}\times\underset{\sim}{k} = \underset{\sim}{0} \tag{2.5.9}$$

and

$$\underset{\sim}{i} \times \underset{\sim}{j} = -\underset{\sim}{j} \times \underset{\sim}{i} = \underset{\sim}{k} \;,\; \underset{\sim}{i} \times \underset{\sim}{k} = -\underset{\sim}{k} \times \underset{\sim}{i} = -\underset{\sim}{j} \tag{2.5.10a}$$

$$\underset{\sim}{j} \times \underset{\sim}{k} = -\underset{\sim}{k} \times \underset{\sim}{j} = \underset{\sim}{i} \tag{2.5.10b}$$

Thus

$$\begin{aligned} \underset{\sim}{1} \times \underset{\sim}{e} = & - w\underset{\sim}{i}\underset{\sim}{j} + v\underset{\sim}{i}\underset{\sim}{k} \\ & + w\underset{\sim}{j}\underset{\sim}{i} - u\underset{\sim}{j}\underset{\sim}{k} \\ & - v\underset{\sim}{k}\underset{\sim}{i} + u\underset{\sim}{k}\underset{\sim}{j} \end{aligned} \tag{2.5.11}$$

Dyadics (2.5.6) and (2.5.7) can be written in matrix form as

$$(\underset{\sim}{e}\underset{\sim}{e}) = \begin{pmatrix} u^2 & uv & uw \\ uv & v^2 & vw \\ uw & vw & w^2 \end{pmatrix}, \quad (\underset{\sim}{1} \times \underset{\sim}{e}) = \begin{pmatrix} 0 & -w & v \\ w & 0 & -u \\ -v & u & 0 \end{pmatrix} \tag{2.5.12}$$

and

$$(\underset{\sim}{1} - \underset{\sim}{e}\underset{\sim}{e}) = \begin{pmatrix} v^2+w^2 & -uv & -uw \\ -uv & u^2+w^2 & -vw \\ -uw & -vw & u^2+v^2 \end{pmatrix} \tag{2.5.13}$$

Substitution of matrices (2.5.12) and (2.5.13) into eq. (2.3.17) leads directly to eq. (2.5.4). This expression of matrix $\underset{\sim}{Q}$ is very useful because it allows one to determine the sign of θ without requiring to compute the image $\underset{\sim}{r}'$ of a vector $\underset{\sim}{r}$ under $\underset{\sim}{Q}$.

Indeed, from eqs. (2.5.5a) and (2.5.5b), it is clear that matrices $\underset{\sim}{R}$ and $\underset{\sim}{I}$ are symmetric, whereas $\underset{\sim}{P}$ is skew symmetric. Hence, and from Theorem 1.7.1, $\underset{\sim}{P}\sin\theta$ can be obtained as

$$\underset{\sim}{P}\sin\theta = \frac{1}{2}(\underset{\sim}{Q} - \underset{\sim}{Q}^T) \tag{2.5.14}$$

i.e. eq. (2.5.4) can be regarded as the cartesian decomposition (see Section 1.7) of matrix $\underset{\sim}{Q}$. Now, calling e_i the ith component of vector $\underset{\sim}{e}$, as given by eq. (2.5.3) and taking definition (2.5.5b) and eq. (2.5.14) into account, one obtains

$$-e_1\sin\theta = \frac{1}{2}(q_{23}-q_{32}) \qquad (2.5.15a)$$

$$e_2\sin\theta = \frac{1}{2}(q_{13}-q_{31}) \qquad (2.5.15b)$$

$$-e_3\sin\theta = \frac{1}{2}(q_{12}-q_{21}) \qquad (2.5.15c)$$

Introducing the alternating tensor ε_{ijk} defined as

$$\varepsilon_{ijk} = \begin{cases} +1, \text{if } i,j \text{ and } k \text{ are in cyclic order} \\ -1, \text{ if } i,j \text{ and } k \text{ are in anticyclic order} \\ 0, \text{ if at one index is repeated} \end{cases}$$

eqs. (2.5.15) can be written as

$$e_i\sin\theta = -\frac{1}{2}\varepsilon_{ijk}(q_{jk}-q_{kj})$$

from which, if e_i does not vanish,

$$\mathrm{sgn}\theta = \varepsilon_{ijk}\ \mathrm{sgn}\left(\frac{q_{kj}-q_{jk}}{e_i}\right)$$

follows directly.

Exercise 2.5.1 Given matrices $\underset{\sim}{T}$ and $\underset{\sim}{P}$, as defined in eqs. (2.5.5a) and (2.5.5b), prove that $\underset{\sim}{T} = -\underset{\sim}{P}^2$ and devise an algorithm to compute $\underset{\sim}{P}$ given $\underset{\sim}{T}$.

Exercise 2.5.2 Use eq. (2.5.16) to determine the sign of θ for the rotation matrix of Example 2.3.1 and verify the result thus obtained with the one obtained previously.

Exercise 2.5.3 (Euler's angles). A rigid body can attain an arbitrary configuration starting from any reference configuration, 0, by means of the composition of three rotations about coordinate axes as follows: Attach axes X_0, Y_0 and Z_0 to the body in the reference configuration and rotate the body through an angle ϕ about Z_0, thus carrying the axes into X_1, Y_1 and $Z_1(=Z_0)$, respectively. Next, rotate the body through an angle θ about axis Y_1, thus carrying the axes into X_2, Y_2 and Z_2, respectively. Finally, rotate the body through an angle ψ about Z_2 so that the axes coincide with their desired final positions, X_3, Y_3, Z_3. Angle ψ is chosen so that axis Z_3 be contained in the plane of Z_0 and X_1, whereas angle θ is chosen so as to carry axis $Z_1(=Z_0)$ into $Z_3(=Z_2)$. Show that the revolute carrying the body from configuration 0 to configuration 3 is

$$(\underset{\sim}{Q}_{03})_0 = \begin{pmatrix} c\theta c\phi c\psi - s\phi s\psi & -c\theta c\phi s\psi - s\phi c\psi & s\theta c\phi \\ c\theta s\phi c\psi + c\phi s\psi & -c\theta s\phi s\psi + c\phi c\psi & s\theta s\phi \\ -s\theta c\psi & s\theta s\psi & c\theta \end{pmatrix} \tag{2.5.16}$$

Moreover, show that the angle of rotation of $\underset{\sim}{Q}_{03}$, α , is given as

$$\cos\frac{\alpha}{2} = \cos\frac{\psi+\phi}{2}\cos\frac{\theta}{2} \tag{2.5.17}$$

2.6 GENERAL MOTION OF A RIGID BODY AND CHASLES' THEOREM

In the previous sections only the motion of a rigid body about a fixed point was discussed. There are rigid-body motions, however, with no fixed point. Such motions are studied in this section.

Consider a motion under which one point is displaced from A to A' and another one is displaced from R to R', as shown in Fig 2.6.1
This motion can take place in any of three different ways, namely i) any pair of points A,R of the body undergo a displacement to A', R', respectively in such a way that line A' R' is parallel to line AR; this motion is referred to as pure translation; ii) a line of the body remains fixed, in which case, according to Euler's Theorem (Theorem 2.3.1), the motion is referred to as pure rotation; iii) no point of the body remains fixed under the motion, in

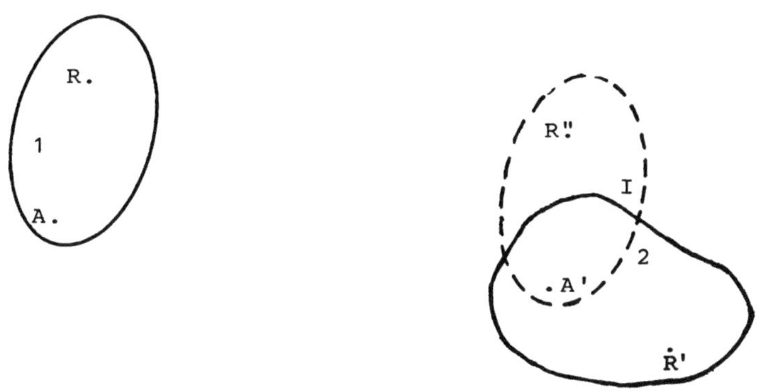

Fig 2.6.1 General motion of a rigid body

which case it is referred to as general motion.

The motion from configuration 1 to configuration 2 can be regarded as the composition of two motions: first the rigid-body is displaced from 1 to I without any rotation. Hence, the lines connecting any pair of points in 1 are parallel to those connecting them in the intermediate configuration I. Since this is a rigid-body motion, the length of each segment remains unchanged. Thus, letting $\underset{\sim}{a}, \underset{\sim}{a}'$, $\underset{\sim}{r}, \underset{\sim}{r}'$ and $\underset{\sim}{r}''$ be the position vectors of points A,A', R,R' and R", respectively,

$$\underset{\sim}{r}''-\underset{\sim}{a}'=\underset{\sim}{r}-\underset{\sim}{a} \tag{2.6.1}$$

Next, to take the body into its final configuration, 2, a rigid-body rotation $\underset{\sim}{Q}$, about point A', must be performed.

Thus,

$$\underset{\sim}{r}'-\underset{\sim}{a}'=\underset{\sim}{Q}(\underset{\sim}{r}''-\underset{\sim}{a}') \tag{2.6.2}$$

Substitution of (2.6.1) into (2.6.2) and rearrangement of the terms yield

$$\underset{\sim}{r}'=\underset{\sim}{a}'+\underset{\sim}{Q}(\underset{\sim}{r}-\underset{\sim}{a}) \tag{2.6.3}$$

which is an expression for the final position R' of any point R of the rigid body in terms of: i) its initial position, $\underset{\sim}{r}$, ii) the initial and the final position of any other point A, and iii) the rotation $\underset{\sim}{Q}$ accompanying the motion. The above expression could have also been obtained considering first a rigid-body rotation about point A from 1 to an intermediate configuration I' in which all lines connecting any pair of points are parallel to the corresponding lines in 2 and, since the motion is rigid, the segments thus defined are of equal lengths; then, perform a pure translation from I' to 2. Summarizing: The general motion of a rigid body is completely defined by the initial and final positions of any one of its points and the rotation involved.

<u>Exercise 2.6.1</u> Obtain eq. (2.6.3) by performing first a rotation and then a translation.

The main result in this section is <u>Chasles'Theorem</u>, which states that, given <u>any</u> rigid-body motion, it can always be obtained as the rotation about a line of the body, known as "the screw axis", followed by a translation parallel* to the axis of rotation. Moreover, the displacements of all points of the body along the screw axis are of minimum magnitude. The <u>displacement vector</u> of a point is defined as the vector between the final and the initial positions of the point, e.g. the displacement of point R in the previous discussion is

$$\underset{\sim}{u}=\underset{\sim}{r}'-\underset{\sim}{r}=\underset{\sim}{a}'+\underset{\sim}{Q}(\underset{\sim}{r}-\underset{\sim}{a})-\underset{\sim}{r} =$$
$$=\underset{\sim}{a}'-\underset{\sim}{Q}\underset{\sim}{a}+(\underset{\sim}{Q}-\underset{\sim}{I})\underset{\sim}{r} \tag{2.6.3a}$$

From eq. (2.6.3) notice that $\underset{\sim}{u}$ is a linear function of one single variable, $\underset{\sim}{r}$. Hence, the norm of $\underset{\sim}{u}$ is a linear function of $\underset{\sim}{r}$ only. The square of this norm is quadratic in $\underset{\sim}{r}$ and is given as

$$\phi(\underset{\sim}{r})=\underset{\sim}{u}^T\underset{\sim}{u}=\underset{\sim}{r}^T(\underset{\sim}{Q}-\underset{\sim}{I})^T(\underset{\sim}{Q}-\underset{\sim}{I})\underset{\sim}{r}+2(\underset{\sim}{a}'-\underset{\sim}{Q}\underset{\sim}{a})^T(\underset{\sim}{Q}-\underset{\sim}{I})\underset{\sim}{r}+(\underset{\sim}{a}'-\underset{\sim}{Q}\underset{\sim}{a})^T(\underset{\sim}{a}'-\underset{\sim}{Q}\underset{\sim}{a}) \tag{2.6.4}$$

The theorem is now proved via the minimization of $\phi(\underset{\sim}{r})$. This function has one extremum at the point $\underset{\sim}{r}_0$ where $\phi'(\underset{\sim}{r}_0)=\underset{\sim}{0}$. The derivative $\phi'(\underset{\sim}{r})$ is next computed, and zeroed at $\underset{\sim}{r}_0$.

Applying the "Chain rule" to ϕ,

$$\phi'(\underset{\sim}{r}) = \left(\frac{\partial \underset{\sim}{u}}{\partial \underset{\sim}{r}}\right)^T \frac{\partial \phi}{\partial u}$$

---

* The direction of a pure translation of a rigid body is understood here as the direction of the displacement vectors of the points of the body.

where, from eq. (2.6.3a),

$$\frac{\partial \mathbf{u}}{\partial \mathbf{r}} = \mathbf{Q}-\mathbf{I} \tag{2.6.5a}$$

and

$$\frac{\partial \phi}{\partial \mathbf{u}} = 2\mathbf{u} \tag{2.6.5b}$$

Thus, letting $\mathbf{u}_0=\mathbf{u}(\mathbf{r}_0)$, the zeroing of the gradient of ϕ at $\mathbf{r}=\mathbf{r}_0$ leads to

$$(\mathbf{Q}-\mathbf{I})^T\mathbf{u}_0=\mathbf{0} \tag{2.6.6}$$

or

$$\mathbf{Q}^T\mathbf{u}_0=\mathbf{u}_0 \tag{2.6.6a}$$

Now, if both sides of eq. (2.6.6) are multiplied by $\mathbf{Q}$, one obtains

$$\mathbf{Q}\mathbf{u}_0=\mathbf{u}_0 \tag{2.6.6b}$$

thereby concluding that the minimum-norm displacement $\mathbf{u}_0$ lies in the real spectral space of $\mathbf{Q}$, i.e., it is parallel to the axis of rotation of $\mathbf{Q}$. What is now left to complete the proof of Chasles' Theorem is to determine the set of points of the rigid body having a displacement vector parallel to the rotation axis. This is done next.

Substituting $\mathbf{u}$, evaluated at $\mathbf{r}_0$, as given by eq. (2.6.3a) into eq. (2.6.6), and rearranging terms leads to

$$(\mathbf{Q}-\mathbf{I})^T(\mathbf{Q}-\mathbf{I})\mathbf{r}_0=(\mathbf{Q}-\mathbf{I})^T(\mathbf{Q}\mathbf{a}-\mathbf{a}') \tag{2.6.6c}$$

from which $\mathbf{r}_0$ cannot be solved for, since $(\mathbf{Q}-\mathbf{I})$, and hence $(\mathbf{Q}-\mathbf{I})^T(\mathbf{Q}-\mathbf{I})$, is singular. In fact, it can be readily proved that this matrix is of rank 2.

<u>Exercise 2.6.1</u> Prove that $(\mathbf{Q}-\mathbf{I})^T(\mathbf{Q}-\mathbf{I})$ is of rank 2, except for $\mathbf{Q}=\mathbf{I}$.

Although $\mathbf{r}_0$ cannot be solved for from the latter equation, interesting results can be derived from it. Indeed, given a point R_0, with position vector $\mathbf{r}_0$, of minimum-magnitude displacement $\mathbf{u}_0$, define a new point S_0, with position vector $\mathbf{s}_0$ given as

$$\mathbf{s}_0 = \mathbf{r}_0 + \alpha\mathbf{e}$$

where $\mathbf{e}$ is the unit vector parallel to the axis of rotation of $\mathbf{Q}$. Multiplying $\mathbf{s}_0$, as given before, times $(\mathbf{Q}-\mathbf{I})^T(\mathbf{Q}-\mathbf{I})$ gives

$$(\mathbf{Q}-\mathbf{I})^T(\mathbf{Q}-\mathbf{I})\mathbf{s}_0=(\mathbf{Q}-\mathbf{I})^T(\mathbf{Q}-\mathbf{I})(\mathbf{r}_0+\alpha\mathbf{e})=(\mathbf{Q}-\mathbf{I})^T(\mathbf{Q}-\mathbf{I})\mathbf{r}_0+ (\mathbf{Q}-\mathbf{I})^T(\mathbf{Q}-\mathbf{I})\mathbf{e}$$

but $\mathbf{e}$, being parallel to the rotation axis of $\mathbf{Q}$, is in the null space of $\mathbf{Q}-\mathbf{I}$; hence, in the null space of $(\mathbf{Q}-\mathbf{I})^T(\mathbf{Q}-\mathbf{I})$. Therefore, the second term on the right-hand side of the latter equation vanishes, the latter equation thus reducing to

$$(\mathbf{Q}-\mathbf{I})^T(\mathbf{Q}-\mathbf{I})\ \mathbf{s}_0 = (\mathbf{Q}-\mathbf{I})^T(\mathbf{Q}\mathbf{a}-\mathbf{a}')$$

i.e. $\mathbf{s}_0$ also satisfies eq. (2.6.6c). In conclusion, <u>all points $\mathbf{s}_0$ of minimum-magnitude displacement, $\mathbf{u}_0$, lie on a line parallel to the axis of rotation of $\mathbf{Q}$.</u>

<u>Exercise. 2.6.2</u> Show that the $\mathbf{r}_0$ satisfying eq. (2.6.6) actually yields a minimum.

From Exercise 2.6.1, if $\mathbf{Q}\neq\mathbf{I}$, the rank of $Q-I$ is exactly 2. Therefore, two of the three scalar equations of (2.6.5) are linearly independent. These two equations can be expressed in matrix form as

$$\mathbf{A}\mathbf{r}_0 = \mathbf{c} \tag{2.6.7}$$

where $\mathbf{A}$ is a 2 x 3-rank-two matrix and vectors $\mathbf{r}_0$ and $\mathbf{c}$ are 3-and 2-dimensional, respectively. Now, since the rank of $\mathbf{A}$ is 2, $\mathbf{A}\mathbf{A}^T$, being 2x2, is nonsingular and hence, the minimum-norm solution to eq. (2.6.7) is (See Section 1.11)

$$\mathbf{r}_0=\mathbf{A}^T(\mathbf{A}\mathbf{A}^T)^{-1}\mathbf{c} \tag{2.6.8}$$

The geometric interpretation of the previous result is that $\mathbf{r}_0$, as given by (2.6.8), is perpendicular to the sought axis. This axis is "<u>the screw</u>

axis" and is totally determined by the axis of rotation, which gives its direction, and the point R_0 whose position vector, $\mathbf{r}_0$, is given by eq. (2.6.8). The name "screw" comes from the fact that the body moves as if it were fastened to the bolt of a screw whose axis were the screw axis. Other facts motivating the name of the screw axis will be shown later. Another method of finding a point on the screw axis is via Rodrigues' formula as it appears in [2.6]. This procedure can be developed as follows: As was pointed out from eq. (2.6.6), the minimum-norm displacement is parallel to the axis of rotation. Hence, the displacement of R_0 must satisfy

$$\mathbf{u}_0=\mathbf{r}_0' -\mathbf{r}_0=\alpha\mathbf{e} \tag{2.6.9}$$

where α is a scalar. Substituting the initial and the final position vectors of R in Rodrigues' formula, eq. (2.5.3),

$$\mathbf{r}_0'-\mathbf{r}_0=\tan\frac{\theta}{2}\,\mathbf{e}\times(\mathbf{r}_0'+\mathbf{r}_0) \tag{2.6.10}$$

which, together with eq. (2.5.3) for vectors $\mathbf{a}$ and $\mathbf{a}'$, denoting the initial and the final positions of point A, yields

$$\mathbf{a}'-\mathbf{r}_0'-(\mathbf{a}-\mathbf{r}_0)=\tan\frac{\theta}{2}\mathbf{e}\times\big((\mathbf{a}'-\mathbf{r}_0')+(\mathbf{a}-\mathbf{r}_0)\big) \tag{2.6.11}$$

From eq. (2.6.9),

$$\mathbf{e}\times(\mathbf{r}_0'-\mathbf{r}_0)=\mathbf{0} \tag{2.6.12}$$

Hence, eq. (2.6.11) becomes

$$\mathbf{a}'-\mathbf{a}-\alpha\mathbf{e}=\tan\frac{\theta}{2}\mathbf{e}\times(\mathbf{a}'+\mathbf{a})-2\tan\frac{\theta}{2}\,\mathbf{e}\times\mathbf{r}_0 \tag{2.6.13}$$

Multiplying both sides of eq. (2.6.13) times $\cot\frac{\theta}{2}\mathbf{e}\times$,

$$\begin{aligned}\cot\frac{\theta}{2}\mathbf{e}\times(\mathbf{a}'-\mathbf{a})&=\mathbf{e}\times\big(\mathbf{e}\times(\mathbf{a}'+\mathbf{a})\big)-2\mathbf{e}\times(\mathbf{e}\times\mathbf{r}_0) =\\ &=\mathbf{e}\times\big(\mathbf{e}\times(\mathbf{a}'+\mathbf{a})\big)-2(\mathbf{e}\cdot\mathbf{r}_0)\mathbf{e}+2(\mathbf{e}\cdot\mathbf{e})\mathbf{r}_0\end{aligned} \tag{2.6.14}$$

To determine $\mathbf{r}_0$ from eq. (2.6.14, it is necessary to impose one extra condition on it, which is done next. Let be the particular point on the screw axis which is closet to the origin; hence,

$$\mathbf{r}_0 \cdot \mathbf{e} = 0$$

and so, substituting this vector into eq. (2.6.14) and solving for $\mathbf{r}_0$ in the same equation, leads to

$$\mathbf{r}_0 = \frac{1}{2}\cot\frac{\theta}{2}\mathbf{e}\times(\mathbf{a}'-\mathbf{a}) - \frac{1}{2}\mathbf{e}\times\left(\mathbf{e}\times(\mathbf{a}'+\mathbf{a})\right) \qquad (2.6.15)$$

which is an alternate expression for $\mathbf{r}_0$. The foregoing result is summarized next.

THEOREM 2.6.1 (CHASLES). *The most general displacement of a rigid body is equivalent to a translation together with a rotation about an axis parallel to the translation.*

Alernatively, Chasles'Theorem can be stated as follows:

"*Given an arbitrary displacement of a rigid body, there exists a set of points of the body, constituting a line, such that all points on that line undergo a displacement parallel to the line, which is of minimum Euclidean norm*"

A property of the screw axis is established in the next theorem.

THEOREM 2.6.2 *The displacement vectors of all the points of a rigid body undergoing an arbitrary motion have the same projection along the screw axis.*

Proof:

Let P be an arbitrary point of a rigid body and S a point on the screw axis; let P' and S' represent the corresponding points after the displacement. From eq. (2.6.3), the displacement of P, $\mathbf{u}_P$, is given in terms of the position vectors of P, S and S', by

$$\mathbf{u}_P = \mathbf{s}' - \mathbf{Q}\mathbf{s} + (\mathbf{Q} - \mathbf{I})\mathbf{p} \qquad (2.6.16)$$

The projection of $\underset{\sim}{u}_P$ onto the screw axis is computed now by obtaining the scalar product of $\underset{\sim}{u}_P$ times $\underset{\sim}{u}_S$. From eq. (2.6.16) this becomes

$$\underset{\sim}{u}_P^T\underset{\sim}{u}_S=(\underset{\sim}{s}'-\underset{\approx}{Q}\underset{\sim}{s})^T\underset{\sim}{u}_S+\underset{\sim}{p}^T(\underset{\approx}{Q}-\underset{\sim}{I})^T\underset{\sim}{u}_S \tag{2.6.17}$$

where the second term on the right hand side vanishes because, as already shown, $\underset{\sim}{u}_S$ is an eigenvector of $\underset{\approx}{Q}$ and $\underset{\approx}{Q}^T$. Thus, eq. (2.6.17) becomes

$$\underset{\sim}{u}_P^T\underset{\sim}{u}_S=(\underset{\sim}{s}')^T\underset{\sim}{u}_S-\underset{\sim}{s}^T\underset{\sim}{u}_S=(\underset{\sim}{s}'-\underset{\sim}{s})^T\underset{\sim}{u}_S=\underset{\sim}{u}_S^T\underset{\sim}{u}_S$$

From the above expression it follows that the projection of $\underset{\sim}{u}_P$ onto the screw axis has lenght $||\underset{\sim}{u}_S||$, q.e.d

Using the same notation as above, the final position vector of a point of a rigid body undergoing an arbitrary motion and its displacement can be expressed as

$$\underset{\sim}{p}'=\underset{\sim}{p}+\underset{\sim}{u}_P \tag{2.6.18}$$

$$\underset{\sim}{u}_P=\underset{\sim}{u}_S+(\underset{\approx}{Q}-\underset{\sim}{I})(\underset{\sim}{p}-\underset{\sim}{s}) \tag{2.6.19}$$

where the two terms of the right hand side are orthogonal.

Hence it is clear that the displacement of any point of the rigid body is known if the following quantities are given:

i) The magnitude of the screw displacement, $||\underset{\sim}{u}_S||$

ii) One point of the screw axis, R_0, whose position vector is $\underset{\sim}{r}_0$

iii) The axis of rotation, $\underset{\sim}{e}$

iV) The angle of rotation, θ

Given the above data, vector $\underset{\sim}{u}_S$ is obtained as

$$\underset{\sim}{u}_S=||\underset{\sim}{u}_S||\underset{\sim}{e} \tag{2.6.20}$$

and matrix $\underset{\sim}{Q}$ is given by eqs.(2.5.1) or (2.5.4). Point R_0 and vector $\underset{\sim}{e}$ completely determine the screw axis, henceforth called L. From Theorem 2.6.2 it is clear that a rigid body undergoing an arbitrary motion, moves as if it were welded to the bolt of a screw whose axis where L and whose pitch were given by

$$\lambda = \frac{2\pi\|\underset{\sim}{u}_S\|}{\theta} \tag{2.6.21}$$

For this reason, the pair $(L,\underset{\sim}{Q})$, which completely determines a rigid-body motion, is called a "screw", and rigid-body motions are thus referred to as "screw motions". It was shown in section 2.3 how to obtain the matrix $\underset{\sim}{Q}$, given a rigid-body motion with a fixed point. Vectors $\underset{\sim}{r}_0$ and $\underset{\sim}{u}_S$, which define L, are obtained from eqs. (2.6.15) or, alternatively, from eq. (2.6.7) and eq. (2.6.20).

The following interesting result is derived immediately from Theorem 2.6.2

Corollary 2.6.1 *A rigid-body motion is a rotation about a fixed point if and only if the orthogonal projection of the displacement of one point of the body on the screw axis vanishes.*

Another useful result is the following

Corollary 2.6.2 *The difference vector of the displacements of any two of the points of a rigid body undergoing an arbitrary motion is perpendicular to the screw axis.*

Exercise 2.6.3 Prove corollaries 2.6.1 and 2.6.2

Clearly, the motion of any one plane of a rigid body completely determines the motion of the body. Futhermore, three noncollinear points determine a plane; thus it follows that the motion of any three noncollinear points of a rigid body determine the motion of the body. In other words, knowing the initial and the final positions of three noncollinear points of a rigid body, one can determine the axis, the displacement and the rotation of the corresponding screw. In the following, formulae are derived to compute the screw parameters of a rigid-body motion in terms of the motion of three noncollinear points of the body. The usually applied formulae require that the displacements of the involved points be noncoplanar. Now, if three points of a rigid body are collinear, their displacements under any motion are coplanar. The converse, however, is not true, for three noncollinear points of a rigid body could have, under special circumstances, coplanar displacements, as is proved in Theorem 2.6.4. To prove this, a previous result is derived in the following

THEOREM 2.6.3 *If the displacements of three noncollinear points of a rigid body are identical, the body undergoes a pure translation.*

Proof:

Let A,B, C be three noncollinear points of a rigid body, and $\underset{\sim}{a},\underset{\sim}{b},\underset{\sim}{c}$ their respective position vectors. Using eq. (2.6.19), the displacements of these points can be written as

$$\underset{\sim}{u}_A=\underset{\sim}{u}_S+(\underset{\sim}{Q}-\underset{\sim}{I})(\underset{\sim}{a}-\underset{\sim}{s})$$

$$\underset{\sim}{u}_B=\underset{\sim}{u}_S+(\underset{\sim}{Q}-\underset{\sim}{I})(\underset{\sim}{b}-\underset{\sim}{s})$$

$$\underset{\sim}{u}_C=\underset{\sim}{u}_S+(\underset{\sim}{Q}-\underset{\sim}{I})(\underset{\sim}{c}-\underset{\sim}{s})$$

where $\underset{\sim}{s}$ is the position vector of a point S on the screw axis. Subtracting the third equation from the first and the second ones, and recalling that the three displacements are identical, one obtains

$$(\underset{\sim}{Q}-\underset{\sim}{I})(\underset{\sim}{a}-\underset{\sim}{c})=\underset{\sim}{0}$$

and

$$(\underset{\sim}{Q}-\underset{\sim}{I})(\underset{\sim}{b}-\underset{\sim}{c})=\underset{\sim}{0}$$

Hence both $\underset{\sim}{a}-\underset{\sim}{c}$ and $\underset{\sim}{b}-\underset{\sim}{c}$ ε $N(\underset{\sim}{Q}-\underset{\sim}{I})$*, i.e. $\underset{\sim}{a}-\underset{\sim}{c}$ and $\underset{\sim}{b}-\underset{\sim}{c}$ lie in the same one-dimensional space spanned by the real eigenvector of $\underset{\sim}{Q}$. This cannot be so

* $N(\underset{\sim}{T})$ and $R(\underset{\sim}{T})$ denote the null space and the range of $\underset{\sim}{T}$, as defined in Section 1.3

because A,B,C were assumed to be noncollinear. Thus, the only possibility for the two latter equations to hold is that $\underset{\sim}{Q}=\underset{\sim}{I}$, i.e. the motion contains no rotation and hence is a pure translation. q.e.d.

THEOREM 2.6.4 *The non-identical* displacements of three points of a rigid body are coplanar if and only if one of the following three conditions is met:*

i) The motion is a pure rotation

ii) The motion is general, but the points are collinear

iii) The motion is general and the points are not collinear, but lie in a plane parallel to the screw axis.

Proof:

(Sufficiency)

i) If the motion is a pure rotation and the origin of coordinates is located along the axis of rotation, the displacement $\underset{\sim}{u}$ of any point of position vector $\underset{\sim}{r}$ is then

$$\underset{\sim}{u}=(\underset{\sim}{Q}-\underset{\sim}{I})\underset{\sim}{r}$$

i.e. $\underset{\sim}{u}\varepsilon R(\underset{\sim}{Q}-\underset{\sim}{I})$. Since $N(\underset{\sim}{Q}-\underset{\sim}{I})$ is of dimension 1, namely the axis of rotation, then from eq. (1.3.1), $R(\underset{\sim}{Q}-\underset{\sim}{I})$ is of dimension 2, namely a plane passing through the origin, normal to the axis of rotation. Thus, all displacements are coplanar, thereby proving this part.

ii) Let A,B and C be the given three collinear points of the rigid body undergoing a general motion. Let $\underset{\sim}{a}$,$\underset{\sim}{b}$ and $\underset{\sim}{c}$ be their respective position vectors. Hence, vectors $\underset{\sim}{c}-\underset{\sim}{a}$ and $\underset{\sim}{b}-\underset{\sim}{a}$ are linearly dependent and they are related by

---

*If the displacements of the three noncollinear points were identical, the motion would be a pure translation, according to Theorem 2.6.3.

$$\mathbf{c}-\mathbf{a}=\alpha(\mathbf{b}-\mathbf{a}) \tag{ii.1}$$

From eq. (2.6.3),

$$\mathbf{u}_C=\mathbf{a}'+\mathbf{Q}(\mathbf{c}-\mathbf{a})-\mathbf{c}$$

$$=\mathbf{a}'-\mathbf{a}+\mathbf{c}+(\mathbf{Q}-\mathbf{I})(\mathbf{c}-\mathbf{a})-\mathbf{c}$$

$$=\mathbf{u}_A+\alpha(\mathbf{Q}-\mathbf{I})(\mathbf{b}-\mathbf{a}) \tag{ii.2}$$

But, also from eq. (2.6.3),

$$(\mathbf{Q}-\mathbf{I})(\mathbf{b}-\mathbf{a})=\mathbf{u}_B-\mathbf{u}_A \tag{ii.3}$$

Hence, eq. (ii.2) can be written as

$$\mathbf{u}_C=(1-\alpha)\mathbf{u}_A+\alpha\mathbf{u}_B$$

thus making evident that the three involved displacements are coplanar.

iii) Using eq. (2.6.19), the displacements of points A, B and C are

$$\mathbf{u}_A=\mathbf{u}_S+(\mathbf{Q}-\mathbf{I})(\mathbf{a}-\mathbf{s})$$

$$\mathbf{u}_B=\mathbf{u}_S+(\mathbf{Q}-\mathbf{I})(\mathbf{b}-\mathbf{s})$$

$$\mathbf{u}_C=\mathbf{u}_S+(\mathbf{Q}-\mathbf{I})(\mathbf{c}-\mathbf{s})$$

Since A, B and C lie in a plane parallel to the screw axis, vectors $\mathbf{b}-\mathbf{a}$, $\mathbf{c}-\mathbf{a}$ and $\mathbf{u}_S$ are coplanar and hence they can be related as

$$\mathbf{c}-\mathbf{a}=\alpha(\mathbf{b}-\mathbf{a})+\beta\mathbf{u}_S$$

or

$$\mathbf{c}=(1-\alpha)\mathbf{a}+\alpha\mathbf{b}+\beta\mathbf{u}_S$$

Substituting the latter expression into $\mathbf{u}_C$, after cancellations and rearrangement of terms,

$$\mathbf{u}_C=\mathbf{u}_A-\alpha(\mathbf{Q}-\mathbf{I})(\mathbf{a}-\mathbf{b})$$

But, from the above expressions for $\mathbf{u}_A$ and $\mathbf{u}_B$,

$$\mathbf{u}_A-\mathbf{u}_B=(\mathbf{Q}-\mathbf{I})(\mathbf{a}-\mathbf{b})$$

and so, from the latter expressions for $\underset{\sim}{u}_C$,

$$\underset{\sim}{u}_C=(1-\alpha)\underset{\sim}{u}_A+\alpha\underset{\sim}{u}_B$$

thereby showing the linear dependence, of the three involved displacements Hence, the said displacements have been shown to be coplanar.

(Necessity)

If $\underset{\sim}{u}_A$, $\underset{\sim}{u}_B$ and $\underset{\sim}{u}_C$ are coplanar, then

$$\det(\underset{\sim}{u}_A,\underset{\sim}{u}_B,\underset{\sim}{u}_C)=0$$

Introducing eq. (2.6.3), $\underset{\sim}{u}_B$ and $\underset{\sim}{u}_C$ can be written as

$$\underset{\sim}{u}_B=\underset{\sim}{u}_A+(\underset{\sim}{Q}-\underset{\sim}{I})(\underset{\sim}{b}-\underset{\sim}{a})$$

$$\underset{\sim}{u}_C=\underset{\sim}{u}_A+(\underset{\sim}{Q}-\underset{\sim}{I})(\underset{\sim}{c}-\underset{\sim}{a})$$

Hence, the above condition, regarding the vanishing of the determinant, can be weitten, after proper simplifications, as

$$\det\left(\underset{\sim}{u}_A,(\underset{\sim}{Q}-\underset{\sim}{I})(\underset{\sim}{b}-\underset{\sim}{a}),(\underset{\sim}{Q}-\underset{\sim}{I})(\underset{\sim}{c}-\underset{\sim}{a})\right) = 0$$

or, in Gibbs notation,

$$(\underset{\sim}{Q}-\underset{\sim}{I})(\underset{\sim}{b}-\underset{\sim}{a})\times(\underset{\sim}{Q}-\underset{\sim}{I})(\underset{\sim}{c}-\underset{\sim}{a})\cdot\underset{\sim}{u}_A=0$$

From (2.6,p.5), the first product can be expressed as

$$(\underset{\sim}{Q}-\underset{\sim}{I})(\underset{\sim}{b}-\underset{\sim}{a})\times(\underset{\sim}{Q}-\underset{\sim}{I})(\underset{\sim}{c}-\underset{\sim}{a})=\alpha\underset{\sim}{e}$$

where

$$\alpha=2(1-\cos\theta)\underset{\sim}{e}\times(\underset{\sim}{b}-\underset{\sim}{a})\cdot(\underset{\sim}{c}-\underset{\sim}{a})$$

θ and $\underset{\sim}{e}$ being the angle of rotation and the unit vector parallel to the axis of this rotation.

<u>Exercise 2.6.4</u> Derive the above expression for α

The double product thus can vanish if any one of the following conditions is met:

i) $\underset{\sim}{e}\cdot\underset{\sim}{u}_A=0$

which, from Corollary 2.6.1, states that the body undergoes a pure rotation

ii) $\alpha=0$

which is satisfied under one of the two following conditions:

ii.1) $1-\cos\theta=0$

which implies $\theta=0$; i.e. the motion reduces to a pure translation. This case, however, has been discarded in the present analysis, for the displacements have been assumed to be non-identical (See Theorem 2.6.3)

ii.2) $\underset{\sim}{e}\underset{\sim}{x}(\underset{\sim}{b}-\underset{\sim}{a}).(\underset{\sim}{c}-\underset{\sim}{a})=0$

which in turn is satisfied under one of the following two conditions:

ii.2.a) $\underset{\sim}{e}$, $\underset{\sim}{b}-\underset{\sim}{a}$ and $\underset{\sim}{c}-\underset{\sim}{a}$ are coplanar, i.e. points A, and C lie in a plane parallel to the axis of rotation (Picture it!)

ii.2.b) $(\underset{\sim}{b}-\underset{\sim}{a})x(\underset{\sim}{c}-\underset{\sim}{a})=0$

which implies that A,B and C are collinear, thus completing the proof.

Corollary 2.6.3 *Assume a rigid body undergoes an arbitrary motion and choose any three noncollinear points of the body, A,B and C. Letting $\underset{\sim}{u}_A$, $\underset{\sim}{u}_B$ and $\underset{\sim}{u}_C$ be the three resulting displacements, then the two difference vectors $\underset{\sim}{u}_A-\underset{\sim}{u}_C$ and $\underset{\sim}{u}_B-\underset{\sim}{u}_C$ are parallel if and only if the points lie in a plane parallel to the screw axis (if the motion is general. If the motion is a pure rotation, these points lie in a plane parallel to the axis of rotation.)*

Exercise 2.6.5 Prove Corollary 2.6.3

Further consequences of Theorem 2.6.4 are next stated.

Corollary 2.6.4. The displacements of any two points of a rigid body cannot be parallel and different, unless the body undergoes a pure rotation

<u>Exercise 2.6.6</u> Prove Corollary 2.6.4. Hint: Use eq. (2.6.19) and the fact that the two terms of its right-hand side are orthogonal.

<u>Corollary 2.6.5</u> If the displacements of any two points of a rigid body are parallel, then either i) the displacements are identical and belong to points lying on a line parallel to the screw axis (or to the axis of rotation, if the motion is one of pure rotation), or ii) the displacements are different, in which case the motion is a pure rotation and the said two points lie on a line intersecting the axis of rotation.

<u>Corollary 2.6.6</u> Keeping the notation introduced in Theorem 2.6.3, let a rigid body motion be defined by $\underset{\sim}{u}_B=\beta\underset{\sim}{u}_A\neq\underset{\sim}{0}$ and $\underset{\sim}{u}_C=\underset{\sim}{0}$, β being a nonzero scalar. The motion is then a pure rotation and its axis is parallel to vector $\underset{\sim}{b}-\underset{\sim}{c}-\beta(\underset{\sim}{a}-\underset{\sim}{c})$.

<u>Exercise 2.6.7</u> Prove Corollaries 2.6.5 and 2.6.6

The procedure enabling one to compute the screw parameters can now be established. This is based upon the "tree-diagram" shown in Fig 2.6.2, which is next described. Given the <u>compatible</u> displacement vectors of three noncollinear points of a rigid body, these can exhibit several different relationships amongst them, all of which are shown in that figure. A set of displacement vectors of points of a rigid body are said to be compatible if they are so defined as to meet the definition of a rigid-body motion given in Sect. 2.2, i.e. if they correspond to a set of displaced configurations of points amongst which the distances are preserved. Even if the distances are preserved, however, the displaced configuration could involve a reflection instead of a rotation. This situation is discarded within this context because the motion is supposed to be defined by three points, and the involved transformation of points is supposed to be a rigid-body motion.

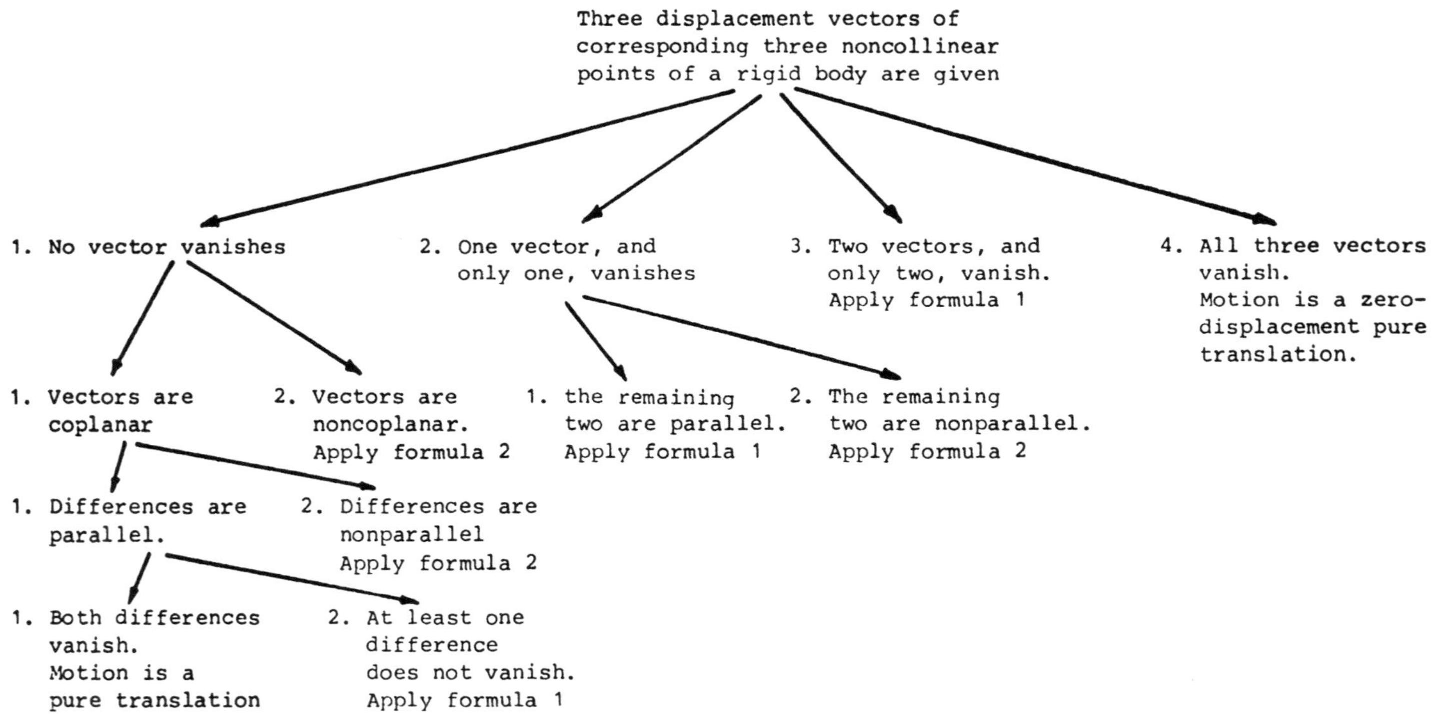

Fig.2.6.2. **Tree diagram** showing the different possible relationships amongst the displacements of three noncollinear points defining a rigid-body motion.

In order to define a reflection, four points are needed. The three displacement vectors can exhibit the following relationships amongst them;

CASE 1. No vector vanishes. This in turn comprises the following cases:

1.1. The given vectors are coplanar, which in turn comprises the following

1.1.1 The difference vectors are parallel, which again comprises two cases, namely

1.1.1.1 Both difference vectors vanish, in which case the motion is a pure translation.

1.1.1.2 At least one diference vector does not vanish

1.1.2 The difference vectors are nonparallel

1.2. The given vectors are noncoplanar

CASE 2. One vector, and only one, vanishes. This, in turn comprises the following subcases.

2.1. The nonvanishing vectors are parallel.

2.2. The nonvanishing vectors are nonparallel.

CASE 3. Two vectors, and only two, vanish.

CASE 4. All three vectors vanish. The motion is a pure displacement with zero displacement.

The crucial step in the present computational procedure is the computation of $\underset{\sim}{e}$, the unit vector parallel to the screw axis. To compute $\underset{\sim}{e}$ is a very simple matter if the displacement vectors observe the relationship of case 2. If they correspond to cases 1 or 3, however, a new motion can be defined arising from the given one,that falls either in case 2.1 or in case 2.2, as will be shown next. Case 4 is of course trivial and thus need not be discussed. In either case 2.1 or 2.2, a particular procedure to compute $\underset{\sim}{e}$ should be followed. Once $\underset{\sim}{e}$ is known, the remaining parameters, i.e. the angle of rotation, θ, the position vector of the point on the screw axis lying closest to the origin, $\mathbf{r}_0$, and the slide displacement, s, defined as the orthogonal projection of the displacement field on the screw axis, can be computed resorting to identical formulae, regardless of the case at hand, except for $\theta = 180°$, in which case $\mathbf{r}_0$ cannot be computed using the general formula. An alternate one is resorted to in this case.

In order to prove that the computation of $\mathbf{e}$ for cases 1 or 3 can be reduced to that of case 2, two previous results will be proved, namely*

<u>*Theorem 2.6.5*</u> *Given a rigid-body motion defined by the displacements of three noncollinear points, such that no displacement vanishes, define a new motion by assigning new displacement vectors* $\mathbf{u}'_A$, $\mathbf{u}'_B$ *and* $\mathbf{u}'_C$ *to points A, B and C, respectively. If these vectors are defined as*

$$\mathbf{u}'_A = \mathbf{u}_A - \mathbf{u}_C, \mathbf{u}'_B = \mathbf{u}_B - \mathbf{u}_C, \mathbf{u}'_C = \mathbf{0}$$

then the matrix describing the rotation involved in the new motion is identical to that in the given one.

<u>Proof</u>:

From eq. (2.6.3a), $\mathbf{u}_A$ can be written as

$$\mathbf{u}_C = \mathbf{a}' + \mathbf{Q}(\mathbf{c}-\mathbf{a}) - \mathbf{c} \tag{2.6.22}$$

where, clearly, instead of $\mathbf{r}$, one has written $\mathbf{c}$. Adding $\mathbf{a}$ to the right-hand side of eq. (2.6.22) and subtracting it from the same side does not alter that equation. Hence,

$$\mathbf{u}_C = \mathbf{a}'-\mathbf{a} + \mathbf{Q}(\mathbf{c}-\mathbf{a})-(\mathbf{c}-\mathbf{a})$$

which can readily be rewritten as

$$\mathbf{u}_C = \mathbf{u}_A + (\mathbf{Q}-\mathbf{I})(\mathbf{c}-\mathbf{a})$$

* The notation introduced in Theorem 2.6.3 still holds.

or, alternatively, as

$$\underset{\sim}{u}_A = \underset{\sim}{u}_C + (\underset{\approx}{Q}-\underset{\sim}{I})(\underset{\sim}{a}-\underset{\sim}{c}) \tag{2.6.23a}$$

Similarly,

$$\underset{\sim}{u}_B = \underset{\sim}{u}_C + (\underset{\approx}{Q}-\underset{\sim}{I})(\underset{\sim}{b}-\underset{\sim}{c}) \tag{2.6.23b}$$

From the latter equations one has immediately

$$\underset{\sim}{u}'_A \equiv \underset{\sim}{u}_A-\underset{\sim}{u}_C = (\underset{\approx}{Q}-\underset{\sim}{I})(\underset{\sim}{a}-\underset{\sim}{c})$$

and

$$\underset{\sim}{u}'_B \equiv \underset{\sim}{u}_B-\underset{\sim}{u}_C = (\underset{\approx}{Q}-\underset{\sim}{I})(\underset{\sim}{b}-\underset{\sim}{c})$$

or, since u'_C has been defined as zero,

$$\underset{\sim}{u}'_A = \underset{\sim}{u}'_C + (\underset{\approx}{Q}-\underset{\sim}{I})(\underset{\sim}{a}-\underset{\sim}{c}) \tag{2.6.24a}$$

$$\underset{\sim}{u}'_B = \underset{\sim}{u}'_C + (\underset{\approx}{Q}-\underset{\sim}{I})(\underset{\sim}{b}-\underset{\sim}{c}) \tag{2.6.24b}$$

which are expressions formally identical to eqs. (2.6.23a and b), with $\underset{\sim}{u}_A$, $\underset{\sim}{u}_B$ and $\underset{\sim}{u}_C$ exchanged for $\underset{\sim}{u}'_A$, $\underset{\sim}{u}'_B$ and $\underset{\sim}{u}'_C$. Vectors $\underset{\sim}{a}$, $\underset{\sim}{b}$ and $\underset{\sim}{c}$ and matrix $\underset{\approx}{Q}$ remain in the second pair of expressions, however, which completes the proof.

<u>*Theorem 2.6.6*</u> *Given a rigid-body motion defined by the displacements of three noncollinear points, such that two of these, and only two, vanish, define a new motion by assigning new displacement vectors $\underset{\sim}{u}'_A$, $\underset{\sim}{u}'_B$, $\underset{\sim}{u}'_C$ to points A, B and C, respectively. Letting $\underset{\sim}{u}_C$ be the unique nonvanishing vector, define $\underset{\sim}{u}'_A = -\underset{\sim}{u}_C$, $\underset{\sim}{u}'_B = -\underset{\sim}{u}_C$, $\underset{\sim}{u}'_C = \underset{\sim}{0}$.*
Then, the matrix describing the rotation involved in the new motion is identical to that in the given one.

<u>Proof:</u>

Since eqs. (2.6.23a and b) are valid regardless of whether any involved displacement vanishes, one can apply them to the case at hand, i.e. for $\underset{\sim}{u}_A=\underset{\sim}{u}_B=\underset{\sim}{0}$, $\underset{\sim}{u}_C \neq \underset{\sim}{0}$. Thus,

$$\underset{\sim}{0} = \underset{\sim}{u}_C + (\underset{\sim}{Q}-\underset{\sim}{I})(\underset{\sim}{a}-\underset{\sim}{c}) \tag{2.6.25a}$$

$$\underset{\sim}{0} = \underset{\sim}{u}_C + (\underset{\sim}{Q}-\underset{\sim}{I})(\underset{\sim}{b}-\underset{\sim}{c}) \tag{2.6.25b}$$

and hence,

$$-\underset{\sim}{u}_C = \underset{\sim}{0} + (\underset{\sim}{Q}-\underset{\sim}{I})(\underset{\sim}{a}-\underset{\sim}{c})$$

$$-\underset{\sim}{u}_C = \underset{\sim}{0} + (\underset{\sim}{Q}-\underset{\sim}{I})(\underset{\sim}{b}-\underset{\sim}{c})$$

Recalling the given definitions of $\underset{\sim}{u}'_A$, $\underset{\sim}{u}'_B$ and $\underset{\sim}{u}'_C$ one obtains readily

$$\underset{\sim}{u}'_A = \underset{\sim}{u}'_C + (\underset{\sim}{Q}-\underset{\sim}{I})(\underset{\sim}{a}-\underset{\sim}{c}) \tag{2.6.26a}$$

$$\underset{\sim}{u}'_B = \underset{\sim}{u}'_C + (\underset{\sim}{Q}-\underset{\sim}{I})(\underset{\sim}{b}-\underset{\sim}{c}) \tag{2.6.26b}$$

and the proof follows by introducing the argument resorted to for proving Theorem 2.6.5.

As an application of the foregoing results, one can compute $\underset{\sim}{e}$ for the motion at hand by first defining new motions leading to Case 2.

Next, the computation of $\underset{\sim}{e}$ for Case 2 is discussed. Since this contains two subcases, each is discussed separately.

CASE 2.1: One vector, and only one, vanishes, the remaining two being parallel. Let $\underset{\sim}{u}_B = \beta\underset{\sim}{u}_A \neq \underset{\sim}{0}$, $\underset{\sim}{u}_C = \underset{\sim}{0}$, $\beta \neq 0$ Then, according to Corollary 2.6.6, the screw axis is parallel to vector

$$\underset{\sim}{w} = \underset{\sim}{b}-\underset{\sim}{c}-\beta(\underset{\sim}{a}-\underset{\sim}{c}) \tag{2.6.27}$$

Vector $\underset{\sim}{e}$ is thus computed as

$$\underset{\sim}{e} = \underset{\sim}{w} / ||\underset{\sim}{w}|| \tag{2.6.28}$$

CASE 2.2: One vector, and only one, vanishes, the remaining two being nonparallel. Let $\underset{\sim}{u}_C = \underset{\sim}{0}$ and $\underset{\sim}{u}_A \times \underset{\sim}{u}_B \neq \underset{\sim}{0}$ Then, the screw axis is perpendicular to both $\underset{\sim}{u}_A$ and $\underset{\sim}{u}_B$ and hence, is given as

$$\underset{\sim}{w} = \underset{\sim}{u}_A \times \underset{\sim}{u}_B \tag{2.6.29}$$

$$\underset{\sim}{e} = \underset{\sim}{w}/||\underset{\sim}{w}|| \tag{2.6.30}$$

From the foregoing results, then follows that Cases 1.1.1.2 and 3 give rise to a new motion falling into case 2.1. Cases 1.1.2 and 1.2 lead to motions falling into case 2.2. Cases 1.1.1.1 and 4 are pure translations, their corresponding screw parameters thus being very simple to compute. Indeed, the direction $\underset{\sim}{e}$ of the screw axis for Case 1.1.1.1 is that of the given identical displacements, the angle of rotation is zero, the slide displacement is the magnitude of the given displacements and the point of the screw axis lying closest to the origin can be arbitrarily defined as the origin itself, since all points of the body possess identical displacements. Once $\underset{\sim}{e}$ is known, θ can be computed directly as the angular displacement of a line of the rigid body undergoing a pure rotation, either falling in case 2.1 or in case 2.2. In fact, considering either of these two cases, and referring to Fig 2.6.3, the orthogonal projections of points A and A' (the displaced position of point A) on the axis of rotation, L, coincide at A". According to that figure, then

$$\theta = \cos^{-1} \frac{\overrightarrow{A''A} \cdot \overrightarrow{A''A'}}{||\overrightarrow{A''A}||^2} \qquad (2.6.31)$$

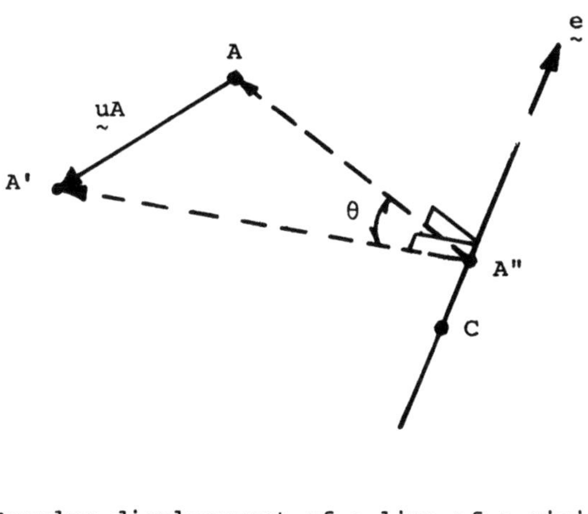

Fig 2.6.3 Angular displacement of a line of a rigid body undergoing a pure rotation.

where

$$\overrightarrow{A''A} = \underset{\sim}{a}-\underset{\sim}{c}- (\underset{\sim}{a}-\underset{\sim}{c}) \cdot \underset{\sim}{e}\underset{\sim}{e} \qquad (2.6.32a)$$

and

$$\overrightarrow{A''A'} = \underset{\sim}{a}'-\underset{\sim}{c}-(\underset{\sim}{a}'-\underset{\sim}{c}).\underset{\sim}{e}\underset{\sim}{e}=\underset{\sim}{a}+\underset{\sim}{u}_A-\underset{\sim}{c}-(\underset{\sim}{a}+\underset{\sim}{u}_A-\underset{\sim}{c}).\underset{\sim}{e}\underset{\sim}{e} \tag{2.6.32b}$$

Notice, however, that eq. (2.6.31) does not provide the sign of θ with respect to $\underset{\sim}{e}$. This is determined according to Theorem 2.3.2 as

$$\mathrm{sgn}(\theta) = \mathrm{sgn}\,(\overrightarrow{A''A}'\mathrm{x}\,\overrightarrow{A''A}'.\underset{\sim}{e}) \tag{2.6.32c}$$

with θ and $\underset{\sim}{e}$ known, the position vector of the point on the screw axis lying closet to the origin, $\underset{\sim}{r}_0$, can be computed by application of eq. (2.6.15), except if $\theta = 0, \pi$, in which case the cotangent function appearing in that equation becomes infinity. The first possibility, $\theta = 0$, is discarded, because this corresponds to a pure translation, which case falls either in case 1.1.1.1 or in case 4. Thus the only case that still has to be dealt with is that corresponding to $\theta = \pi$. In order to treat this case one needs the following result:

Theorem 2.6.7 *Given a rigid motion involving a 180°-rotation, let S be a point of the screw axis and let A be an arbitrary point of the body outside the screw axis. Moreover, let* $\underset{\sim}{s}, \underset{\sim}{a}, \underset{\sim}{u}_S$ *and* $\underset{\sim}{u}_A$ *be the respective position and displacement vectors of those points. Then, a second point on the screw axis is located by vector* $\underset{\sim}{a}+\frac{1}{2}\underset{\sim}{u}_A$

Proof:

If the point located by $\underset{\sim}{a} + \frac{1}{2}\underset{\sim}{u}_A$ is on the screw axis, then vector $\underset{\sim}{a} + \frac{1}{2}\underset{\sim}{u}_A-\underset{\sim}{s}$ is parallel to this axis. Thus, denoting by $\underset{\sim}{Q}$ the involved rotation matrix, the following equality holds:

$$\underset{\sim}{Q}(\mathbf{a} + \tfrac{1}{2}\underset{\sim}{u}_A-\underset{\sim}{s}) = \underset{\sim}{a}+ \tfrac{1}{2}\underset{\sim}{u}_A-\underset{\sim}{s} \tag{2.6.33}$$

Thus, all that need be proved is eq. (2.6.33). This follows from the fact that, according to eq. (2.6.19), $\underset{\sim}{u}_A$ can be written as

$$\underset{\sim}{u}_A = \underset{\sim}{u}_S + (\underset{\sim}{Q}-\underset{\sim}{I})(\underset{\sim}{a}-\underset{\sim}{s}) \tag{2.6.34}$$

Now, $(\tilde{Q}-\tilde{I})(\tilde{a}-\tilde{s})$ lies naturally in the range of $(\tilde{Q}-\tilde{I})$ (by definition) and hence is perpendicular to its null space, i.e., to $\tilde{u}_S$. Then, if $(\tilde{Q}-\tilde{I})(\tilde{a}-\tilde{s})$ is a vector perpendicular to $\tilde{u}_S$ or equivalently, to the screw axis, its image under $\tilde{Q}$ must be its negative, because, by assumption, $\theta=\pi$. Thus

$$\tilde{Q}(\tilde{Q}-\tilde{I})(\tilde{a}-\tilde{s})=-(\tilde{Q}-\tilde{I})(\tilde{a}-\tilde{s}) \tag{2.6.35}$$

On the other hand, making use of eq. (2.6.34),

$$\begin{aligned}\tilde{Q}(\tilde{a}+\frac{1}{2}\tilde{u}_A-\tilde{s})&=\tilde{Q}\left(\tilde{a}+\frac{1}{2}\tilde{u}_S+\frac{1}{2}(\tilde{Q}-\tilde{I})(\tilde{a}-\tilde{s})-\tilde{s}\right)=\\&=\frac{1}{2}\tilde{u}_S+\tilde{Q}\left((\tilde{a}-\tilde{s})+\frac{1}{2}(\tilde{Q}-\tilde{I}))\tilde{a}-\tilde{s})\right)=\\&=\frac{1}{2}\tilde{u}_S+\tilde{Q}\ (\tilde{a}-\tilde{s})+\frac{1}{2}\tilde{Q}(\tilde{Q}-\tilde{I})(\tilde{a}-\tilde{s})\end{aligned}$$

By virtue of eq. (2.6.35) the latter equation becomes

$$\tilde{Q}(\tilde{a}+\frac{1}{2}\tilde{u}_A-\tilde{s})=\frac{1}{2}\tilde{u}_S+\tilde{Q}(\tilde{a}-\tilde{s})-\frac{1}{2}(\tilde{Q}-\tilde{I})(\tilde{a}-\tilde{s})$$

Adding and subtracting $\tilde{a}-\tilde{s}$ to the left-hand side of the latter equation leads to

$$\begin{aligned}\tilde{Q}(a+\frac{1}{2}u_A-s)&=\frac{1}{2}u_S+a-s+(\tilde{Q}-I)(a-s)-\frac{1}{2}(\tilde{Q}-I)(a-s)\\&=\tilde{a}+\frac{1}{2}\left(\tilde{u}_S+(\tilde{Q}-\tilde{I})(\tilde{a}-\tilde{s})\right)-\tilde{s}=\\&=\tilde{a}+\frac{1}{2}\tilde{u}_A-\tilde{s},\ \text{q.e.d.}\end{aligned}$$

Now, if vector $\tilde{a}+\frac{1}{2}\tilde{u}_A$ is located on the screw axis, the point of that axis lying closest to the origin, R_0, is the one whose position vector is perpendicular to $\tilde{e}$. Thus, the position vector, $\tilde{r}_0$, of R_0, is computed as

$$\tilde{r}_0=\tilde{a}+\frac{1}{2}\tilde{u}_A-(\tilde{a}+\frac{1}{2}\tilde{u}_A)\cdot\tilde{e}\tilde{e} \tag{2.6.36}$$

which is the formula that is to be used to compute $\tilde{r}_0$ when $\theta=\pi$.

To complete the computation of the screw parameters, only the slide displacement, s, need be computed. This is easily done as

$$s=\tilde{u}_A\cdot\tilde{e} \tag{2.6.37}$$

thereby completing the computation of the screw parameters of a rigid body, considering all possible particular cases that could arise within this context.

Example 2.6.1 Determine the screw parameters of the motion under which the cube ABCDEFGH of Fig 2.6.4 is carried into configuration A'B' C'D'E'F'G'H'.

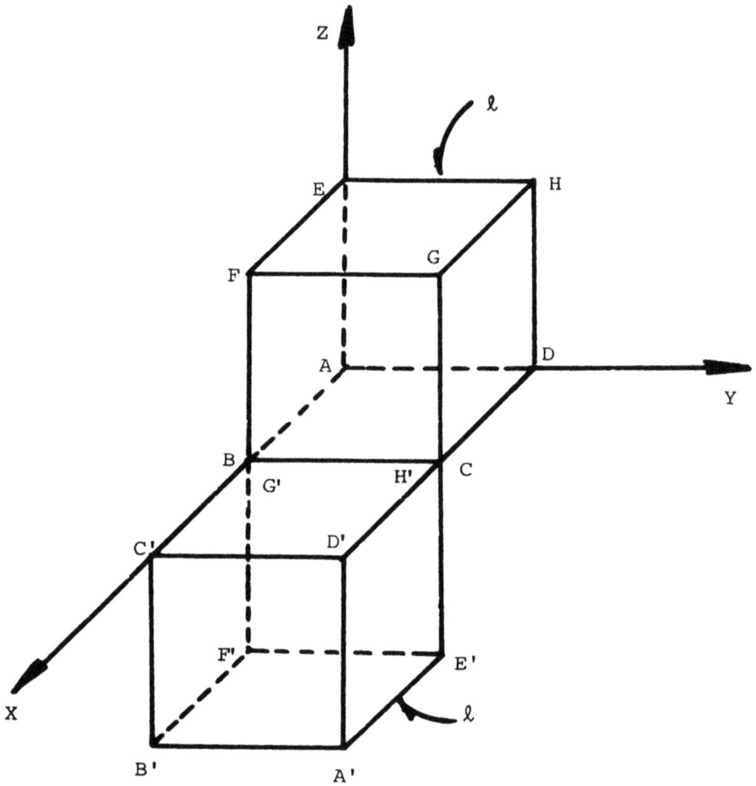

Fig 2.6. 4 Motion of a cube

Following the foregoing algorithm, the motion is first classified under one of the different cases appearing in Fig 2.6.2. To this end, choose the displacements of any three points, namely B, F and H. These are

$$u_B = \ell(i-k), \; u_F = -2\ell k, \; u_H = \ell(i-k)$$

i.e. no displacement vanishes and two are parallel. Hence, the said description of the motion leads to case 2.1. To compute vector e a new motion is defined as follows:

$$u'_B = u_B - u_F = \ell(i+k), \; u'_F = 0, \; u'_H = \ell(i+k)$$

One has then $u'_B = u'_H$, i.e. $\beta = 1$. Thus, the screw axis is computed by application of eqs. (2.6.27) and (2.6.28), i.e.

$$w = h - f - (b-f) = h - b = \ell(-i+j+k)$$

$$e = w/||w|| = \frac{\sqrt{3}}{3}(-i+j+k)$$

Angle θ is now computed by first finding the projection of point, say B, on the screw axis, B", and then computing vectors $\overrightarrow{B''B}$ and $\overrightarrow{B''B'}$, following formulae (2.6.32a and b), i.e.

$$\overrightarrow{B''B} = b - f - (b-f).ee$$

$$\overrightarrow{B''B'} = b' - f - (b'-f).ee$$

thus obtaining

$$\overrightarrow{B''B} = \frac{\ell}{3}(-i\;\; j - 2k)$$

$$\overrightarrow{B''B'} = \ell(j-k)$$

Hence, from eq. (2.6.31)

$$\cos\theta = \frac{\ell^2}{2\ell^2} = \frac{1}{2}$$

Thus θ could be either 60° or -60°. Its sign is determined according to eq. (2.6.32c) as

$$\text{sgn}(\theta) = \text{sgn}\,(\overrightarrow{B''B}\times\overrightarrow{B''B'}\cdot\underset{\sim}{e}) =$$

$$=\text{sgn}\left(\det \frac{\ell^2\sqrt{3}}{9}\begin{pmatrix} 1 & -1 & -4 \\ 0 & 1 & -1 \\ -1 & 1 & 1 \end{pmatrix}\right) =$$

$$=\text{sgn}\left(-\frac{\sqrt{3}}{9}\ell^2\right) = -1$$

Hence,

$$\theta = -60°$$

The position vector $\underset{\sim}{r}_0$ of the point on the screw axis lying closest to the origin can thus be computed by application of eq. (2.6.15), using point F as base point. Thus,

$$\underset{\sim}{r}_0 = \frac{1}{2}\cot\frac{\theta}{2}\,\underset{\sim}{e}\times(\underset{\sim}{f}'-\underset{\sim}{f}) - \frac{1}{2}\,\underset{\sim}{e}\times\left(\underset{\sim}{e}\times(\underset{\sim}{f}'+\underset{\sim}{f})\right)$$

Performing the above computations, one obtains

$$\underset{\sim}{r}_0 = \frac{1}{3}\ell(3\underset{\sim}{i}+2\underset{\sim}{j}+\underset{\sim}{k})$$

and

$$s = \underset{\sim}{u}_F\cdot\underset{\sim}{e} = -\frac{2\sqrt{3}}{3}\ell$$

SUBROUTINE SCREW, whose listing appears in Fig 2.6.5, implements the foregoing algorithm. This subroutine makes use of the subsidiary subroutines EXCHGE, CROSS and SCAL, listings of which appear in Figs 2.6.6, 2.6.7 and 2.6.8.

Exercise 2.6.8. In a manufacturing process it is required to position the workpiece of Fig 2.6.9 in configuration 2 starting from configuration 1, by means of an arm fastened to the bolt of a screw. Determine the location of the axis of this screw as well as its pitch. If the operation is to take place in n screw revolutions plus a fraction, what is the value of this fraction?

```
      SUBROUTINE SCREW(AIN,BIN,CIN,AFIN,BFIN,CFIN,E,RHO,THETA,DISPL)
C
C  THIS SUBROUTINE COMPUTES THE SCREW-PARAMETERS OF A RIGID BODY MOTION.
C
C
C  INPUT:
C   THE X, Y AND Z-COORDINATES OF THREE NONCOLLINEAR POINTS OF THE
C   RIGID BODY IN BOTH ITS INITIAL (AIN,BIN,CIN 3-DIMENSIONAL VECTORS)
C   AND  FINAL (AFIN,BFIN,CFIN 3-DIMENSIONAL VECTORS) CONFIGURATIONS.
C
C  OUTPUT:
C   1.  THE DIRECTION E, OF THE SCREW AXIS, A 3-DIMENSIONAL VECTOR.
C   2.  THE LOCATION RHO OF THE POINT ON THE SCREW AXIS LYING CLOSEST
C       TO THE ORIGIN.
C   3.  THE ANGLE OF ROTATION (SIGN WITH RESPECT TO E INCLUDED), THETA.
C   4.  THE SCALAR DISPLACEMENT DISPL, ALONG E (SIGN WITH RESPECT TO
C       E INCLUDED).
C
C  SUBSIDIARY SUBROUTINES:
C
C  CROSS(A,B,C)     COMPUTES THE CROSS PRODUCT OF VECTORS A AND B, IN THIS
C                   ORDER, AND STORES THE PRODUCT IN VECTOR C.
C  SCAL(A,B,S)      COMPUTES THE SCALAR PRODUCT OF VECTORS A AND B AND
C                   STORES THE PRODUCT IN THE SCALAR S.
C  EXCHGE(A,B)      EXCHANGES FIELDS A AND B.
C
C  3-DIMENSIONAL VECTORS A, B, C ARE AUXILIARY FIELDS.
C
      REAL AIN(3),BIN(3),CIN(3),AFIN(3),BFIN(3),CFIN(3),UA(3),UB(3)
     -     ,UC(3),A(3),B(3),C(3),E(3),RHO(3)
      LOGICAL LO(3)
      COMMON ZERO
C
C  COLLINEARITY OF GIVEN POINTS IS VERIFIED. WHEN POINTS ARE COLLINEAR,
C  ZERO IS SET EQUAL TO -1 AND SUBROUTINE RETURNS TO MAIN PROGRAM.
      DO 10 I=1,3
            LO(I)=.FALSE.
            A(I)=AIN(I)-CIN(I)
   10 B(I)=BIN(I)-CIN(I)
      CALL CROSS(A,B,C)
      CALL SCAL(C,C,S)
      S=SQRT(S)
      IF(S-ZERO) 20,20,30
   20 ZERO=-1.
      WRITE(6,1000)
      RETURN
C
C  DONE
C  COMPABILITY IS VERIFIED. IF THIS IS NOT MET, THEN ZERO IS SET EQUAL
C  TO -2, -3, OR -4, DEPENDING UPON WHETHER DISTANCE AC, BC,  OR AB DOES
C  NOT REMAIN CONSTANT THROUGHOUT THE MOTION.
   30 DO 40 I=1,3
            C(I)=AFIN(I)-CFIN(I)
   40 CONTINUE
#
```

Fig. 2.6.5 Listing of SUBROUTINE SCREW (first part)

```
      CALL SCAL(A,A,S1)
      S1=SQRT(S1)
      CALL SCAL(C,C,S2)
      S2=SQRT(S2)
      IF(ABS(S1-S2).LE.ZERO) GO TO 50
      ZERO=-2.
      WRITE(6,1010)
      RETURN
   50 DO 60 I=1,3
            C(I)=BFIN(I)-CFIN(I)
   60 CONTINUE
      CALL SCAL(B,B,S1)
      CALL SCAL(C,C,S2)
      S1=SQRT(S1)
      S2=SQRT(S2)
      IF(ABS(S1-S2).LE.ZERO) GO TO 70
      ZERO=-3
      WRITE(6,1020)
      RETURN
   70 DO 80 I=1,3
            A(I)=AIN(I)-BIN(I)
   80 B(I)=AFIN(I)-BFIN(I)
      CALL SCAL(A,A,S1)
      CALL SCAL(B,B,S2)
      S1=SQRT(S1)
      S2=SQRT(S2)
      IF(ABS(S1-S2).LE.ZERO) GO TO 90
      ZERO=-4.
      WRITE(6,1030)
      RETURN
C
C  DONE
C  DISPLACEMENT VECTORS ARE COMPUTED
   90 DO 100 I=1,3
            UA(I)=AFIN(I)-AIN(I)
            UB(I)=BFIN(I)-BIN(I)
  100 UC(I)=CFIN(I)-CIN(I)
C
C  DONE
C  NUMBER OF VANISHING DISPLACEMENTS IS DETERMINED AND STORED IN NUZE.
C  DISPLACEMENT MAGNITUDES ARE STORED TEMPORARILY IN E. IF NUZE.EQ.0
C  THEN NUZE IS SET EQUAL TO 4.
      CALL SCAL(UA,UA,E(1))
      CALL SCAL(UB,UB,E(2))
      CALL SCAL(UC,UC,E(3))
      NUZE=0
      DO 110 I=1,3
            E(I)=SQRT(E(I))
            IF(E(I).GT.ZERO) GO TO 110
            NUZE=NUZE+1
            LO(I)=.TRUE.
  110 CONTINUE
*
```

Fig. 2.6.5 Listing of SUBROUTINE SCREW (second part)

```
C  DONE
      WRITE(6,1060) NUZE
      IF(NUZE.EQ.0)NUZE=4
C
C  EACH CASE (NUZE=1,2,3 AND 0) IS NOW INVESTIGATED.
      GO TO(111,211,311,411) NUZE
C
C  IF ONE, AND ONLY ONE, VECTOR VANISHES, THE SUBROUTINE INVESTIGATES
C  WHETHER THE REMAINING ONES ARE PARALLEL OR NOT. FIRST, THE
C  VANISHING VECTOR IS DETECTED BY SETTING THE INTEGER VARIABLE IN
C  EQUAL TO 1,2 OR 3, DEPENDING UPON WHETHER UA,UB OR UC, RESPECTIVELY,
C  IS THAT WHICH VANISHES.
  111 DO 120 I=1,3
            IF(LO(I))IN=I
  120 CONTINUE
      GO TO (121,131,141),IN
  121 CALL EXCHGE(UA,UC)
      CALL EXCHGE(AIN,CIN)
      CALL EXCHGE(AFIN,CFIN)
      GO TO 141
  131 CALL EXCHGE(UB,UC)
      CALL EXCHGE(BIN,CIN)
      CALL EXCHGE(BFIN,CFIN)
  141 CALL CROSS(UA,UB,C)
      CALL SCAL(C,C,S)
      S=SQRT(S)
C
C  THE FORMULA TO COMPUTE THE DIRECTION OF THE SCREW AXIS IS
C  CHOSEN DEPENDING UPON WHETHER THE TWO NONVANISHING VECTORS
C  ARE PARALLEL OR NOT.
      IF(S.GT.ZERO) GO TO 150
      WRITE(6,1070)
      GO TO 509
  150 WRITE(6,1080)
      GO TO 600
C
C  IF EXACTLY TWO VECTORS VANISH, IN IS SET EQUAL TO 1,2 OR 3, DEPEND-
C  ING UPON WHETHER UA, UB OR UC, RESPECTIVELY, IS THE NONVANISHING
C  VECTOR.
  211 DO 221 I=1,3
            IF(LO(I))GO TO 221
            IN=I
  221 CONTINUE
C
C  ALL VECTORS ARE RELABELLED SO THAT THE NONVANISHING VECTOR IS
C  STORED IN FIELD UC. VECTOR E IS THEN COMPUTED USING THE
C  FORMULA REALISED IN STATEMENTS 501-520.
      GO TO (231,232,233),IN
  231 CALL EXCHGE(AIN,CIN)
      CALL EXCHGE(AFIN,CFIN)
      CALL EXCHGE(UA,UC)
      GO TO 233
  232 CALL EXCHGE(BIN,CIN)
      CALL EXCHGE(BFIN,CFIN)
      CALL EXCHGE(UB,UC)
```

Fig. 2.6.5 Listing of SUBROUTINE SCREW (third part)

```
C  UA-UC AND UB-UC ARE STORED IN UA AND UB, RESPECTIVELY.
  233 DO 240 I=1,3
              UA(I)=UA(I)-UC(I)
  240 UB(I)=UB(I)-UC(1)
      GO TO 509
C
C  IF ALL THREE VECTORS VANISH,A MESSAGE IS WRITEN AND SCREW RE-
C  TURNS TO MAIN PROGRAM.
  311 WRITE(6,1040)
      ZERO=-5
      RETURN
C
C  DONE
C  ONE-, TWO- AND THREE- VANISHING-DISPLACEMENT CASES WERE ALREADY
C  DEALT WITH. NO-VANISHING-DISPLACEMENT CASE IS NEXT INVESTIGATED.
C  FIRST IT IS INVESTIGATED WHETHER THE GIVEN VECTORS ARE
C  COPLANAR AND, IF SO, WHETHER THEIR DIFFERENCES ARE PARALLEL.
C  THE FORMULA EMPLOYED TO COMPUTE VECTOR E IS DECIDED UPON
C  ACCORDINGLY.
  411 CALL CROSS(UA,UB,C)
      CALL SCAL(UC,C,S1)
C
C  UA-UC AND UB-UC ARE STORED IN UA AND UB RESPECTIVELY.
      DO 420 I=1,3
              UA(I)=UA(I)-UC(I)
  420 UB(I)=UB(I)-UC(I)
      CALL CROSS (UA,UB,C)
      CALL SCAL(C,C,S2)
      S2=SQRT(S2)
      JUMP=1
      IF(ABS(S1).LT.ZERO)JUMP=JUMP+1
      IF(S2.LT.ZERO)JUMP=JUMP+1
      GO TO (421,422,423),JUMP
  421 WRITE(6,1090)
      GO TO 600
  422 WRITE(6,1100)
      GO TO 600
C
C  IF VECTORS ARE COPLANAR AND DIFFERENCES ARE PARALLEL, INVESTIGATES
C  WHETHER MOTION IS A PURE TRANSLATION.
  423 CALL SCAL(UA,UA,UAUA)
      CALL SCAL(UB,UB,UBUB)
      UAUA=SQRT(UAUA)
      UBUB=SQRT(UBUB)
      IF(UAUA.LE.ZERO.AND.UBUB.LE.ZERO)GO TO 900
      WRITE(6,1110)
C
C  COMPUTES VECTOR E. FIRST THE SCALAR FACTOR BETA RELATING
C  PARALLEL VECTORS UA AND UB IS COMPUTED. VECTOR E IS STORED
C  TEMPORARILY IN C.
  509 CALL SCAL(UA,UA,UAMAG)
      CALL SCAL(UB,UB,UBMAG)
      CALL SCAL(UA,UB,UAUB)
```

Fig. 2.6.5 Listing of SUBROUTINE SCREW (fourth part)

```
      UAMAG=SQRT(UAMAG)
      UBMAG=SQRT(UBMAG)
      IF(UAMAG.LE.ZERO) WRITE(6,1120)
      BETA=UBMAG/UAMAG
      IF(UAUB.LE.0)BETA=-BETA
      DO 510 I=1,3
  510 E(I)=BIN(I)-CIN(I)-BETA*(AIN(I)-CIN(I))
      GO TO 605
  600 CALL CROSS(UA,UB,E)
  605 CALL SCAL(E,E,EE)
      EE=SQRT(EE)
      DO 610 I=1,3
  610 E(I)=E(I)/EE
C
C  DONE
C  THE ANGLE OF ROTATION IS COMPUTED.AUXILIARY VECTORS A AND B
C  ARE INTRODUCED.
      DO 710 I=1,3
               A(I)=AIN(I)-CIN(I)
               C(I)=AIN(I)+UA(I)
  710 B(I)=C(I)-CIN(I)
      CALL SCAL(A,E,APROY)
      CALL SCAL(B,E,BPROY)
      DO 720 I=1,3
               A(I)=A(I)-APROY*E(I)
  720 B(I)=B(I)-BPROY*E(I)
      CALL SCAL(A,B,AB)
      CALL SCAL(A,A,AA)
      THETA=ARCOS(AB/AA)
      CALL CROSS(A,B,C)
      CALL SCAL(C,E,CE)
      IF(CE.LE.0)THETA=-THETA
C
C DONE
C  COMPUTATION OF RHO AND DISPL.
      COTH=COS(THETA)
      DIF=1+COSH
      DIF=ABS(DIF)
      IF(DIF.LE.ZERO) GO TO 830
      DO 810 I=1,3
               UA(I)=AFIN(I)-AIN(I)
  810 C(I)=AFIN(I)+AIN(I)
      CALL CROSS(E,C,A)
      CALL CROSS(E,A,B)
      CALL CROSS(E,UA,RHO)
      DO 820 I=1,3
               RHO(I)=COTAN(THETA/2.)*RHO(I)
  820 RHO(I)=(RHO(I)-B(I))/2.
      GO TO 840
  830 DO 850 I=1,3
  850 RHO(I)=AIN(I)+UA(I)*0.5
      CALL SCAL(RHO,E,S)
      DO 860 I=1,3
  860 RHO(I)=RHO(I)-S*E(I)
#
```

Fig. 2.6.5 Listing of SUBROUTINE SCREW (fifth part)

```
 240 CALL SCAL(UA,E,DISPL)
     RETURN
 900 DO 905 I=1,3
 905 UA(I)=AFIN(I)-AIN(I)
     WRITE(6,1050)(UA(I),I=1,3)
     ZERO=-6
     RETURN
1000 FORMAT(5X,"POINTS ARE COLLINEAR.MOTION IS UNDEFINED"/)
1010 FORMAT(5X,"MOTION IS NOT RIGID.LENGTH AC DOES NOT REMAIN",
    -       " CONSTANT."/)
1020 FORMAT(5X,"MOTION IS NOT RIGID.LENGTH BC DOES NOT REMAIN",
    -       " CONSTANT."/)
1030 FORMAT(5X,"MOTION IS NOT RIGID.LENGTH AB DOES NOT REMAIN",
    -       " CONSTANT."/)
1040 FORMAT(5X,"NO MOTION.ALL THREE DISPLACEMENTS VECTORS ARE",
    -       " ZERO."/)
1050 FORMAT(9X,"VECTORS DIFFERENCES ARE BOTH ZERO "/
    -       9X,"THE MOTION IS PURE TRANSLATION. "//15X,"THE ",
    -       "DISPLACEMENT HAS THE FOLLOWING X-,Y-AND Z COMPONENTS :"
    -       /15X,F12.5,5X,F12.5,5X,F12.5,/)
1060 FORMAT(9X,"NUMBER OF VANISHING DISPLACEMENTS IS",I5)
1070 FORMAT(9X,"DIFFERENCES ARE PARALLEL"/)
1080 FORMAT(9X,"DIFFERENCES ARE NONPARALLEL"/)
1090 FORMAT(9X,"VECTORS ARE NONCOPLANAR"/)
1100 FORMAT(9X,"VECTORS ARE COPLANAR AND DIFFERENCES ARE NONPARALLEL"/)
1110 FORMAT(9X,"VECTOR DIFFERENCES ARE PARALLEL"/)
1120 FORMAT(1X,"THE MAGNITUD OF THE NONVANISING VECTOR UA WAS "
    -      "COMPUTED AS ZERO BETWEEN STATEMENS 509 AND 510",/)
     END
```

Fig. 2.6.5 Listing of SUBROUTINE SCREW (sixth and last part)

```
      SUBROUTINE EXCHGE (A,B)
C
C  THIS SUBROUTINE EXCHANGES THE FIELDS OF A AND B, I.E., IT
C  RETURNS B AS A AND A AS B.
C
      REAL A(3),B(3),AUX(3)
      DO 10 I=1,3
            AUX(I)=A(I)
            A(I)=B(I)
            B(I)=AUX(I)
   10 CONTINUE
      RETURN
      END
```

Fig. 2.6.6 Listing of SUBROUTINE EXCHGE

```
      SUBROUTINE CROSS (A,B,C)
C
C  THIS SUBROUTINE PERFORMS THE CROSS PRODUCT OF A AND B, IN THIS ORDER,
C  AND STORES THIS PRODUCT IN C.
C
      REAL A(3),B(3),C(3)
      DO 10 K=1,3
            C(K)=0.
            DO 10 L=1,3
                  DO 10 M=1,3
                  N=(L-K)*(M-L)*(K-M)
                  C(K)=C(K)-N*A(L)*B(M)/2.
   10 CONTINUE
      RETURN
      END
#
```

Fig. 2.6.7 Listing of SUBROUTINE CROSS

```
      SUBROUTINE SCAL (A,B,S)
C
C  THIS SUBROUTINE PERFORMS THE SCALAR PRODUCT OF VECTORS A AND B
C  AND STORES THIS PRODUCT IN S.
C
      REAL A(3),B(3)
      S=0.
      DO 10 I=1,3
            S=S+A(I)*B(I)
   10 CONTINUE
      RETURN
      END
#
```

Fig. 2.6.8 Listing of SUBROUTINE SCAL

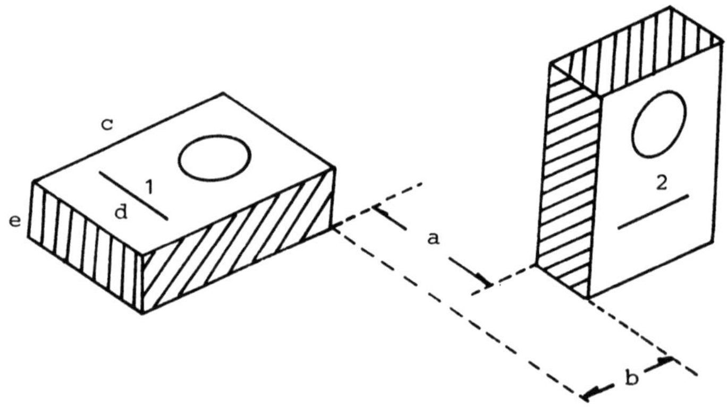

Fig 2.6.9. Motion of a workpiece

Exercise 2.6.9 (Taken from [2.7]) Let $\underset{\sim}{a}, \underset{\sim}{r}, \underset{\sim}{a}'$ and $\underset{\sim}{r}'$ be the position vectors of the initial and the displaced positions of points A and R of a rigid body under the screw motion

$$\underset{\sim}{r}' = \underset{\sim}{a}' + \underset{\sim}{Q}(\underset{\sim}{r} - \underset{\sim}{a})$$

$\underset{\sim}{Q}$ being the rotation of the screw. Show that the set of points of the body that, under the given motion remain equidistant from a fixed point P, lie in a plane.

Exercise 2.6.10 The rectangular plate shown in Fig 2.6.10 is displaced from configuration 1 to configuration 2. Determine the locus of the points of the plate that undergo a displacement of minimum magnitude from 1 to 2. What is the value of this displacement?

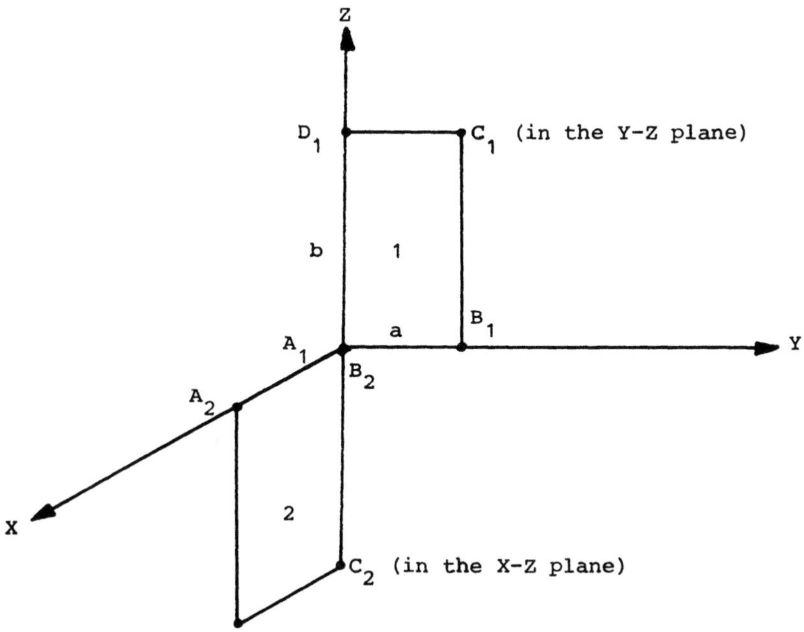

Fig. 2.6.10. Rigid plate undergoing a general displacement.

2.7 VELOCITY OF A POINT OF A RIGID BODY ROTATING ABOUT A FIXED POINT.

In the previous sections the motion of a rigid body when moving between two finitely-separated configurations was analysed. In this section and the following ones, the motion of a rigid body between two infinitesimally-separated configurations is analysed. The variables involved in the body motion are considered to be functions of time and results concerning their time derivatives are obtained.

Let $\underset{\sim}{y}(t)$ be the image of vector $\underset{\sim}{x}$ under a pure rotation $\underset{\sim}{Q}(t)$. Clearly,

$\underset{\sim}{x}$ is an independent variable; however, its image, $\underset{\sim}{y}(t)$, is a function of time. If the origin of coordinates is placed at the fixed point, then

$$\underset{\sim}{y}(t)=\underset{\sim}{Q}(t)\underset{\sim}{x} \tag{2.7.1}$$

Differentiating the above equation with respect to time, one obtains

$$\underset{\sim}{y}(t)=\dot{\underset{\sim}{Q}}(t)\underset{\sim}{x} \tag{2.7.2}$$

which is an expression for the velocity of the point located by vector $\underset{\sim}{x}$ in its initial configuration, at time t. Expression (2.7.2), however, is not practical to compute the velocity of the said point, for it requires knowledge of the point position in its initial configuration. Solving for $\underset{\sim}{x}$ in eq. (2.7.1) and introducing the corresponding value in eq. (2.7.2) yields

$$\underset{\sim}{v}(t)=\dot{\underset{\sim}{y}}(t)=\dot{\underset{\sim}{Q}}(t)\underset{\sim}{Q}^T(t)\underset{\sim}{y}(t) \tag{2.7.3}$$

which is an expression for the velocity of a point of a rigid body moving about a fixed point, in terms of the current position vector of the moving point. The matrix poduct, $\dot{\underset{\sim}{Q}}(t)\underset{\sim}{Q}^T(t)$, called the angular velocity* of the rigid body, represented by $\underset{\sim}{\Omega}(t)$, is a skew symmetric matrix. Then , the velocity $\underset{\sim}{v}(t)$ can be expressed as

$$\underset{\sim}{v}(t)=\underset{\sim}{\Omega}(t)\underset{\sim}{y}(t) \tag{2.7.4a}$$

where

$$\underset{\sim}{\Omega}(t)=\dot{\underset{\sim}{Q}}(t)\underset{\sim}{Q}^T(t)$$

Exercise 2.7.1 Show that, if $\underset{\sim}{Q}(t)$ is orthogonal, then $\dot{\underset{\sim}{Q}}(t)\underset{\sim}{Q}^T(t)$ is skew symmetric.

---

* Truesdell [2.8] prefers to call it "the spin" and so it is found also under this name in the literature

<u>Exercise 2.7.2</u> Show that the velocity of a point of a rigid body moving about a fixed point is perpendicular to its position vector (directed from the fixed point).

Since $\underset{\sim}{\Omega}(t)$ is skew symmetric and 3x3 it is totally determined by three independent scalars, thus being isomorphic to a cartesian vector, $\underset{\sim}{\omega}(t)$, called also <u>the angular velocity</u> of the rigid body. Using cartesian vector notation, the velocity $\underset{\sim}{v}(t)$ then can be expressed as

$$\underset{\sim}{v}(t)=\underset{\sim}{\omega}(t)\times\underset{\sim}{r}(t) \qquad (2.7.5)$$

where the components ω_i of $\underset{\sim}{\omega}$ and Ω_{ij} of $\underset{\sim}{\Omega}$ are related by $\omega_i=-\varepsilon_{ijk}\Omega_{jk}$, the alternating symbol ε_{ijk} being that defined in Section 2.5.

Since $\underset{\sim}{\Omega}(t)$ is skew symmetric and 3x3, it has one zero eigenvalue, as is shown below. Furthermore, its other two eigenvalues are complex (and conjugate, of course). Indeed, assume $\underset{\sim}{Q}(t)$ is in its canonical form, i.e.

$$(\underset{\sim}{Q}(t))=\begin{pmatrix}\cos\theta & -\sin\theta & 0\\ \sin\theta & \cos\theta & 0\\ \sin\theta & 0 & 1\end{pmatrix} \qquad (2.7.6)$$

$$(\underset{\sim}{\dot{Q}}(t))=\begin{pmatrix}-\sin\theta & -\cos\theta & 0\\ \cos\theta & -\sin\theta & 0\\ 0 & 0 & 0\end{pmatrix}\dot{\theta} \qquad (2.7.7)$$

From the above expressions,

$$\underset{\sim}{\dot{Q}}(t)\underset{\sim}{Q}^T(t) = \begin{pmatrix}0 & -1 & 0\\ 1 & 0 & 0\\ 0 & 0 & 0\end{pmatrix}\dot{\theta} \qquad (2.7.8)$$

which makes evident that all vectors of the form $(\underset{\sim}{e}) = (0,0,\alpha)^T$, α being any scalar, correspond to a zero eigenvalue. The other two eigenvalues are readily found to be

$$\lambda_1 = i, \lambda_2 = -i \tag{2.7.9}$$

where i is the imaginary unity, $\sqrt{-1}$.

The null space of $\underset{\sim}{\Omega}(t)$ (see Sec. 1.3) is, from the foregoing discussion, of dimension 1, i.e. a line. All the points lying on that line have zero velocity, the line thus being called "the instantaneous axis of rotation" of the rigid body. A cone rolling without slipping on a plane is a simple example of a rigid body rotating about a fixed point, its apex; its instantaneous axis of rotation is clearly, the element of the cone touching instantaneously the plane. Another example would be a sphere rotating on a plane in such a way that the contact point remains fixed; the instant axis of rotation of the sphere is thus the diameter passing through the point of contact.

Exercise 2.7.3 A cone of revolution rolls on a conic surface, also of revolution, without slipping, in such a way that both appices are coincident. What is the instantaneous axis of rotation of the cones in motion?.

Exercise 2.7.4 Show that the spin matrix $\underset{\sim}{\Omega}$ can be written as

$$\underset{\sim}{\Omega} = \underset{\sim}{A}\dot{\theta}$$

where $\underset{\sim}{A}$ is a constant matrix and $\dot{\theta}$ is the time derivative of θ , the angle of rotation.

Exercise 2.7.5 If the revolute is given in terms of Euler's angles, as appearing in eq. (2.5.16), show that vector $\underset{\sim}{\omega}$ is given as

$$(\underset{\sim}{\omega})_0 = \begin{bmatrix} 0 & c\phi & s\phi s\theta \\ 0 & s\phi & c\phi s\theta \\ 1 & 0 & c\theta \end{bmatrix} \begin{bmatrix} \dot{\phi} \\ \dot{\theta} \\ \dot{\psi} \end{bmatrix} \qquad (2.7.9a)$$

or, alternatively, as

$$(\underset{\sim}{\omega})_3 = \begin{bmatrix} s\theta s\psi & c\psi & 0 \\ s\theta c\psi & -s\psi & 0 \\ c\psi & 0 & 1 \end{bmatrix} \begin{bmatrix} \dot{\phi} \\ \dot{\theta} \\ \dot{\psi} \end{bmatrix} \qquad (2.7.9b)$$

where reference frames 0 and 3 were described in Exercise 2.5.3

Additivity of the angular velocity

In what follows it will be shown that the angular velocity, as opposed to the rotation, is additive. That is to say, if a rigid body undergoes a sequence of rotations, the overall rotation matrix is obtained not as the sum of individual matrices, but as their product. If the rigid body, however, rotates in such a way that its angular velocity is the composition of rotations about certain given axes, the matrix of the overall angular velocity is the sum of the individual matrices of angular velocities about those axes. Mathematically this difference can be stated as: Whereas rotation matrices form a group under multiplications, angular velocities do under additions.

Indeed, let a rigid body rotate from configuration 0 to configuration 1 about a given axis of rotation L_0, which exists by virtue of Euler's Theorem. Furthermore, let $\underset{\sim}{Q}_{01}$ be the rotation matrix carrying the body from configuration 0 to configuration 1. Next assume the body undergoes a second rotation, carrying it from configuration 1 to configuration 2, about axis L_1, the corresponding rotation matrix being $\underset{\sim}{Q}_{12}$. From the results of Section 2.4, the overall rotation matrix is given as

$$\left(\underset{\sim}{Q}_{02}\right)_0 = \left(\underset{\sim}{Q}_{12}\right)_0 \left(\underset{\sim}{Q}_{01}\right)_0 \tag{2.7.11}$$

But usually $\underset{\sim}{Q}_{12}$ is not readily described in reference frame 0, but in that labelled 1. Thus, the following similarity transformation (See Section 1.5) has to be introduced in eq. (2.7.11)

$$\left(\underset{\sim}{Q}_{12}\right)_0 = \left(\underset{\sim}{Q}_{01}\right)_0 \left(\underset{\sim}{Q}_{12}\right)_1 \left(\underset{\sim}{Q}^T_{01}\right)_0 \tag{2.7.12}$$

Substituting eq. (2.7.12) into eq. (2.7.11), one obtains

$$\left(\underset{\sim}{Q}_{02}\right)_0 = \left(\underset{\sim}{Q}_{01}\right)_0 \left(\underset{\sim}{Q}_{12}\right)_1 \tag{2.7.13}$$

The overall angular velocity of the body, $\underset{\sim}{\Omega}$, is then

$$\left(\underset{\sim}{\Omega}\right)_0 = \left(\dot{\underset{\sim}{Q}}_{02}\right)_0 \left(\underset{\sim}{Q}^T_{02}\right)_0 = \left(\left(\dot{\underset{\sim}{Q}}_{01}\right)_0 \left(\underset{\sim}{Q}_{12}\right)_1 + \left(\underset{\sim}{Q}_{01}\right)_0 \left(\dot{\underset{\sim}{Q}}_{12}\right)_1\right) \left(\underset{\sim}{Q}^T_{12}\right)_1 \left(\underset{\sim}{Q}^T_{01}\right)_0 =$$

$$= \left(\dot{\underset{\sim}{Q}}_{01}\right)_0 \left(\underset{\sim}{Q}_{12}\right)_1 \left(\underset{\sim}{Q}^T_{12}\right)_1 \left(\underset{\sim}{Q}^T_{01}\right)_0 + \left(\underset{\sim}{Q}_{01}\right)_0 \left(\dot{\underset{\sim}{Q}}_{12}\right)_1 \left(\underset{\sim}{Q}^T_{12}\right)_1 \left(\underset{\sim}{Q}^T_{01}\right)_0 \tag{2.7.14}$$

Now, let $\underset{\sim}{\Omega}_0$ and $\underset{\sim}{\Omega}_1$ be the angular-velocity matrices of the rotations about axes L_0 and L_1, respectively, i.e.

$$\underset{\sim}{\Omega}_0 = \dot{\underset{\sim}{Q}}_{01}\underset{\sim}{Q}^T_{01} \text{ and } \underset{\sim}{\Omega}_1 = \dot{\underset{\sim}{Q}}_{12}\underset{\sim}{Q}^T_{12} \tag{2.7.15}$$

Substitution of eqs. (2.7.15) into eq. (2.7.14) yields

$$\left(\Omega\right)_0 = \left(\Omega_0\right)_0 + \left(\underset{\sim}{Q}_{01}\right)_0 \left(\Omega_1\right)_1 \left(\underset{\sim}{Q}^T_{01}\right)_0 \tag{2.7.16a}$$

or, alternatively,

$$\left(\underset{\sim}{\Omega}\right)_0 = \left(\underset{\sim}{\Omega}_0 + \underset{\sim}{\Omega}_1\right)_0 \tag{2.7.16b}$$

or, in invariant form,

$$\underset{\sim}{\Omega} = \underset{\sim}{\Omega}_0 + \underset{\sim}{\Omega}_1 \tag{2.7.16c}$$

Now it is a simple matter to prove the result contained in the following

Exercise 2.7.7 Show that the angular-velocity matrices constitute a group under the operation of addition.

2.8 VELOCITY OF A MOVING POINT REFERRED TO A MOVING OBSERVER.

In what follows, an observer will be understood to be a set of coordinate axes provided with a clock (2.8,p.26). Assume a point P_o, located by vector $\underset{\sim}{x}_o$, is the origin of a coordinate system in motion with respect to another coordinate system, which will be arbitrarily referred to as "fixed".

The latter system constitutes a fixed observer, whereas the first, a moving one.

Let $\underset{\sim}{x}$ be the position vector of a point P, in motion with respect to both observers. Vectors and matrices expressed with respect to the fixed observer will be indexed with letter F, whereas those expressed with respect to the moving one will be indexed with letter M. Let $\underset{\sim}{\xi}$ be the position vector of P in the moving observer and $\underset{\sim}{Q}$ the rotation dyadic from the fixed observer to the moving one.

Hence,

$$(\underset{\sim}{x})_F = (\underset{\sim}{x}_o)_F + (\underset{\sim}{\xi})_F \tag{2.8.1}$$

where it is understood that all three vectors are functions of time.

The velocity of P is obtained differentiating both sides of eq. (2.8.1) with respect to time, i.e.

$$(\underset{\sim}{v})_F = (\underset{\sim}{v}_o)_F + (\dot{\underset{\sim}{\xi}})_F \tag{2.8.2}$$

where v_o is the velocity of point P_o and, since

$$(\underset{\sim}{\xi})_F = (\underset{\sim}{Q})_F(\underset{\sim}{\xi})_M \tag{2.8.3}$$

then

$$(\dot{\underset{\sim}{\xi}})_F = (\dot{\underset{\sim}{Q}})_F(\underset{\sim}{\xi})_M + (\underset{\sim}{Q})_F(\dot{\underset{\sim}{\xi}})_M = (\dot{\underset{\sim}{Q}})_F(\underset{\sim}{Q}^T)_F(\underset{\sim}{Q})_F(\underset{\sim}{\xi})_M + (\underset{\sim}{Q})_F(\dot{\underset{\sim}{\xi}}_M) =$$

$$= (\underset{\sim}{\Omega})_F(\underset{\sim}{\xi})_F + (\underset{\sim}{Q})_F(\dot{\underset{\sim}{\xi}})_M \tag{2.8.4}$$

Thus, eq. (2.8.2) becomes

$$(\underset{\sim}{v})_F = (\underset{\sim}{v}_o)_F + (\underset{\sim}{\Omega})_F(\underset{\sim}{\xi})_F + (\underset{\sim}{Q})_F(\dot{\underset{\sim}{\xi}})_M =$$

$$= (v_o)_F + (\Omega)_F(Q)_F(\xi)_M + (Q)_F(\dot{\xi})_M \tag{2.8.5}$$

where the first two terms of the right hand side represent the velocity of P as if it were one point of the rigid body defined by the moving observer*

---

* See Section 2.9

and the last term is the velocity of P as measured by the moving observer. Matrix $(\underset{\sim}{Q})_F$ transfers the description of velocity $(\dot{\underset{\sim}{\xi}})_M$ to the fixed observer. Thus, eq. (2.8.5) states that the velocity of point P equals that of point P as if it were fixed to the moving coordinate axes, plus the velocity of point P as if the moving observer were fixed.

Given any two points, P_1 and P_2, moving with velocities $\underset{\sim}{v}_1$ and $\underset{\sim}{v}_2$, respectively, 'the relative velocity of P_2 with respect to P_1 is defined as

$$\underset{\sim}{v}_{2/1}=\underset{\sim}{v}_2-\underset{\sim}{v}_1 \qquad (2.8.6)$$

Similarly, the relative angular velocity of body 2, moving with angular velocity $\underset{\sim}{\Omega}_2$, with respect to body 1, moving with angular velocity $\underset{\sim}{\Omega}_1$, is defined as

$$\underset{\sim}{\Omega}_{2/1}=\underset{\sim}{\Omega}_2-\underset{\sim}{\Omega}_1 \qquad (2.8.7)$$

or, alternatively,

$$\underset{\sim}{\omega}_{2/1}=\underset{\sim}{\omega}_2-\underset{\sim}{\omega}_1 \qquad (2.8.7)$$

2.9 GENERAL MOTION OF A RIGID BODY

Let a rigid body B undergo the most general motion, i.e., in general, no point of B remains fixed. Let $\underset{\sim}{v}_P$ be the velocity of one of its points, P, with position vector $\underset{\sim}{x}_P$ and angular velocity $\underset{\sim}{\Omega}$.

Thus, the relative velocity, $\underset{\sim}{v}-\underset{\sim}{v}_P$, of any other point R(located by $\underset{\sim}{x}$) with respect to P, is given by

$$\underset{\sim}{v}-\underset{\sim}{v}_P=\underset{\sim}{\Omega}(\underset{\sim}{x}-\underset{\sim}{x}_P) \qquad (2.9.1)$$

for $\underset{\sim}{v}-\underset{\sim}{v}_P$ is the velocity that R would have if P were fixed. From eq. (2.9.1),

$$\underset{\sim}{v}=\underset{\sim}{v}_P+\underset{\sim}{\Omega}(\underset{\sim}{x}-\underset{\sim}{x}_P) \qquad (2.9.2)$$

is the velocity of point R, and is given in terms of the velocity of another point, P, the angular velocity $\underset{\sim}{\Omega}$ and the position vector of R with respect to P.

Given two rigid bodies in motion, body 1 rolls without slipping with respect to body 2 if, and only if, there exists a set of points on both 1 and 2 such that the relative velocity of points on that set is zero. Regarding the velocity $\underset{\sim}{v}$, as given by eq. (2.9.2) as the relative velocity of one point R of body B with respect to body C, the fixed observer, the condition for B to roll without slipping on C is that there exists a set of points, whose position vector is given by $\underset{\sim}{x}$, for which $\underset{\sim}{v}=0$. But for this to happen, the condition is

$$\underset{\sim}{v}_P = -\underset{\sim}{\Omega}(\underset{\sim}{x}-\underset{\sim}{x}_P) \tag{2.9.3}$$

which states that $\underset{\sim}{v}_P$ is in the range (See section 1.3) of $\underset{\sim}{\Omega}$. However, it was shown in Section 2.8 that the null space of $\underset{\sim}{\Omega}$ if of dimension 1; hence -eq. (1.3.1)- the range of $\underset{\sim}{\Omega}$ is of dimension 2, thereby existing vectors in E^3 not belonging to the range of $\underset{\sim}{\Omega}$. If $\underset{\sim}{v}_P$ happens to lie outside the range of $\underset{\sim}{\Omega}$, it is impossible to find a vector $\underset{\sim}{x}$ for which eq. (2.9.3) is satisfied. Those vectors lying outside the range of $\underset{\sim}{\Omega}$ lie necessarily on its null space, i.e., on a line parallel to the eignevector of $\underset{\sim}{\Omega}$ corresponding to its zero eigenvalue or, equivalently, are parallel to the vector $\underset{\sim}{\omega}$ associated with $\underset{\sim}{\Omega}$. In case $\underset{\sim}{v}_P$ has a nonzero component along the null space of $\underset{\sim}{\Omega}$, body B is said to slide on body C, which is the case of the worm-gear or of the hypoid gear couplings. In these couplings there are power losses due to the involved sliding and, since the dissipated power is proportional to the sliding velocity, the contact between the two mechanical elements under consideration should take place along points of minimum magnitude of sliding velocity. For hypoid gears this set of points lie on a line which, paralleling Chasles' Theorem, is called <u>the instantaneous axis of the screw motion of body B with respect to C</u> or, for short, <u>the instantaneous axis</u> . Indeed, let the sliding velocity be given by eq. (2.9.2). Finding the points of minimum magnitude of sliding velocity corresponds to finding the vectors $\underset{\sim}{x}$ of expression (2.8.2) which minimize the quadratic form $\phi(\underset{\sim}{x})=\underset{\sim}{v}^T\underset{\sim}{v}$, which has a stationary value (Section 1.10) when $\phi'(\underset{\sim}{x})$ vanishes.

Applying the "chain rule" to $\phi(\underset{\sim}{x})$,

$$\phi'(x)=2\left(\frac{\partial \underset{\sim}{v}}{\partial \underset{\sim}{x}}\right)^T\underset{\sim}{v} \qquad (2.9.4a)$$

where, from eq. (2.9.2),

$$\frac{\partial \underset{\sim}{v}}{\partial \underset{\sim}{x}} = \underset{\sim}{\Omega} \qquad (2.9.4b)$$

Thus, points $\underset{\sim}{x}_0$ yielding an extremum of $\phi(\underset{\sim}{x})$ satisfy the equation

$$\underset{\sim}{\Omega}^T\underset{\sim}{v} =- \underset{\sim}{\Omega}\underset{\sim}{v}=\underset{\sim}{0} \qquad (2.9.5)$$

<u>Exercise 2.9.1</u> Show that the gradient of $\phi(\underset{\sim}{x})$ is twice the left hand side of eq. (2.9.5).

Since $\underset{\sim}{\Omega}$ has one zero eigenvalue (and only one), eq. (2.9.5) states that the minimum-magnitude velocity, given by

$$\underset{\sim}{v}_0 = \underset{\sim}{v}_P + \underset{\sim}{\Omega}(\underset{\sim}{x}_0-\underset{\sim}{x}_P) \qquad (2.9.6)$$

is in the null space of $\underset{\sim}{\Omega}$, i.e. is a vector parallel to $\underset{\sim}{\omega}$. Notice that eq. (2.9.6) does not yield a unique vector $\underset{\sim}{x}_0$ for, if any vector $\alpha\underset{\sim}{\omega}$ (in the null space of $\underset{\sim}{\Omega}$) is added to $\underset{\sim}{x}_0$, the velocity $\underset{\sim}{v}_N$ of the new point, is given by

$$\underset{\sim}{v}_N=\underset{\sim}{v}_P+\underset{\sim}{\Omega}(x_0+\alpha\underset{\sim}{\omega}-\underset{\sim}{x}_P)=\underset{\sim}{v}_P+\underset{\sim}{\Omega}(\underset{\sim}{x}_0-\underset{\sim}{x}_P)$$

Hence, the points of minimum-magnitude velocity lie on a line passing through one point $\underset{\sim}{x}_0$, in the direction of $\underset{\sim}{\omega}$, this line being <u>the instanta-</u>

neous screw axis of the motion under study. The particular point P_0 on the screw axis, located by $\underset{\sim}{x}_0$, is chosen such that $\underset{\sim}{x}_0$ be normal to the screw axis; thus, $\underset{\sim}{x}_0$ happens to be the minimum-norm vector satisfying (2.9.6). This vector can be found in a similar way as vector $\underset{\sim}{r}_0$ of eq. (2.6.8) was found, namely, choose two linearly independent equations out of eq (2.9.5) and form the system

$$\underset{\sim}{A}\underset{\sim}{x}_0=\underset{\sim}{b} \tag{2.9.7}$$

where $\underset{\sim}{A}$ is a 2x3 matrix and $\underset{\sim}{b}$ is a two-dimensional vector. Hence, the minimum-norm solution $\underset{\sim}{x}_0$ is given as

$$\underset{\sim}{x}_0=\underset{\sim}{A}^T(\underset{\sim}{A}\underset{\sim}{A}^T)^{-1}\underset{\sim}{b} \tag{2.9.8}$$

An alternative way of finding $\underset{\sim}{x}_0$ is now presented, expressing eq (2.9.5) in cartesian vector form, namely,

$$-\underset{\sim}{\omega}\times\left(\underset{\sim}{v}_P+\underset{\sim}{\omega}\times(\underset{\sim}{r}_0-\underset{\sim}{r}_P)\right)=\underset{\sim}{0} \tag{2.9.9}$$

which can be expanded as

$$\underset{\sim}{\omega}\times\underset{\sim}{v}_P+\underset{\sim}{\omega}\times\left(\underset{\sim}{\omega}\times(\underset{\sim}{r}_0-\underset{\sim}{r}_P)\right)=\underset{\sim}{0}$$

or, expanding the double cross product*,

$$\underset{\sim}{\omega}\times\underset{\sim}{v}_P+\left(\underset{\sim}{\omega}\cdot(\underset{\sim}{r}_0-\underset{\sim}{r}_P)\right)\underset{\sim}{\omega}-\underset{\sim}{\omega}^2(\underset{\sim}{r}_0-\underset{\sim}{r}_P)=\underset{\sim}{0}$$

If now $\underset{\sim}{r}_0$ is specified to be normal to $\underset{\sim}{\omega}$, i.e., to be the minimum-norm vector of all those satifying eq. (2.9.9), then it can be obtained from the above equation as

$$\underset{\sim}{r}_0=\underset{\sim}{r}_P+\frac{1}{\underset{\sim}{\omega}^2}\left(\underset{\sim}{\omega}\times\underset{\sim}{v}_P-(\underset{\sim}{\omega}\cdot\underset{\sim}{r}_P)\underset{\sim}{\omega}\right) \tag{2.9.10}$$

which is an expression similar to that appearing in eq. (2.6.15).

---

* $\underset{\sim}{\omega}^2\equiv\underset{\sim}{\omega}\cdot\underset{\sim}{\omega}$

The foregoing results are summarized in a theorem similar to that of Chasles'.

THEOREM 2.9.1 *Any rigid-body motion is equivalent to a screw motion, composed of a velocity* $\underset{\sim}{v}_0$ *and a spin about an axis parallel to* $\underset{\sim}{v}_0$ *. The points whose velocity is* $\underset{\sim}{v}_0$ *are located on a line parallel to* $\underset{\sim}{v}_0$*, called the axis of the instantaneous screw and* $\underset{\sim}{v}_0$ *is of minimum magnitude. The screw axis passes through point* P_0 *whose position vector is given either by eq. (2.9.8) or by eq. (2.9.10)*

The counterpart of Theorem 2.6.2 now follows:

THEOREM 2.9.2 *The velocities of all the points of a rigid body undergoing an arbitrary motion have identical projections along the axis of the instantaneous screw.*

Proof:

The velocity of any point of the rigid body can be written as

$$\underset{\sim}{v}=\underset{\sim}{v}_P+\underset{\sim}{\omega}\times(\underset{\sim}{r}-\underset{\sim}{r}_P)$$

Dot multiplying both sides of the above equation times $\underset{\sim}{\omega}$ (a vector parallel to the screw axis) yields

$$\underset{\sim}{v}\cdot\underset{\sim}{\omega}=\underset{\sim}{v}_P\cdot\underset{\sim}{\omega}+\underset{\sim}{\omega}\times(\underset{\sim}{r}-\underset{\sim}{r}_P)\cdot\underset{\sim}{\omega}$$

But the second term of the right hand side clearly vanishes. Hence

$$\underset{\sim}{v}\cdot\underset{\sim}{\omega}=\underset{\sim}{v}_P\cdot\underset{\sim}{\omega} \qquad \text{q.e.d.}$$

By virtue of the latter result, the projection of the velocities of all the points of a rigid body in motion along the screw axis is given by $||\underset{\sim}{v}_0||$, which is called the sliding . The pitch of the instant aneous screw is given by

$$\lambda = \frac{2\pi||\underset{\sim}{v}_0||}{||\underset{\sim}{\omega}||} \tag{2.9.11}$$

which is the counterpart of eq. (2.6.21a)

The sliding velocity, $\underset{\sim}{v}_0$, can then be obtained as

$$\underset{\sim}{v}_0 = (\underset{\sim}{v}_A \cdot \underset{\sim}{e})\underset{\sim}{e} \tag{2.9.12}$$

To completely determine the instantaneous screw, only the angular velocity need be computed. This is done in what follows, after deriving some results similar to those of Section 2.6. In fact, within the algorithm derived next, to compute all the parameters of the instantaneous screw, the angular velocity $\underset{\sim}{\omega}$ is first computed, as will be shown in what follows.

Corollary 2.9.1 *If the velocity of at least one point of a rigid body has a vanishing orthogonal projection on its angular velocity, the body undergoes a pure rotation.*

The foregoing Corollary is a direct consequence of Theorem 2.9.2 and so its proof is left as an exercise for the reader.

Exercise 2.9.2 Prove Corollary 2.9.1

THEOREM 2.9.3 *The difference vector of the velocities of any two points of a rigid body undergoing an arbitrary motion is perpendicular to the axis of the instantaneous screw.*

<u>Proof</u>

Let $\mathbf{v}_A$ and $\mathbf{v}_B$ be the velocities of two points, A and B, of a rigid body. From eq. (2.9.2) these are related by

$$\mathbf{v}_B=\mathbf{v}_A+\boldsymbol{\Omega}(\mathbf{b}-\mathbf{a})$$

Hence, the difference, $\mathbf{d}$, is given by

$$\mathbf{d} = \mathbf{v}_B-\mathbf{v}_A=\boldsymbol{\Omega}(\mathbf{b}-\mathbf{a})$$

which makes clear that $\mathbf{d}$ lies in the range (See Section 1.3) of $\boldsymbol{\Omega}$, thereby being normal to the null space of $\boldsymbol{\Omega}$, i.e., normal to the screw axis. Alternately this result can be proved resorting to Gibbs' notation. This way, $\mathbf{d}$ can be written as

$$\mathbf{d} = \boldsymbol{\omega}\times(\mathbf{b}-\mathbf{a})$$

and hence

$$\mathbf{d}\cdot\boldsymbol{\omega}= \boldsymbol{\omega}\times(\mathbf{b}-\mathbf{a})\cdot\boldsymbol{\omega}$$

which vanishes because the double product at the right hand side contains two identical vectors, q.e.d.

<u>THEOREM 2.9.4</u> *If the velocities of three noncollinear points of a rigid body are identical, the body undergoes a pure translation.*

<u>Proof</u>

Let $\mathbf{v}_A$, $\mathbf{v}_B$ and $\mathbf{v}_C$ be the respective velocities of points A,B and C. Referring these to the velocity of an arbitrary point P, one obtains

$$\mathbf{v}_A=\mathbf{v}_P+\boldsymbol{\Omega}(\mathbf{a}-\mathbf{p})$$

$$\mathbf{v}_B=\mathbf{v}_P+\boldsymbol{\Omega}(\mathbf{b}-\mathbf{p})$$

$$\mathbf{v}_C=\mathbf{v}_P+\boldsymbol{\Omega}(\mathbf{c}-\mathbf{p})$$

Subtracting the third equation from the first two yields

$$\mathbf{v}_A-\mathbf{v}_C=\boldsymbol{\Omega}(\mathbf{a}-\mathbf{c})=\mathbf{0}$$

$$\mathbf{v}_B-\mathbf{v}_C=\boldsymbol{\Omega}(\mathbf{b}-\mathbf{c})=\mathbf{0}$$

which implies that both $\underset{\sim}{a}-\underset{\sim}{c}$ and $\underset{\sim}{b}-\underset{\sim}{c}$ lie the null space of $\underset{\sim}{\Omega}$. This space, however, is of dimension 1, as discussed in Section 2.7. Since points A, B and C are noncollinear, vectors $\underset{\sim}{a}-\underset{\sim}{c}$ and $\underset{\sim}{b}-\underset{\sim}{c}$ are linearly independent and hence cannot be simultaneously in the null space of $\underset{\sim}{\Omega}$, unless $\underset{\sim}{\Omega}=\underset{\sim}{0}$, the motion thus reducing to a pure translation, q.e.d.

THEOREM 2.9.5 *The nonidentical velocities of three points of a rigid body are coplanar if and only if one of the following conditions is met:*

i) The motion is a pure rotation

ii) The motion is general, but the points are collinear

iii) The motion is general and the points are not collinear, but lie in a plane parallel to the screw axis.

Proof

(Sufficiency)

i) If the motion is a pure rotation, the velocity of any point with position vector $\underset{\sim}{r}$ is given by

$$\underset{\sim}{v} = \underset{\sim}{\Omega}\underset{\sim}{r}$$

which states that $\underset{\sim}{v}$ lies in the range (See Section 1. 3) of $\underset{\sim}{\Omega}$, which is of dimension 2, as was discussed in Section 2.7. This means that all velocity vectors lie in a plane perpendicular to the null space of $\underset{\sim}{\Omega}$, i.e. they are perpendicular to the axis of rotation, thereby showing that these velocities are coplanar.

ii) Let A, B and C be three collinear points of the rigid body and let $\underset{\sim}{a}$, $\underset{\sim}{b}$ and $\underset{\sim}{c}$ be their respective position vectors. The velocities of these points, referred to an arbitrary point with position vector $\underset{\sim}{p}$ are

$$\underset{\sim}{v}_A=\underset{\sim}{v}_P+\underset{\sim}{\Omega}(\underset{\sim}{a}-\underset{\sim}{p})$$

$$\underset{\sim}{v}_B=\underset{\sim}{v}_P+\underset{\sim}{\Omega}(\underset{\sim}{b}-\underset{\sim}{p})$$

$$\underset{\sim}{v}_C=\underset{\sim}{v}_P+\underset{\sim}{\Omega}(\underset{\sim}{c}-\underset{\sim}{p})$$

Since the points are collinear, their position vectors are related by

$$\underset{\sim}{c}-\underset{\sim}{a}=\alpha(\underset{\sim}{b}-\underset{\sim}{a})$$

Now, adding $\underset{\sim}{\Omega}\underset{\sim}{a}$ to $\underset{\sim}{v}_C$ and subtracting it simultaneously from the same expression, one obtains

$$\underset{\sim}{v}_C=\underset{\sim}{v}_P+\underset{\sim}{\Omega}(\underset{\sim}{a}-\underset{\sim}{p})+\underset{\sim}{\Omega}(\underset{\sim}{c}-\underset{\sim}{a})$$

whose first two terms can be readily identified with $\underset{\sim}{v}_A$. Moreover, substituting $\underset{\sim}{c}-\underset{\sim}{a}$ in the third term of the latter equation by $\alpha(\underset{\sim}{b}-\underset{\sim}{a})$, as given above, leads to

$$\underset{\sim}{v}_C=\underset{\sim}{v}_A+\alpha\underset{\sim}{\Omega}(\underset{\sim}{b}-\underset{\sim}{a})$$

But

$$\underset{\sim}{\Omega}(\underset{\sim}{b}-\underset{\sim}{a})=\underset{\sim}{v}_B-\underset{\sim}{v}_A$$

Hence, the expression for $\underset{\sim}{v}_C$ is transformed into

$$\underset{\sim}{v}_C=\underset{\sim}{v}_A+\alpha(\underset{\sim}{v}_B-\underset{\sim}{v}_A)$$

or, equivalently,

$$\underset{\sim}{v}_C=(1-\alpha)\underset{\sim}{v}_A+\alpha\underset{\sim}{v}_B$$

thereby proving the linear dependence of $\underset{\sim}{v}_A$, $\underset{\sim}{v}_B$ and $\underset{\sim}{v}_C$. Hence these vectors are coplanar.

iii) The velocities of the three given points, A,B and C, are referred to that of a point P on the screw axis. These velocities take on the form appearing in ii). Thus, the velocity of P, $\underset{\sim}{v}_P$, is parallel to the screw axis. On the other hand, the fact that A, B and C lie in a plane parallel to the screw axis allows one to establish the following relationship

$$\underset{\sim}{c}-\underset{\sim}{a}=\alpha(\underset{\sim}{b}-\underset{\sim}{a})+\beta\underset{\sim}{v}_P$$

or, equivalently,

$$\underset{\sim}{c}=(1-\alpha)\underset{\sim}{a}+\alpha\underset{\sim}{b}+\beta\underset{\sim}{v}_P$$

Substituting the latter expression in $\underset{\sim}{v}_C$ as given in ii). leads to

$$\underset{\sim}{v}_C=\underset{\sim}{v}_P+\underset{\sim}{\Omega}(\underset{\sim}{c}-\underset{\sim}{p})=$$

$$=\underset{\sim}{v}_P+\underset{\sim}{\Omega}(\underset{\sim}{a}-\underset{\sim}{p})-\alpha\underset{\sim}{\Omega}(\underset{\sim}{b}-\underset{\sim}{a})+\beta\underset{\sim}{\Omega}\underset{\sim}{v}_P$$

whose two first terms can be readily identified as $\underset{\sim}{v}_A$, its fourth term vanishing because it lies in the null space of $\underset{\sim}{\Omega}$. Hence

$$\underset{\sim}{v}_C=\underset{\sim}{v}_A-\alpha\underset{\sim}{\Omega}(\underset{\sim}{b}-\underset{\sim}{a})$$

But

$$\underset{\sim}{\Omega}(\underset{\sim}{b}-\underset{\sim}{a})=\underset{\sim}{v}_B-\underset{\sim}{v}_A$$

Thus, the latter expression for $\underset{\sim}{v}_C$ is transformed into

$$\underset{\sim}{v}_C=\underset{\sim}{v}_A-\alpha\underset{\sim}{\Omega}(\underset{\sim}{v}_B-\underset{\sim}{v}_A)$$

which shows the linear dependence of the three given velocity vectors, i.e. that they are coplanar.

(Necessity)

Assuming that the velocities $\underset{\sim}{v}_A$, $\underset{\sim}{v}_B$ and $\underset{\sim}{v}_C$ of three given points A,B and C are coplanar, the following relationship holds

$$\det(\underset{\sim}{v}_A,\ \underset{\sim}{v}_B,\ \underset{\sim}{v}_C)=0$$

Referring $\underset{\sim}{v}_B$ and $\underset{\sim}{v}_C$ to $\underset{\sim}{v}_A$ one has

$$\underset{\sim}{v}_B=\underset{\sim}{v}_A+\underset{\sim}{\Omega}(\underset{\sim}{b}-\underset{\sim}{a})$$

$$\underset{\sim}{v}_C=\underset{\sim}{v}_A+\underset{\sim}{\Omega}(\underset{\sim}{c}-\underset{\sim}{a})$$

Thus, the above expression for the determinant becomes

$$\det\left(\underset{\sim}{v}_A,\ \underset{\sim}{v}_A+\underset{\sim}{\Omega}(\underset{\sim}{b}-\underset{\sim}{a}),\ \underset{\sim}{v}_A+\underset{\sim}{\Omega}(\underset{\sim}{c}-\underset{\sim}{a})\right)=0$$

Subtracting the first column of this determinant from the remaining ones does not change the value of the determinant. Hence

$$\det(\underset{\sim}{v}_A,\ \underset{\sim}{\Omega}(\underset{\sim}{b}-\underset{\sim}{a}),\ \underset{\sim}{\Omega}(\underset{\sim}{c}-\underset{\sim}{a}))=0$$

which is equivalent to

$$\underset{\sim}{\Omega}(\underset{\sim}{b}-\underset{\sim}{a})\times\underset{\sim}{\Omega}(\underset{\sim}{c}-\underset{\sim}{a})\cdot\underset{\sim}{v}_A=0$$

Introducing Gibbs' notation, and expanding the resulting expression,

$$\underset{\sim}{\Omega}(\underset{\sim}{b}-\underset{\sim}{a})\times\underset{\sim}{\Omega}(\underset{\sim}{c}-\underset{\sim}{a})=\left(\underset{\sim}{\omega}\times(\underset{\sim}{b}-\underset{\sim}{a})\right)\times\left(\underset{\sim}{\omega}\times(\underset{\sim}{c}-\underset{\sim}{a})\right)=$$

$$=\left(\underset{\sim}{\omega}\times(\underset{\sim}{b}-\underset{\sim}{a})\cdot(\underset{\sim}{c}-\underset{\sim}{a})\right)\underset{\sim}{\omega}-\left(\underset{\sim}{\omega}\times(\underset{\sim}{b}-\underset{\sim}{a})\cdot\underset{\sim}{\omega}\right)(\underset{\sim}{c}-\underset{\sim}{a})$$

where the expression in brackets in the second term of the rightmost-hand side clearly vanishes. Hence

$$\underset{\sim}{\Omega}(\underset{\sim}{b}-\underset{\sim}{a})\times\underset{\sim}{\Omega}(\underset{\sim}{c}-\underset{\sim}{a})\cdot\underset{\sim}{v}_A=\left(\underset{\sim}{\omega}\times(\underset{\sim}{b}-\underset{\sim}{a})\cdot(\underset{\sim}{c}-\underset{\sim}{a})\right)\underset{\sim}{\omega}\cdot\underset{\sim}{v}_A$$

which vanishes under one of the following conditions:

i) $\underset{\sim}{\omega}\cdot\underset{\sim}{v}_A = 0$

which implies, under Corollary 2.9.1, that the motion is a pure rotation

ii) $(\underset{\sim}{b}-\underset{\sim}{a})\times(\underset{\sim}{c}-\underset{\sim}{a})=\underset{\sim}{0}$

which means that points A, B and C are collinear

iii) $\underset{\sim}{\omega}\times(\underset{\sim}{b}-\underset{\sim}{a})\cdot(\underset{\sim}{c}-\underset{\sim}{a})=0$

which indicates that vectors $\underset{\sim}{\omega}$, $\underset{\sim}{b}-\underset{\sim}{a}$ and $\underset{\sim}{c}-\underset{\sim}{a}$ are coplanar, q.e.d.

A direct consequence of the foregoing result is the following

Corollary 2.9.2 *Assume a rigid body under motion and choose any three noncollinear points A, B and C of the body. Letting $\underset{\sim}{v}_A$, $\underset{\sim}{v}_B$ and $\underset{\sim}{v}_C$ be the three involved velocities, then the difference vectors $\underset{\sim}{v}_A-\underset{\sim}{v}_C$ and $\underset{\sim}{v}_B-\underset{\sim}{v}_C$ (and, consequently, $\underset{\sim}{v}_A-\underset{\sim}{v}_B$) are parallel if and only if the points lie in a plane parallel to the screw axis (if the motion is general. If the motion reduces to a pure rotation, then the said plane is parallel to the axis of rotation.)*

Exercise 2.9.3 Prove Corollary 2.9.2

More results connected with the present discussion are the following

Corollary 2.9.3 The velocities of any two points of a rigid body cannot be parallel and different, unless the body undergoes a pure rotation

Corollary 2.9.4. If the velocities of any two points of a rigid body are parallel, then either i) the velocities are identical and belong to points lying on a line parallel to the axis of the instantaneous screw or ii) those velocities are different, in which case the motion is a pure rotation and the points lie on a line intersecting the axis of rotation.

Corollary 2.9.5 Given three noncollinear points, A, B and C, of a rigid body in motion, such that $\underset{\sim}{v}_C=\underset{\sim}{0}$ and $\underset{\sim}{v}_A$ and $\underset{\sim}{v}_B$ are parallel, i.e. $\underset{\sim}{v}_B=\beta\underset{\sim}{v}_A$, then the body undergoes a pure rotation and its axis passes through C and is parallel to vector $\underset{\sim}{b}-\underset{\sim}{c}-\beta(\underset{\sim}{a}-\underset{\sim}{c})$, $\underset{\sim}{a},\underset{\sim}{b}$ and $\underset{\sim}{c}$ being the position vectors of A, B and C, respectively.

The computation of the parameters of the instantaneous screw is next dealt with. These parameters are: i) the angular velocity $\underset{\sim}{\omega}$, ii) the position vector of the point of the instantaneous screw lying closest to the origin, $\underset{\sim}{r}_0$, and iii) the sliding velocity, v, i.e. the orthogonal projection of the velocity field on the screw axis.

The procedure enabling one to compute the aforementioned parameters is now established. This is based upon the tree diagram appearing in Fig. 2.9.1. This diagram is equal to that of Fig 2.6.2, except that instead of containing displacement vectors, it contains, velocity vectors. Thus, all cases shown in the "branches" of that tree appear in Fig 2.9.1. It is assumed that the given points are noncollinear and that the three velocity vectors are compatible. By virtue of Theorem 2.10.1, the orthogonal projections of the velocities of two points of a rigid body on the line joining these points are identical.

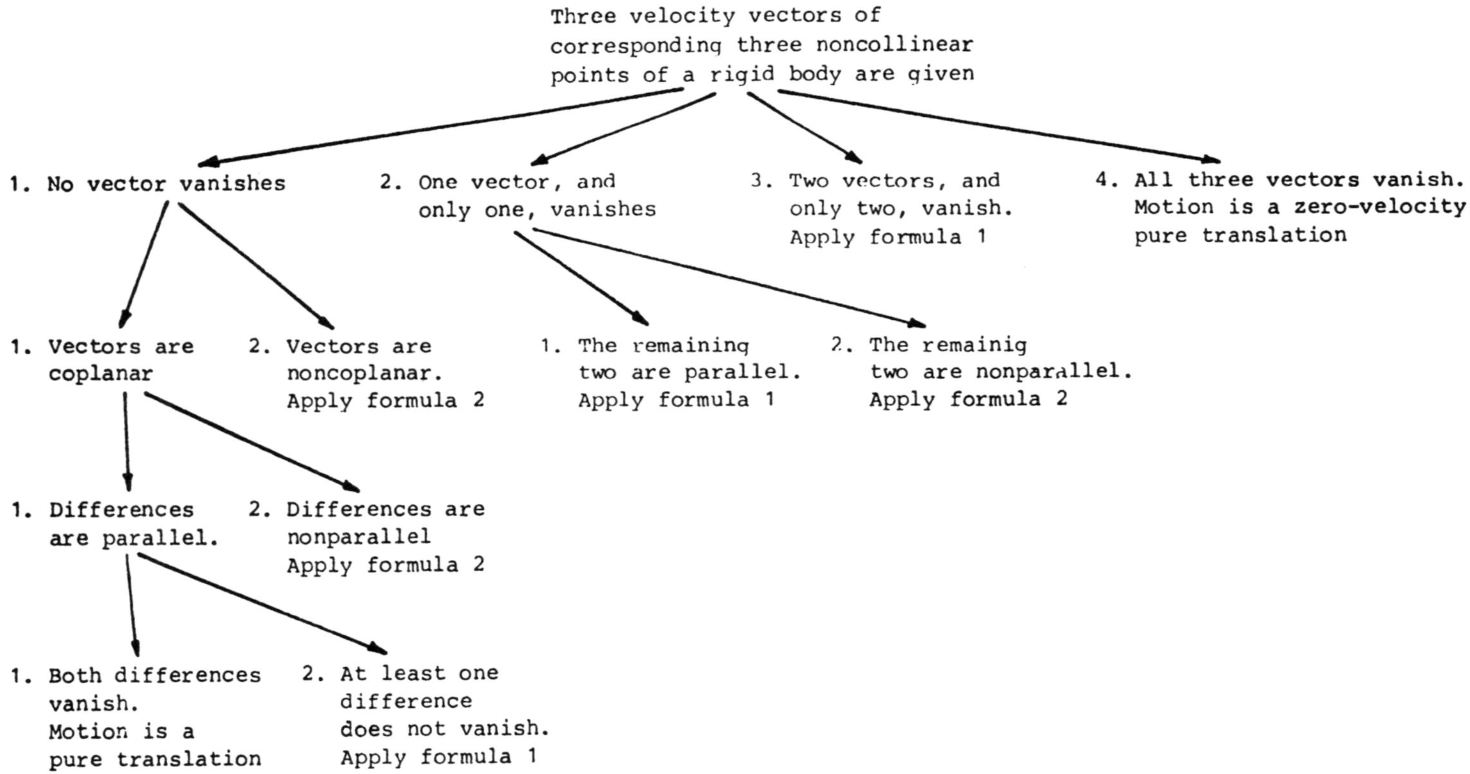

Fig 2.9.1. Tree diagram showing the different possible relationships amongst the velocities of three noncollinear points defining a rigid-body motion.

The crucial step in the present computational procedure is the computation of $\underset{\sim}{\omega}$. To compute this vector is a simple matter if the velocity vectors observe the relationship of case 2. If they correspond to cases 1 or 3, however, a new motion can be defined arising from the given one, that falls either in case 2.1 or in case 2.2, as will be shown next. Case 4, being trivial, need not be further discussed. In either case, 2.1 or 2.2, a particular procedure to compute $\underset{\sim}{\omega}$ should be followed. Once $\underset{\sim}{\omega}$ is known, the remaining parameters, i.e. $\underset{\sim}{r}_0$ and v, as described before, can be computed resorting to identical formulae, regardless of the case at hand.

In order to prove that the computation of $\underset{\sim}{\omega}$ for cases 1 or 3 can be reduced to that of case 2, two previous results are needed, namely:

THEOREM 2.9.6. *Given a rigid-body motion defined by the velocities of three noncollinear points, such that no velocity vanishes, define a new motion by assigning new velocity vectors* $\underset{\sim}{v}'_A$, $\underset{\sim}{v}'_B$ *and* $\underset{\sim}{v}'_C$ *to points A,B and C, respectively. If these vectors are defined as*

$$\underset{\sim}{v}'_A = \underset{\sim}{v}_A - \underset{\sim}{v}_C, \quad \underset{\sim}{v}'_B = \underset{\sim}{v}_B - \underset{\sim}{v}_C, \quad \underset{\sim}{v}'_C = \underset{\sim}{0}$$

then the angular velocity of the new motion is identical to the given one.

Proof:

From eq. (2.9.2), $\underset{\sim}{v}_A$ and $\underset{\sim}{v}_B$ can be written as

$$\underset{\sim}{v}_A = \underset{\sim}{v}_C + \underset{\sim}{\Omega}(\underset{\sim}{a}-\underset{\sim}{c}) \tag{2.9.13a}$$

$$\underset{\sim}{v}_B = \underset{\sim}{v}_C + \underset{\sim}{\Omega}(\underset{\sim}{b}-\underset{\sim}{c}) \tag{2.9.13b}$$

Hence,

$$\underset{\sim}{v}_A - \underset{\sim}{v}_C = \underset{\sim}{0} + \underset{\sim}{\Omega}(\underset{\sim}{a}-\underset{\sim}{c})$$

and

$$\underset{\sim}{v}_B - \underset{\sim}{v}_C = \underset{\sim}{0} + \underset{\sim}{\Omega}(\underset{\sim}{b}-\underset{\sim}{c})$$

and recalling the definitions of $\underset{\sim}{v}'_A$, $\underset{\sim}{v}'_B$ and $\underset{\sim}{v}'_C$, the latter equations become, then,

$$\underset{\sim}{v}'_A = \underset{\sim}{v}'_C + \underset{\sim}{\Omega}(\underset{\sim}{a}-\underset{\sim}{c}) \tag{2.9.14a}$$

$$\underset{\sim}{v}'_B = \underset{\sim}{v}'_C + \underset{\sim}{\Omega}(\underset{\sim}{b}-\underset{\sim}{c}) \tag{2.9.14b}$$

which are expressions analogous to those of eqs. (2.9.13a and b), except that $\underset{\sim}{v}'_A$, $\underset{\sim}{v}'_B$ and $\underset{\sim}{v}'_C$ have been placed instead of $\underset{\sim}{v}_A$, $\underset{\sim}{v}_B$ and $\underset{\sim}{v}_C$, respectively. $\underset{\sim}{\Omega}$, $\underset{\sim}{a}$, $\underset{\sim}{b}$ and $\underset{\sim}{c}$ remain, however, thereby completing the proof.

THEOREM 2.9.7. *Given a rigid-body motion defined by the velocities of three noncollinear points, such that two of these, and only two, vanish, define a new motion by assigning new velocity vectors* $\underset{\sim}{v}'_A$, $\underset{\sim}{v}'_B$ *and* $\underset{\sim}{v}'_C$ *to points A, B and C, respectively. Letting* $\underset{\sim}{v}_C$ *be the unique nonvanishing velocity, define*

$$\underset{\sim}{v}'_A = -\underset{\sim}{v}_C,\quad \underset{\sim}{v}'_B = -\underset{\sim}{v}_C,\quad \underset{\sim}{v}'_C = \underset{\sim}{0}$$

then the angular velocity of the new motion is identical to that of the given one.

Proof:

Since eqs. (2.9.13a and b) are valid regardless of whether any involved velocity vanishes, one can apply them to the case at hand, i.e. for $\underset{\sim}{v}_A = \underset{\sim}{v}_B = \underset{\sim}{0}$ and $\underset{\sim}{v}_C \neq \underset{\sim}{0}$. Thus,

$$\underset{\sim}{0} = \underset{\sim}{v}_C + \underset{\sim}{\Omega}(\underset{\sim}{a}-\underset{\sim}{c}) \tag{2.9.15a}$$

$$\underset{\sim}{0} = \underset{\sim}{v}_C + \underset{\sim}{\Omega}(\underset{\sim}{b}-\underset{\sim}{c}) \tag{2.9.15b}$$

Hence,

$$-\underset{\sim}{v}_C = \underset{\sim}{0} + \underset{\sim}{\Omega}(\underset{\sim}{a}-\underset{\sim}{c})$$

and

$$-\underset{\sim}{v}_C = \underset{\sim}{0} + \underset{\sim}{\Omega}(\underset{\sim}{b}-\underset{\sim}{c})$$

Recalling the definition of v'_A, v'_B and v'_C, the latter equations become

$$\underset{\sim}{v}'_A = \underset{\sim}{v}'_C + \underset{\sim}{\Omega}(\underset{\sim}{a}-\underset{\sim}{c}) \tag{2.9.16a}$$

$$\mathbf{v}'_B = \mathbf{v}'_C + \boldsymbol{\Omega}(\mathbf{b}-\mathbf{c}) \tag{2.9.16b}$$

and the proof follows by introducing the argument resorted to for proving Theorem 2.9.6, q.e.d.

As an application of the foregoing results one can compute $\boldsymbol{\omega}$ for the given motion by first defining motions leading to case 2 of Fig 2.9.1. Next the computation of $\boldsymbol{\omega}$ for that case is discussed. Since this contains two subcases, each is discussed separately.

CASE 2.1 One vector, and only one, vanishes, the remaining ones being parallel. Let

$$\mathbf{v}_B = \beta\mathbf{v}_A, \ \mathbf{v}_C = \mathbf{0} \tag{2.9.17}$$

Then, according to Corollary 2.9.5, $\boldsymbol{\omega}$ is given as

$$\boldsymbol{\omega} = \alpha\left((\mathbf{b}-\mathbf{c}) - \beta(\mathbf{a}-\mathbf{c})\right) \tag{2.9.18}$$

Since the axis of rotation passes through C, $\mathbf{v}_A$ can be written as

$$\mathbf{v}_A = \boldsymbol{\omega} \times (\mathbf{a}-\mathbf{c}) \tag{2.9.19}$$

Upon substitution of $\boldsymbol{\omega}$ as given in (2.9), eq. (2.9.19) is transformed into

$$\mathbf{v}_A = \alpha(b-c)\times(a-c)$$

and hence

$$\alpha = \frac{||\mathbf{v}_A||}{||(\mathbf{b}-\mathbf{c})\times(\mathbf{a}-\mathbf{c})||}\operatorname{sgn}\left((\mathbf{b}-\mathbf{c})\times(\mathbf{a}-\mathbf{c})\cdot\mathbf{v}_A\right) \tag{2.9.10}$$

CASE 2.2 One vector, and only one, vanishes, the remaining ones being non-parallel. Let

$$\mathbf{v}_C = \mathbf{0}, \ \mathbf{v}_A \times \mathbf{v}_B \neq \mathbf{0} \tag{2.9.21}$$

Since the axis of rotation passes through C, $\mathbf{v}_A$ can be written as

$$\mathbf{v}_A = \boldsymbol{\omega} \times (\mathbf{a}-\mathbf{c}) \tag{2.9.22}$$

and hence

$$\underset{\sim}{v}_B \times \underset{\sim}{v}_A = \left(\underset{\sim}{v}_B \cdot (\underset{\sim}{a}-\underset{\sim}{c})\ \underset{\sim}{\omega} - (\underset{\sim}{v}_B \cdot \underset{\sim}{\omega})(\underset{\sim}{a}-\underset{\sim}{c})\right) \tag{2.9.22}$$

where the scalar product $\underset{\sim}{v}_B \cdot \underset{\sim}{\omega}$ vanishes, due to the fact that the motion is a pure rotation.

Thus, one can solve for $\underset{\sim}{\omega}$ from eq. (2.9.22) as

$$\underset{\sim}{\omega} = \frac{\underset{\sim}{v}_B \times \underset{\sim}{v}_A}{\underset{\sim}{v}_B \cdot (\underset{\sim}{a}-\underset{\sim}{c})} \tag{2.9.23}$$

where the denominator does not vanish by virtue of Corollary 4 and the nature of this case. Indeed, the denominator can be written as

$$\underset{\sim}{v}_B \cdot (\underset{\sim}{a}-\underset{\sim}{c}) = \underset{\sim}{\omega} \times (\underset{\sim}{b}-\underset{\sim}{c}) \cdot (\underset{\sim}{a}-\underset{\sim}{c})$$

which vanishes only if the given points lie in a plane parallel to the axis of the instantaneous screw. Since the differences $\underset{\sim}{v}_A - \underset{\sim}{v}_C (= \underset{\sim}{v}_A)$ and $\underset{\sim}{v}_B - \underset{\sim}{v}_C (= \underset{\sim}{v}_B)$ are nonparallel, by virtue of Corollary 4, the points lie in a plane not parallel to the axis of the instantaneous screw, and the denominator does not vanish.

Once $\underset{\sim}{\omega}$ is known, the position vector, $\underset{\sim}{r}_0$, of the point on the axis of the instantaneous screw lying closest to the origin, is computed by application of eq. (2.9.10). The sliding velocity v is computed simply as the orthogonal projection of the velocity field on the screw axis, i.e. as

$$\underset{\sim}{v} = \underset{\sim}{v}_A \cdot \underset{\sim}{\omega} / ||\underset{\sim}{\omega}|| \tag{2.9.24}$$

thereby completing the computation of all the screw parameters defining a rigid-body motion for infinitesimally separated positions.

The foregoing algorithm is implemented by SUBROUTINE INSCRU, whose listing appears in Fig 2.9.2. This subroutine uses EXCHGE, CROSS and SCAL as subsidiary subroutines. Listings of the latter are shown in Figs 2.6.5, 2.6.6 and 2.6.7.

```
      $
      SUBROUTINE INSCRU(A,B,C,VA,VB,VC,RHO,SPIN,SPEED)
C
C  THIS SUBROUTINE COMPUTES THE PARAMETERS OF THE INSTANTANEOUS
C  SCREW OF A RIGID-BODY MOTION.
C  INPUT:
C  THE X,Y AND Z COORDINATES OF THREE NONCOLINEAR POINTS,
C  A,B AND C OF A RIGID BODY AND THEIR CORRESPONDING
C  VELOCITIES, VA,VB AND VC. FIELDS A,B,C,VA,VB AND VC ARE
C  THREE DIMENSIONAL, THEIR ELEMENTS BEING THE X,Y AND Z
C  COMPONENTS OF THE CARTESIAN VECTORS STORED IN THEM.
C  OUTPUT:
C  1.- THE POSITION VECTOR RHO OF THE POINT ON THE INSTANTANEOUS
C      -SCREW AXIS LYING CLOSEST TO THE ORIGIN.
C  2.- THE SPIN. THIS VARIABLE IS LABELED 'SPIN'.
C  3.- THE SLIDING SPEED, I.E THE PROJECTION OF THE VELOCITY FIELD
C      ON THE AXIS OF THE INSTANTANEOUS SCREW. THIS VARIABLE IS
C      LABELED 'SPEED'.
C  SUBSIDIARY SUBROUTINES:
C  EXCHGE(A,B) -EXCHANGES FIELDS A AND B.
C  CROSS(U,V,W)-COMPUTES THE CROSS PRODUCT  OF VECTOR U TIMES
C              VECTOR V, IN THIS ORDER, AND STORES THE PRODUCT
C              IN VECTOR W.
C  SCAL(U,V,S) -COMPUTES THE SCALAR PRODUCT OF VECTORS U AND
C              V AND STORES THE PRODUCT IN THE SCALAR S.
C
      REAL A(3),B(3),C(3),VA(3),VB(3),VC(3),RHO(3),SPIN(3),AU(3),BU(3),
     -     CU(3)
      LOGICAL LO(3)
      COMMON ZERO
C
C  IT IS VERIFIED WHETHER THE GIVEN POINTS ARE COLLINEAR. IF
C  THEY ARE, ZERO IS SET EQUAL TO -1 AND SUBROUTINE RETURNS
C  TO MAIN PROGRAM. DIFFERENCE VECTORS ARE TEMPORARILY STORED
C  IN THE SPIN AND RHO FIELDS.
C
      DO 10 I=1,3
            SPIN(I)=A(I)-C(I)
10    RHO(I)=B(I)-C(I)
      CALL CROSS(SPIN,RHO,CU)
      CALL SCAL(CU,CU,S)
      S=SQRT(S)
      IF(S.GT.ZERO) GO TO 30
      ZERO=-1
      WRITE(6,1000)
      RETURN
C
C  DONE
C  COMPATIBILITY IS VERIFIED. IF THIS IS NOT MET, ZERO IS
C  SET EQUAL TO -2.,-3.,-4., DEPENDING UPON WHETHER DISTANCE
C  AC, BC OR AB HAS A NONZERO TIME RATE OF CHANGE.
C
*
```

Fig. 2.9.2 Listing of SUBROUTINE INSCRU (first part)

```
        $
30      CALL SCAL(VA,SPIN,SA)
        CALL SCAL(VC,SPIN,SC)
        IF(ABS(SA-SC).LE.ZERO) GO TO 40
        ZERO=-2.
        WRITE(6,1010)
        RETURN
40      CALL SCAL(VB,RHO,SB)
        CALL SCAL(VC,RHO,SC)
        IF(ABS(SB-SC).LE.ZERO) GO TO 50
        ZERO=-3.
        WRITE(6,1020)
        RETURN
50      DO 60 I=1,3
60      AU(I)=B(I)-A(I)
        CALL SCAL(VA,AU,SA)
        CALL SCAL(VB,AU,SB)
        IF(ABS(SA-SB).LE.ZERO) GO TO 70
        ZERO=-4.
        WRITE(6,1030)
        RETURN
C
C  DONE
C  NUMBER OF VANISHING-VELOCITY VECTORS IS DETERMINED AND
C  STORED IN NUZE. DISPLACEMENT MAGNITUDES ARE TEMPORARILY
C  STORED IN AU. IF NUZE.EQ.0, THEN NUZE IS SET EQUAL TO 4.
C
70      CALL SCAL(VA,VA,AU(1))
        CALL SCAL(VB,VB,AU(2))
        CALL SCAL(VC,VC,AU(3))
        NUZE=0
        DO 80 I=1,3
              LO(I)=.FALSE.
              AU(I)=SQRT(AU(I))
              IF(AU(I).GT.ZERO) GO TO 80
              NUZE=NUZE+1
              LO(I)=.TRUE.
80      CONTINUE
        WRITE(6,1040)NUZE
        IF(NUZE.EQ.0) NUZE=4
C
C  DONE
C  EACH CASE (NUZE=1,2,3,0) IS NOW INVESTIGATED. PARALLELISM AND
C  EQUALITY OF VECTORS ARE THEN VERIFIED.
C
        GO TO(100,200,300,400),NUZE
C
C  IF ONE, AND ONLY ONE VELOCITY VECTOR VANISHES, THE
C  SUBROUTINE INVESTIGATES WHETHER THE REMAINING ONES ARE
C  PARALLEL OR NOT.
C
*
```

Fig. 2.9.2 Listing of SUBROUTINE INSCRU (second part)

```
        $
 100      DO 101 I=1,3
                 IF(LO(I)) IN=I
 101      CONTINUE
          GO TO (110,120,130),IN
 110      CALL EXCHGE(VA,VC)
          CALL EXCHGE(A,C)
          GO TO 130
 120      CALL EXCHGE(VB,VC)
          CALL EXCHGE(B,C)
 130      CALL CROSS(VA,VB,AU)
          CALL SCAL(AU,AU,S)
          S=SQRT(S)
          DO 140 I=1,3
                 AU(I)=VA(I)
                 BU(I)=VB(I)
 140      RHO(I)=B(I)-C(I)
C
C  THE FORMULA TO COMPUTE THE DIRECTION OF THE SCREW AXIS IS
C  CHOSEN DEPENDING UPON WHETHER THE TWO NONVANISHING VECTORS
C  ARE PARALLEL OR NOT.
C
          IF(S.GT.ZERO) GO TO 150
          WRITE(6,1050)
          GO TO 500
 150      WRITE(6,1060)
          GO TO 600
C
C  IF EXACTLY TWO VECTORS VANISH, IN IS SET EQUAL TO 1,2 OR 3
C  DEPENDING UPON WHETHER VA,VB OR VC, RESPECTIVELY, IS THE
C  NONVANISHING VECTOR.
C
 200      DO 202 I=1,3
                 IF(LO(I)) GO TO 202
                 IN=I
 202      CONTINUE
C
C  ALL VECTORS ARE RELABELLED SO THAT THE NONVANISHING
C  VELOCITY VECTOR IS STORED IN VC.
C
          GO TO(210,220,230),IN
 210      CALL EXCHGE(A,C)
          CALL EXCHGE(VA,VC)
          GO TO 230
 220      CALL EXCHGE(B,C)
          CALL EXCHGE(VB,VC)
C
C  VA-VC AND VB-VC ARE STORED IN AU AND BU RESPECTIVELY.
C
 230      DO 240 I=1,3
                 AU(I)=VA(I)-VC(I)
*
```

Fig. 2.9.2 Listing of SUBROUTINE INSCRU (third part)

```
        $
240     BU(I)=VB(I)-VC(I)
        GO TO 500
C
C  IF ALL THREE VELOCITIES VANISH, A MESSAGE IS WRITTEN AND INSCRU
C  RETURNS TO THE MAIN PROGRAM.
C
300     WRITE(6,1070)
        ZERO=-5
        RETURN
C
C  DONE
C  ONE, TWO AND THREE-VANISHING-VELOCITY CASES WERE ALREADY DEALT
C  WITH. NO-VANISHING DISPLACEMENT CASE IS NOW INVESTIGATED. FIRST
C  IT IS INVESTIGATED WHETHER THE GIVEN VECTORS ARE COPLANAR AND
C  IF SO, WHETHER THEIR DIFFERENCES ARE PARALLEL.
C
400     CALL CROSS(VA,VB,CU)
        CALL SCAL(VC,CU,S1)
C
C  VA-VC AND VB-VC ARE STORED IN AU AND BU RESPECTIVELY.
C
        DO 402 I=1,3
               AU(I)=VA(I)-VC(I)
402     BU(I)=VB(I)-VC(I)
        CALL CROSS(AU,BU,CU)
        CALL SCAL(CU,CU,S2)
        S2=SQRT(S2)
        IND=1
        IF(ABS(S1).LT.ZERO)IND=IND+1
        IF(S2.LT.ZERO)IND=IND+1
        GO TO (410,420,430),IND
410     WRITE(6,1080)
        GO TO 600
420     WRITE(6,1090)
        GO TO 600
C
C  IF VECTORS ARE COPLANAR AND DIFERENCES ARE PARALLEL, INVESTIGATES
C  WHETHER THE MOTION IS A PURE TRANSLATION.
C
430     CALL SCAL(AU,AU,VA2)
        CALL SCAL(BU,BU,VB2)
        VA2=SQRT(VA2)
        VB2=SQRT(VB2)
        IF(VA2.LE.ZERO.AND.VB2.LE.ZERO) GO TO 700
        WRITE(6,1100)
C
C  COMPUTES VECTOR SPIN. FIRST THE SCALAR FACTOR ALFA IS COMPUTED.
C
500     DO 502 I=1,3
               RHO(I)=B(I)-C(I)
*
```

Fig. 2.9.2 Listing of SUBROUTINE INSCRU (fourth part)

```
        $
502     SPIN(I)=A(I)-C(I)
        CALL SCAL(AU,AU,VA2)
        CALL CROSS(RHO,SPIN,CU)
        CALL SCAL(CU,CU,ABC)
        VA2=SQRT(VA2)
        ABC=SQRT(ABC)
        ALFA=VA2/ABC
        CALL SCAL(AU,CU,SG)
        IF(SG.LT.0)ALFA=-ALFA
C
C  DONE
C  THE SCALAR FACTOR BETA IS COMPUTED
C
        CALL SCAL(BU,BU,SB)
        CALL SCAL(AU,AU,SA)
        CALL SCAL(AU,BU,SAB)
        SB=SQRT(SB)
        SA=SQRT(SA)
        BETA=SB/SA
        IF(SAB.LT.0)BETA=-BETA
C
C  DONE
C
        DO 504 I=1,3
              SPIN(I)=RHO(I)-BETA*SPIN(I)
504     SPIN(I)=SPIN(I)*ALFA
C
C  DONE
C  RHO AND SPIN ARE NEXT COMPUTED
C
        GO TO 610
600     CALL CROSS(AU,BU,CU)
        CALL SCAL(AU,RHO,DEN)
        DO 602 I=1,3
602     SPIN(I)=CU(I)/DEN
610     CALL SCAL(SPIN,SPIN,W2)
        CALL SCAL(SPIN,A,COFW)
        CALL CROSS(SPIN,VA,RHO)
        DO 612 I=1,3
612     RHO(I)=A(I)+(RHO(I)-COFW*SPIN(I))/W2
        CALL SCAL(VA,SPIN,SPEED)
        W2=SQRT(W2)
        SPEED=SPEED/W2
C
C  DONE
C
        RETURN
700     WRITE(6,1110)(VA(I),I=1,3)
        ZERO=-6
        RETURN
*
```

Fig. 2.9.2 Listing of SUBROUTINE INSCRU (fifth part)

```
      $
1000      FORMAT(5X,'POINTS ARE COLINEAR. MOTION IS UNDEFINED'/)
1010      FORMAT(5X,'MOTION IS NOT RIGID. LENGTH AC DOES NOT REMAIN',
     -              ' CONSTANT'/)
1020      FORMAT(5X,'MOTION IS NOT RIGID. LENGTH BC DOES NOT REMAIN',
     -              ' CONSTANT'/)
1030      FORMAT(5X,'MOTION IS NOT RIGID. LENGTH AB DOES NOT REMAIN',
     -              ' CONSTANT'/)
1040      FORMAT(5X,'NUMBER OF VANISHING VELOCITIES IS :',I5/)
1050      FORMAT(5X,'VELOCITIES ARE PARALLEL'/)
1060      FORMAT(5X,'VELOCITIES ARE NONPARALLEL'/)
1070      FORMAT(5X,'NO MOTION. ALL THREE VELOCITY VECTORS ARE ZERO'/)
1080      FORMAT(5X,'VELOCITIES ARE NONCOPLANAR'/)
1090      FORMAT(5X,'VELOCITIES ARE COPLANAR AND DIFFERENCES ARE NON',
     -              'PARALLEL'/)
1100      FORMAT(5X,'VELOCITIES ARE COPLANAR AND DIFERENCES ARE ',
     -              'PARALLEL'/)
1110      FORMAT(5X,'MOTION IS PURE TRANSLATION'/5X,'THE VELOCITIES ',
     -              'HAVE THE FOLLOWING X-,Y- AND Z- COORDINATES:'/
     -               3(5X,F12.5)/)
          END
*
```

Fig. 2.9.2 Listing of SUBROUTINE INSCRU (sixth and last part)

2.10 THEOREMS RELATED TO THE VELOCITY DISTRIBUTION IN A MOVING RIGID BODY.

Some results concerning the velocity field in a rigid body in motion are now obtained, the main result of this section being the Aronhold-Kennedy Theorem. First a very useful result is proved.

THEOREM 2.10.1 *The velocities of two points of a rigid body have identical components along the line connecting them.*

Proof:

Let $\underset{\sim}{a}$ and $\underset{\sim}{b}$ be the position vectors of two points, A and B, of a rigid body in motion. Thus, for any configuration,

$$(\underset{\sim}{b}-\underset{\sim}{a})\cdot(\underset{\sim}{b}-\underset{\sim}{a})=\text{const} \qquad (2.10.1)$$

from the rigidity condition. Differentiating both sides of eq. (2.10.1),

$$(\dot{\underset{\sim}{b}}-\dot{\underset{\sim}{a}})\cdot(\underset{\sim}{b}-\underset{\sim}{a})=0$$

or, alternatively,

$$\underset{\sim}{v}_B\cdot(\underset{\sim}{b}-\underset{\sim}{a})=\underset{\sim}{v}_A\cdot(\underset{\sim}{b}-\underset{\sim}{a}),\text{q.e.d.} \qquad (2.10.2)$$

This theorem is used to check the compatibility of the given velocities of a rigid body in subroutine INSCRU of Sect. 2.9.

Exercise 2.10.1 The triangular plate of Fig 2.10.1 is constrained to move in such a way that vertex C remains on the Z-axis, while vertex A remains on the X-axis and side AB remains on the X-Y plane. Vertex C has a velocity $\underset{\sim}{v}_C=-5\underset{\sim}{e}_z$ m/sec.

i) Determine the velocity of vertices A and B

ii) Determine the angular velocity of the plate

iii) Locate the instantaneous screw axis of the motion of the plate, and compute the pitch of its screw.

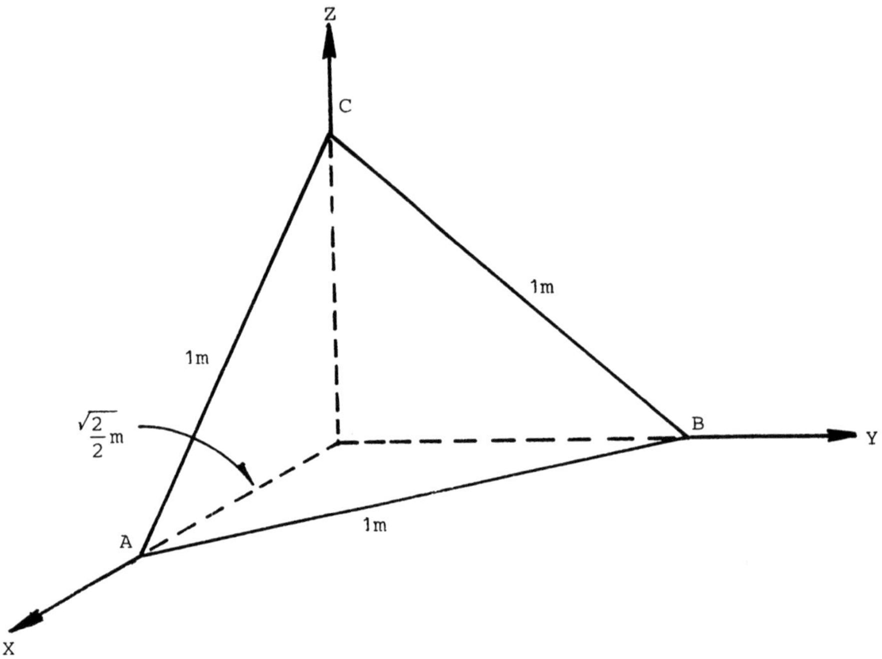

Fig 2.10.1 Triangular plate in constrained motion

THEOREM 2.10.2 *(Aronhold-Kennedy). Given three rigid bodies in motion, the resulting three instant screw axes have one common normal intersecting all three axes.*

Proof:

Referring to Fig 2.10.2 let S_B and S_C be the instant aneous screw axes of bodies B and C, respectively, with respect to body A; let $\underset{\sim}{v}_B$ and $\underset{\sim}{v}_C$ be the relative sliding velocities of the instant screws S_B and S_C, with respect to A. Finally, let $\underset{\sim}{\Omega}_B$ and $\underset{\sim}{\Omega}_C$ be the relative angular velocities of bodies B and C, respectively, with respect to body A, and $\underset{\sim}{c}$, the common normal to S_B and S_C, joining both axes. It will be shown that the third instant screw axis, $S_{B/C}$, passes through the common normal B*C*.

Let P be any point of the three-dimensional Euclidean space, with position vector $\underset{\sim}{r}$. Points P_A, P_B and P_C of bodies A, B and C, coincide at P. Let

$\underset{\sim}{v}_{PA}$, $\underset{\sim}{v}_{PB}$ and $\underset{\sim}{v}_{PC}$ be the velocities of each of these points. Furthermore, let $\underset{\sim}{v}$ be the relative velocity of P_B with respect to P_C and let B* and C* be the points in which the common normal intersects S_B and S_C.

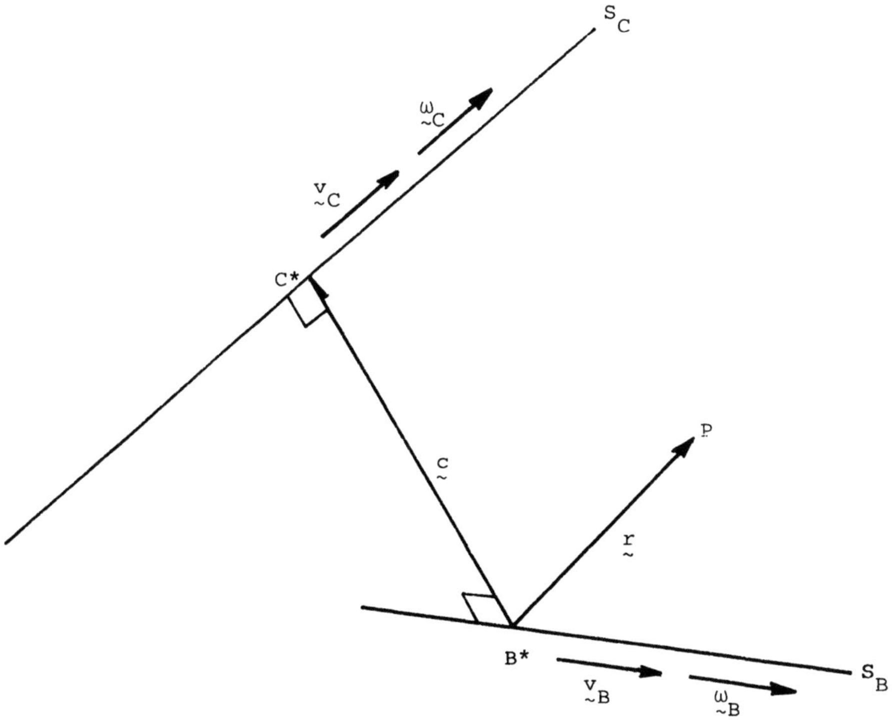

Fig. 2.10.2 Instantaneous screws of two bodies in motion with respect to a third one.

Thus,

$$
\begin{aligned}
&\underset{\sim}{v}=\underset{\sim}{v}_{PB}-\underset{\sim}{v}_{PC}= \\
&=(\underset{\sim}{v}_B+\underset{\sim}{\Omega}_B\underset{\sim}{r})-\left(\underset{\sim}{v}_C+\underset{\sim}{\Omega}_C(\underset{\sim}{r}-\underset{\sim}{c})\right)= \qquad (2.10.3)^* \\
&=\underset{\sim}{v}_{B/C}+\underset{\sim}{\Omega}_{B/C}\underset{\sim}{r}+\underset{\sim}{\Omega}_C\underset{\sim}{c}
\end{aligned}
$$

It is next shown that, if P is a point of the relative instant aneous screw axis $S_{B/C}$, then it lies on the line defined by points B* and C*. This is done

---

* $\underset{\sim}{v}_{B/C}$ is to be interpreted as the relative velocity of B* with respect to C*.

via the minimization of the quadratic form

$$\phi(\underset{\sim}{r})=\underset{\sim}{v}^T\underset{\sim}{v} \tag{2.10.4}$$

$\phi(\underset{\sim}{r})$ has an extremum at a point $\underset{\sim}{r}_0$ where its gradient vanishes. The said gradient is, applying the "chain rule" again,

$$\phi'(\underset{\sim}{r})=2\underset{\sim}{\Omega}_{B/C}^T\underset{\sim}{v} \tag{2.10.5a}$$

Thus, at point $\underset{\sim}{r}_0$,

$$\underset{\sim}{\Omega}_{B/C}(\underset{\sim}{v}_{B/C}+\underset{\sim}{\Omega}_C c+\underset{\sim}{\Omega}_{B/C}\underset{\sim}{r}_0)=\underset{\sim}{0} \tag{2.10.5b}$$

from which $\underset{\sim}{r}_0$ cannot be solved for, since $\underset{\sim}{\Omega}_{B/C}$ is singular, of rank two. One possible way to find $\underset{\sim}{r}_0$ is imposing on it the minimum-norm condition, as done previously in similar instances. Another possible way is to write eq. (2.10.5) in Gibbs' notation as

$$\underset{\sim}{\omega}_{B/C}\times(\underset{\sim}{\omega}_{B/C}\times\underset{\sim}{r}_0)+\underset{\sim}{\omega}_{B/C}\times(\underset{\sim}{v}_{B/C}+\ {}_C\times\underset{\sim}{c})=\underset{\sim}{0}$$

Expanding the first term and imposing the condition that $\underset{\sim}{\omega}_{B/C}\cdot\underset{\sim}{r}_0$ be zero, one obtains

$$-\underset{\sim}{\omega}_{B/C}^2\underset{\sim}{r}_0+\underset{\sim}{\omega}_{B/C}\times(\underset{\sim}{v}_{B/C}+\underset{\sim}{\omega}_C\times\underset{\sim}{c})=\underset{\sim}{0}$$

from which,

$$\underset{\sim}{r}_0=-\frac{1}{\underset{\sim}{\omega}_{B/C}^2}(\underset{\sim}{v}_{B/C}+\underset{\sim}{\omega}_C\times\underset{\sim}{c})\times\underset{\sim}{\omega}_{B/C} \tag{2.10.6}$$

which is parallel to vector $\underset{\sim}{c}$. Since $\underset{\sim}{r}_0$ is parallel to vector $\underset{\sim}{c}$, the common normal to axes S_B and S_C, then $S_{B/C}$ passes through line B*C*, q.e.d.

<u>Exercise 2.10.3</u> Show that $\underset{\sim}{r}_0$, as given by eq. (2.10.6), is parallel to $\underset{\sim}{c}$.

One application of the Aronhold-Kennedy Theorem arises in the pitch-surface synthesis of the coupling of two bodies whose relative motion is the composition of sliding and rotation, as is the case in hypoid gears. This is shown in the following example.

<u>Example 2.10.1</u> Let L_1 and L_2 be the non-intersecting non-parallel axes of two shafts required to be coupled. These axes are shown in Fig 2.10.3.

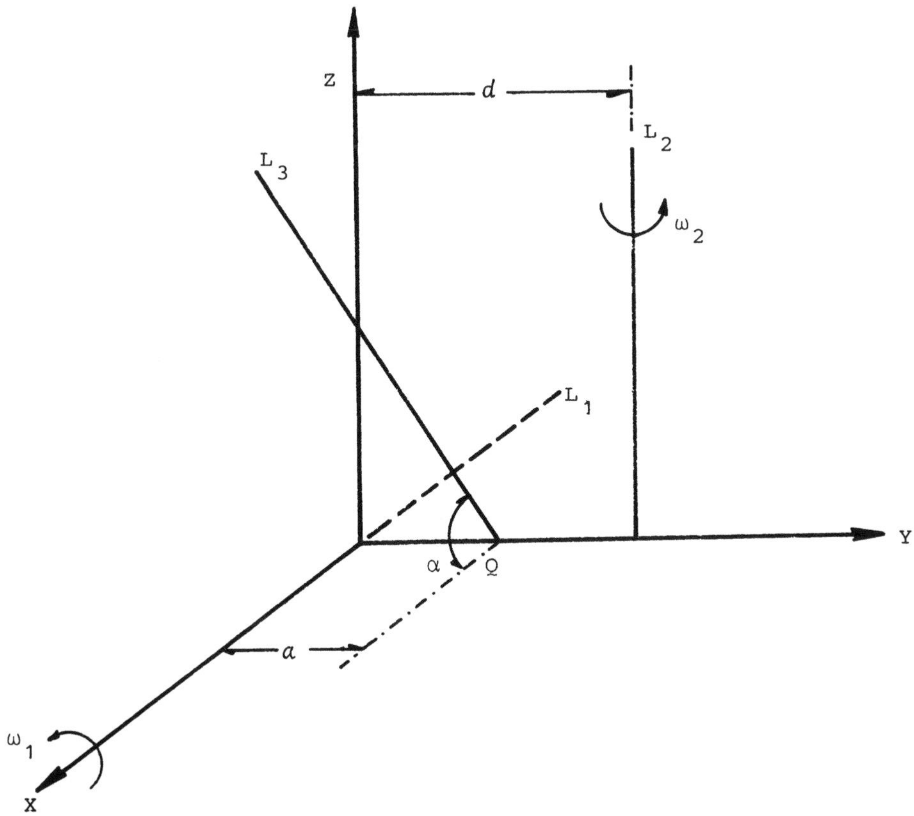

Fig 2.10.3 An application of the Aronhold-Kennedy Theorem

In order to have the most efficient coupling, it is required that this takes place along points of minimum-magnitude relative velocity, i.e. on the axis of the instantaneous screw of shaft 2 with respect to shaft 1.

From Theorem 2.10.2, that set of points constitutes line L_3, normal to the Y-axis, a distance a from L_1. Hence, line L_3 is determined by distance a and angle α. Point Q, the intersection of line L_3 and the Y-axis is found from the minimum condition on the relative velocity magnitude. Let $\underset{\sim}{v}_{Q1}$

and $\underset{\sim}{v}_{Q2}$ be the velocities of points Q1 and Q2, with respect to the fixed axes X, Y, Z.

Thus,

$$\underset{\sim}{v}_{Q1}=a\omega_1\underset{\sim}{e}_z \qquad (2.10.7a)$$

$$\underset{\sim}{v}_{Q2}=(d-a)\omega_2\underset{\sim}{e}_x \qquad (2.10.7b)$$

Assuming a required reduction m, i.e.,

$$\omega_2=m\omega_1 \qquad (2.10.8)$$

eq. (2.10.7b) can be rewritten as

$$\underset{\sim}{v}_{Q2}=(d-a)m\omega_1\underset{\sim}{e}_x \qquad (2.10.7c)$$

Next the quadratic form $\phi(a)$, obtained squaring the relative velocity magnitude, is minimized.

$$\phi(a)=(\underset{\sim}{v}_{Q2}-\underset{\sim}{v}_{Q1})\cdot(\underset{\sim}{v}_{Q2}-\underset{\sim}{v}_{Q1})=$$

$$=\omega_1^2\left(m^2(d-a)^2+a^2\right) \qquad (2.10.9)$$

$\phi(a)$ has an extremum when $\phi'(a)$ becomes zero, i.e.

$$\phi'(a)=\omega_1^2\left(-2m^2(d-a)+2a\right)=0 \qquad (2.10.10)$$

from which the minimizing value of a is obtained as

$$a = \frac{m^2 d}{1+m^2} \qquad (2.10.11)$$

Angle α is now obtained from the relationship

$$\cos\alpha=\frac{|\underset{\sim}{\omega}_{2/1}\cdot\underset{\sim}{e}_x|}{||\underset{\sim}{\omega}_{2/1}||}$$

Thus

$$\cos\alpha=\frac{1}{\sqrt{1+m^2}} \qquad (2.10.12)$$

Summarizing, the pitch surface (on 1) is a ruled surface whose elements are lines a distance a form the X axis, making an angle α with this axis.

This is a one-fold hyperboloid of revolution. Hence the name "hypoid" given to such gears.

One very important consequence of the Theorem 2.10.2 now follows.

Corollary 2.10.2 *Given three rigid bodies in motion, A, B and C, there exists an instantaneous axis of pure rotation (i.a.p.r.) of B with respect to C if, and only if, there exist i.a's.p.r. of both B and C with respect to A and these intersect, the i.a.p.r. of B with respect to C passing through the said intersection. Furthermore, all three axes are coplanar.*

Exercise 2.10.4 Prove Corollary 2.10.2.

As an application of Corollary 2.10.2, solve the following problem.

Example 2.10.2 (Kane (2.9))*. A shaft, terminating in a truncated cone C of semivertical angle θ, see Fig 2.10.4, is supported by a thrust bearing consisting of a fixed race R and four identical spheres S of radius r. When the shaft rotates about its axis, S rolls on R at both of its points of contact with R, and C rolls on S.

Proper choice of the dimension b allows to obtain pure rolling of C on S. Determine b

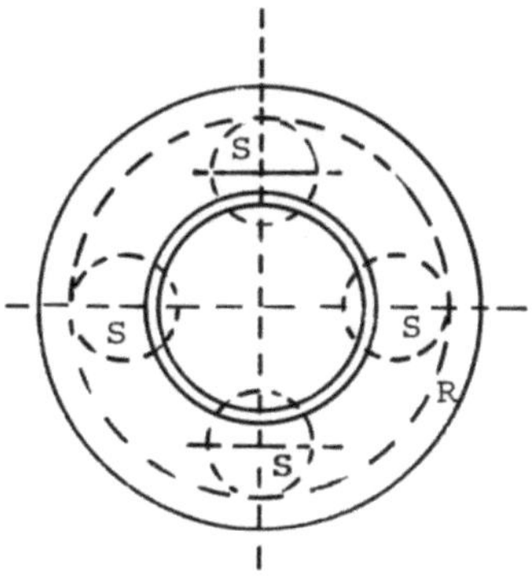

Fig 2.10.4 Shaft rotating on thrust bearings.

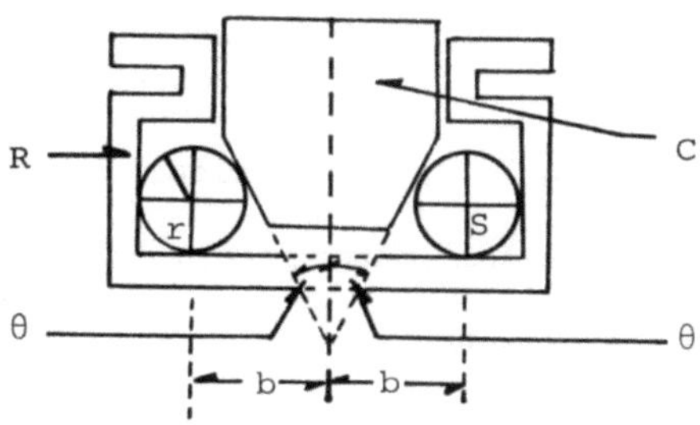

* Taken from: Kane, T.R., Dynamics,

Dr. Thomas Kane
817 Lathrop Drive
Stanford, CA - 94305

Solution:

From Corollary 2.10.2, if all C, S and R move with pure rolling relative motion, then the i.a's.p.r. all coincide at one common point. Clearly, the i.a.p.r. of C with respect to S is the cone element passing through the contact point (between C and S), whereas the i.a.p.r. of C with respect to R is the symmetry axis of C. The intersection of those two axes is the cone apex, which henceforth is referred to as point O. Length b is now determined by the condition that the i.a.p.r. of C with respect to R passes through O.

Now, two points of this axis are already known, namely, the two points of contact of S on R, henceforth referred to as points P_1 and P_2. Then, the geometry of Fig 2.10.5 follows

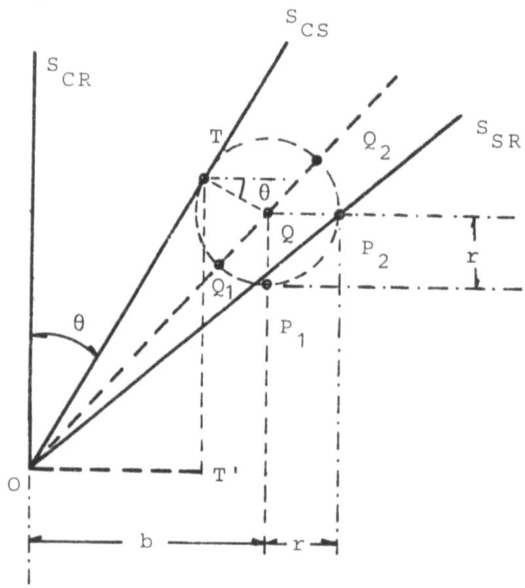

Fig 2.10.5 Instantaneous axes of pure rotation of bodies C, S and R of Fig 2.10.4.

From Fig 2.10.5 it is clear that axis S_{SR} makes a 45° angle with axis S_{CR}. Let T be the contact point between C and S. From a well-known theorem of plane geometry,

$$\overline{OT}^2 = \overline{OP}_1\ \overline{OP}_2 \qquad (2.10.13)$$

Applying the Pythagorean Theorem to triangle OT'T,

$$\overline{OT}^2 = \overline{OT'}^2 + \overline{T'T}^2 \qquad (2.10.14)$$

But

$$\overline{OT'} = b - r\cos\theta \qquad (2.10.15a)$$

and

$$\overline{T'T} = b + r + r\sin\theta \qquad (2.10.15b)$$

Hence,

$$\overline{OT}^2 = 2r^2(1+\sin\theta)+2br(1+\sin\theta-\cos\theta)+2b^2 \qquad (2.10.16)$$

Also,

$$\overline{OP}_1 = \sqrt{2}\ b \qquad (2.10.17a)$$

$$\overline{OP}_2 = \sqrt{2}(b+r) \qquad (2.10.17b)$$

Substitution of eqs. (2.10.16) and (2.10.17 a and b) into eq. (2.10.13) yields

$$r(1+\sin\theta)+b(\sin\theta-\cos\theta)=0$$

from which,

$$b=r\frac{1 + \sin\theta}{\cos\theta-\sin\theta} \qquad (2.10.18)$$

One more consequence of Theorem 2.10.2 is summarized in the following

Corollary 2.10.3 (Three center Theorem). In plane motion the three instantaneous axes (centers in this context) of three rigid bodies in motion lie on a line (2.10)

2.11 ACCELERATION DISTRIBUTION IN A RIGID BODY MOVING ABOUT A FIXED POINT

It was shown in Section 2.7 that the velocity of a point of a rigid body moving about a fixed point is given by

$$\underset{\sim}{v}(t)=\underset{\sim}{\Omega}(t)\underset{\sim}{y}(t) \qquad (2.11.1)$$

where $\underset{\sim}{\Omega}(t)$ is the rigid-body angular velocity and $\underset{\sim}{y}(t)$ is the current position vector of the point under consideration.

The acceleration $\underset{\sim}{a}(t)$ of the said point is now obtained differentiating both sides of eq. (2.11.1) with respect to time; thus

$$\underset{\sim}{a}(t)=\dot{\underset{\sim}{\Omega}}(t)\underset{\sim}{y}(t)+\underset{\sim}{\Omega}(t)\dot{\underset{\sim}{y}}(t)$$

But $\dot{\underset{\sim}{y}}(t)$ is $v(t)$, the above equation taking on the form

$$\underset{\sim}{a}(t)=\left(\dot{\underset{\sim}{\Omega}}(t)+\underset{\sim}{\Omega}^2(t)\right)\underset{\sim}{y}(t) \qquad (2.11.2)$$

The matrix in brackets appearing in eq. (2.11.2) is referred to, by analogy with eq. (2.11.1), as <u>the angular acceleration matrix</u>. The acceleration of the point under study is formed by two components, as appearing in eq. (2.11.2), namely, the <u>tangential acceleration</u>, $\dot{\underset{\sim}{\Omega}}(t)y(\underset{\sim}{t})$, and the <u>normal acceleration</u>, $\underset{\sim}{\Omega}^2(t)y(\underset{\sim}{t})$, the former being tangential and the latter being normal to the velocity.

<u>Exercise 2.11.1</u> Show that $\dot{\underset{\sim}{\Omega}}(t)\underset{\sim}{y}(t)$ and $\underset{\sim}{\Omega}^2(t)\underset{\sim}{y}(t)$ are, respectively, parallel and normal to the velocity.

There is one implicit fact in the above result, namely, in the square matrix vector space, one scalar product (See Section 1.7) can be defined as $Tr(\underset{\sim}{A}\underset{\sim}{B}^T)$, $\underset{\sim}{A}$ and $\underset{\sim}{B}$ being matrices of the same space.

In this context, matrices $\dot{\underset{\sim}{\Omega}}(t)$ and $\underset{\sim}{\Omega}^2(t)$ are orthogonal, i.e., its scalar product vanishes.

<u>Exercise 2.11.2</u> Show that $Tr(\dot{\underset{\sim}{\Omega}}\underset{\sim}{\Omega}^2)=0$

Result (2.11.2) can be expressed, in Gibbs' notation as

$$\underset{\sim}{a}(t)=\dot{\underset{\sim}{\omega}}(t)x\underset{\sim}{r}(t)+\underset{\sim}{\omega}(t)x\left(\underset{\sim}{\omega}(t)x\underset{\sim}{r}(t)\right) \qquad (2.11.3)$$

thereby making the result of Exercise 2.11.1 apparent

2.12 ACCELERATION DISTRIBUTION IN A RIGID BODY UNDER GENERAL MOTION.

Consider now the most general case of rigid-body motion, in which none of the points of the body remains fixed.

From eq. (2.9.2), the velocity of a point-whose position vector is $\mathbf{y}(t)$- of a rigid body under general motion is

$$\mathbf{v}(t)=\mathbf{v}_P(t)+\mathbf{\Omega}(t)\left(\mathbf{y}(t)-\mathbf{y}_P(t)\right) \qquad (2.12.1)$$

where $\mathbf{y}_P(t)$ and $\mathbf{v}_P(t)$ are the position vector and the velocity, both known, of a given point P of the rigid body. The acceleration of a general point, $\mathbf{a}(t)$, of the body under consideration is next obtained differentiating both sides of eq. (2.12.1) with respect to time, i.e.

$$\mathbf{a}(t)=\mathbf{a}_P(t)+\dot{\mathbf{\Omega}}(t)\left(\mathbf{y}(t)-\mathbf{y}_P(t)\right)+\mathbf{\Omega}(t)\left(\dot{\mathbf{y}}(t)-\dot{\mathbf{y}}_P(t)\right) \qquad (2.12.2)$$

and, from eq. (2.12.1),

$$\dot{\mathbf{y}}(t)-\dot{\mathbf{y}}_P(t)= \mathbf{v}(t)-\mathbf{v}_P(t)=\mathbf{\Omega}(t)\left(\mathbf{y}(t)-\mathbf{y}_P(t)\right) \qquad (2.12.3)$$

which, when substituted in eq. (2.12.2), leads to

$$\mathbf{a}(t)=\mathbf{a}_P(t)+\left(\dot{\mathbf{\Omega}}(t)+\mathbf{\Omega}^2(t)\right)\left(\mathbf{y}(t)-\mathbf{y}_P(t)\right) \qquad (2.12.4)$$

which, except for the term $\mathbf{a}_P(t)$, is identical to eq. (2.11.2), with $\mathbf{y}(t)-\mathbf{y}_P(t)$ instead of $\mathbf{y}(t)$ of that equation.

The relative acceleration, $\mathbf{a}(t)-\mathbf{a}_P(t)$, of the general point with respect to P is clearly given as

$$\mathbf{a}(t)-\mathbf{a}_P(t)=\left(\dot{\mathbf{\Omega}}(t)+\mathbf{\Omega}^2(t)\right)\left(\mathbf{y}(t)-\mathbf{y}_P(t)\right) \qquad (2.12.5)$$

which again, is seen to be composed of both a tangential and a normal component.

Paralelling previous sections, the set of points of minimum-magnitude acceleration is now determined. Thus, the function ϕ defined as

$$\phi(\mathbf{y})=\mathbf{a}^T\mathbf{a} \qquad (2.12.6)$$

is now minimized over $\underset{\sim}{y}$. Applying the "chain rule" to it,

$$\phi'(\underset{\sim}{y}) = 2\left(\frac{\partial \underset{\sim}{a}}{\partial \underset{\sim}{y}}\right)^T \underset{\sim}{a} \tag{2.12.7}$$

where, from eq. (2.12.4),

$$\frac{\partial \underset{\sim}{a}}{\partial \underset{\sim}{y}} = \dot{\underset{\sim}{\Omega}} + \underset{\sim}{\Omega}^2 \tag{2.12.8}$$

Hence, the minimum-magnitude acceleration satisfies

$$(-\dot{\underset{\sim}{\Omega}} + \underset{\sim}{\Omega}^2)\underset{\sim}{a} = \underset{\sim}{0} \tag{2.12.9}$$

i.e. the minimum-magnitude acceleration is in the null space of $\dot{\underset{\sim}{\Omega}}+\underset{\sim}{\Omega}^2$. If both $\dot{\underset{\sim}{\Omega}}$ and $\underset{\sim}{\Omega}^2$ have the same null space, then that minimum-magnitude acceleration lies in that space. Since both $\dot{\underset{\sim}{\Omega}}$ and $\underset{\sim}{\Omega}$ are skew symmetric, vectors $\dot{\underset{\sim}{\omega}}$ and $\underset{\sim}{\omega}$ lying in their null space, can be defined in such a way that, for any vector $\underset{\sim}{r}$,

$$\underset{\sim}{\Omega}\underset{\sim}{r} = \underset{\sim}{\omega} \times \underset{\sim}{r}, \quad \dot{\underset{\sim}{\Omega}}\underset{\sim}{r} = \dot{\underset{\sim}{\omega}} \times \underset{\sim}{r} \tag{2.12.10}$$

Hence it becomes clear that for $\dot{\underset{\sim}{\Omega}}$ and $\underset{\sim}{\Omega}$ to have the same null space, $\underset{\sim}{\omega}$ and $\dot{\underset{\sim}{\omega}}$ should be parallel. Furthermore, $\underset{\sim}{\Omega}^2$ and $\underset{\sim}{\Omega}$ have the same null space (prove it) and hence, for $\dot{\underset{\sim}{\Omega}}$ and $\underset{\sim}{\Omega}^2$ to have the same null space, $\underset{\sim}{\omega}$ and $\dot{\underset{\sim}{\omega}}$ should be parallel. A simple case for which $\dot{\underset{\sim}{\Omega}}$ and $\underset{\sim}{\Omega}^2$ have the same null space is that for which the rotation axis, $\underset{\sim}{e}$, has a constant direction. In fact, if this is so, then

$$\underset{\sim}{\Omega}\underset{\sim}{e} = \underset{\sim}{0} \tag{2.12.11}$$

for all time t. Differentiating the latter expresion with respect to time yields

$$\dot{\underset{\sim}{\Omega}}\underset{\sim}{e} + \underset{\sim}{\Omega}\dot{\underset{\sim}{e}} = \underset{\sim}{0}$$

But, since the magnitude of e is unity and its direction is constant, $\dot{\underset{\sim}{e}}=\underset{\sim}{0}$ and hence, the latter equation leads to

$$\dot{\underset{\sim}{\Omega}}\underset{\sim}{e} = \underset{\sim}{0} \tag{2.12.12}$$

thereby showing that, under the conditions stated, if $\underset{\sim}{e}$ is in the null space of $\underset{\sim}{\Omega}$, then it is also in the null space of $\dot{\underset{\sim}{\Omega}}$. Hence, when the instant screw axis has a contant direction, the minimum-magnitude acceleration is parallel to that direction. If the involved matrices do not have a common null space, then the only possibility of eq. (2.12.9) to hold is if $\underset{\sim}{a}=\underset{\sim}{0}$. The latter condition is equivalent to

$$\underset{\sim}{a}_P+(\dot{\underset{\sim}{\Omega}}+\underset{\sim}{\Omega}^2)(\underset{\sim}{y}_0-\underset{\sim}{y}_P)=\underset{\sim}{0}$$

or

$$(\dot{\underset{\sim}{\Omega}}+\underset{\sim}{\Omega}^2)\underset{\sim}{y}_0=(\dot{\underset{\sim}{\Omega}}+\underset{\sim}{\Omega}^2)\underset{\sim}{y}_P-\underset{\sim}{a}_P \tag{2.12.13}$$

which has a unique solution if $\dot{\underset{\sim}{\Omega}}$ and $\underset{\sim}{\Omega}^2$ do not have a common null space, for then, the sum of them becomes of full rank, i.e. in that case $\text{rank}(\dot{\underset{\sim}{\Omega}}+\underset{\sim}{\Omega}^2)=3$, and hence this sum is nonsigular. In this case, then, one single point of the body, located by the position vector $\underset{\sim}{y}_0$, has a zero acceleration. This point is called an *acceleration pole* and its position vector is given as

$$\underset{\sim}{y}_0=\underset{\sim}{y}_P-(\dot{\underset{\sim}{\Omega}}+\underset{\sim}{\Omega}^2)^{-1}\underset{\sim}{a}_P \tag{2.12.14}$$

Example 2.12.1 For a rigid-circular cone rolling without slipping on a plane, its acceleration pole is its apex (prove it)

Exercise 2.12.2 The system shown in Fig 2.12.1 is an inversion of the worm-gear mechanism and is composed of a rigid arm OA of lenght b that can rotate freely about axis EE', this axis being normal to the plane of motion of OA. A rigid wheel is coupled to OA at A in such a way that the wheel can rotate freely about axis FF' passing though A; this axis is perpendicular to both OA and EE'. If OA rotates at a constant rate p and the wheel rotates about FF' at a constant rate q, show that the point of the disk on OA, a distance d from O, has zero acceleration, the distance d being given by

$$d = \frac{q^2}{p^2+q^2} b$$

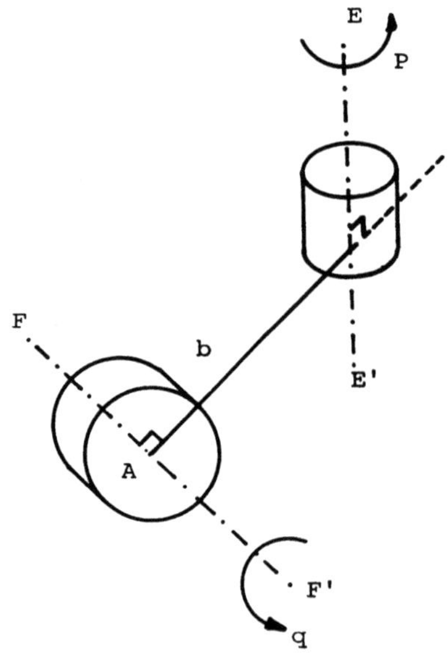

Fig 2.12.1 Inversion of the worm-gear mechanism

An extensive account of this topic is presented in (2.11)

2.13. ACCELERATION OF A MOVING POINT REFERRED TO A MOVING OBSERVER. CORIOLIS' THEOREM.

In Section 2.18 it was shown that the velocity $(\underset{\sim}{v})^{*}_{F}$ of a moving point, referred to a fixed observer, given in terms of its position vector, $(\underset{\sim}{\xi})_{M}$, referred to a moving observer is given by

$$(\underset{\sim}{v})_F = (\underset{\sim}{v}_o)_F+(\underset{\sim}{\Omega})_F(\underset{\sim}{\xi})_F+(\underset{\sim}{Q})_F(\dot{\underset{\sim}{\xi}})_M \tag{2.13.1}$$

where $\underset{\sim}{Q}$ and $\underset{\sim}{\Omega}$ are the rotation and the angular velocity matrices, respectively, of the moving axes with respect to the fixed ones.

The accelaration, $(\underset{\sim}{a})_F$, of the moving point, in the fixed observer, is now obtained differentiating eq. (2.13.1) with respect to time, namely

$$(\underset{\sim}{a})_F=(\underset{\sim}{a}_o)_F+(\dot{\underset{\sim}{\Omega}})_F(\underset{\sim}{\xi})_F+(\underset{\sim}{\Omega})_F(\dot{\underset{\sim}{\xi}})_F+(\dot{\underset{\sim}{Q}})_F(\dot{\underset{\sim}{\xi}})_M+(\underset{\sim}{Q})_F(\ddot{\underset{\sim}{\xi}})_M \tag{2.13.2}$$

where

$$(\underset{\sim}{a}_o)_F=(\dot{\underset{\sim}{v}}_o)_F \tag{2.13.3}$$

But

$$(\underset{\sim}{\xi})_F=(\underset{\sim}{Q})_F(\underset{\sim}{\xi})_M \tag{2.13.4}$$

Hence

$$(\dot{\underset{\sim}{\xi}})_F=(\dot{\underset{\sim}{Q}})_F(\underset{\sim}{\xi})_M+(\underset{\sim}{Q})_F(\dot{\underset{\sim}{\xi}})_M \tag{2.13.5}$$

Substitution of eqs. (2.13.4) and (2.13.5) into eq. (2.13.2) yields

$$(\underset{\sim}{a})_F=(\underset{\sim}{a}_o)_F+(\dot{\underset{\sim}{\Omega}})_F(\underset{\sim}{Q})_F(\underset{\sim}{\xi})_M+(\underset{\sim}{\Omega})_F((\dot{\underset{\sim}{Q}})_F(\underset{\sim}{\xi})_M+(\underset{\sim}{Q})_F(\dot{\underset{\sim}{\xi}})_M)+(\dot{\underset{\sim}{Q}})_F(\dot{\underset{\sim}{\xi}})_M+(\underset{\sim}{Q})_F(\ddot{\underset{\sim}{\xi}})_M \tag{2.13.6}$$

But

$$(\dot{\underset{\sim}{Q}})_F(\underset{\sim}{\xi})_M=(\dot{\underset{\sim}{Q}})_F(\underset{\sim}{Q}^T)_F(\underset{\sim}{Q})_F(\underset{\sim}{\xi})_M=(\underset{\sim}{\Omega})_F(\underset{\sim}{Q})_F(\underset{\sim}{\xi})_M$$

and, similarly,

$$(\dot{\underset{\sim}{Q}})_F(\dot{\underset{\sim}{\xi}})_M=(\underset{\sim}{\Omega})_F(\underset{\sim}{Q})_F(\dot{\underset{\sim}{\xi}})_M$$

---

* All vectors and matrices appearing in this section are functions of time, but for simplicity the argument (t) has been dropped.

and hence,

$$(\underset{\sim}{\Omega})_F(\dot{\underset{\sim}{Q}})_F(\underset{\sim}{\xi})_M = (\underset{\sim}{\Omega})_F(\underset{\sim}{\Omega})_F(\underset{\sim}{Q})_F \ (\underset{\sim}{\xi})_M = (\underset{\sim}{\Omega}^2)_F(\underset{\sim}{Q})_F(\underset{\sim}{\xi})_M$$

Substituting the two latter expressions into eq. (2.13.6) one obtains

$$(\underset{\sim}{a})_F = (\underset{\sim}{a}_o)_F + (\dot{\underset{\sim}{\Omega}} + \underset{\sim}{\Omega}^2)_F(\underset{\sim}{Q})_F(\underset{\sim}{\xi})_M + (\underset{\sim}{Q})_F(\ddot{\underset{\sim}{\xi}})_M + 2(\underset{\sim}{\Omega})_F(\underset{\sim}{Q})_F \ (\dot{\underset{\sim}{\xi}})_M \qquad (2.13.7)$$

which is an expression for the acceleration of a point in terms of measurements of its position, velocity and acceleration, taken by a moving observer. The first two terms of eq. (2.13.7) are identical to the right hand side of eq. (2.12.4) with $\underset{\sim}{y}-\underset{\sim}{y}_P$ substituted for $\underset{\sim}{\xi}$; hence, the two said terms constitute the acceleration of a point fixed in the moving observer, coincident with the moving point under study, at a particular time. The third term stands for the acceleration of the moving point, as measured by the moving observer, and the fourth term is an acceleration term arising from the rotation of the moving observer, as is apparent from eq. (2.13.7); this term is known as <u>Coriolis' acceleration</u> . Equation (2.13.7) constitutes, then, the Theorem of Coriolis. (2.11)

<u>Exercise 2.13.1</u> The mechanism shown in Fig 2.13.1 is a component of a quick-return mechanism used in a crank shaper. Assuming that disk 2 rotates at a constant angular velocity ω_2=1800 rpm, determine graphically the angular acceleration of link 3, for the given configuration.

Hint: Two points, B2 and B3, coincide at B. Find the acceleration of B3 via eq. (2.13.7), referred to an observer fixed in 2.

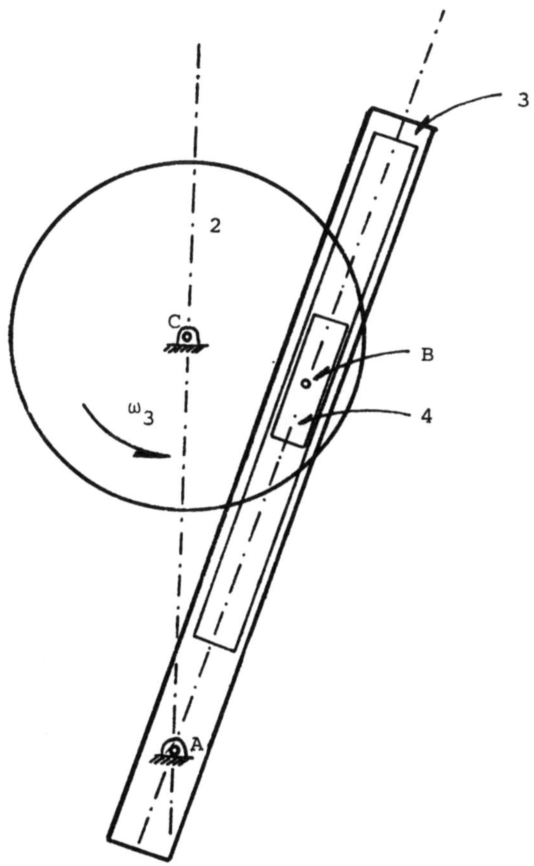

Fig 2.13.1 Driving system of a quick-return mechanism

REFERENCES

2.1 Euler L., "Du movement de rotation des corps solides autour d'un axe variable", Mémoirs de l'Académie de Sciences de Berlin (15), 1759, pp. 210-240

2.2 Synge J.L., "Classical Dynamics", in Flügge S., ed., Encyclopedia of Physics, Vol. III/1, Springer-Verlag, Berlin, 1960, pp. 18-19.

2.3. Rodrigues O., "Des lois géométriques qui régissent les déplacements d'un système solide dans l'éspace, et de la variation des coordonnées provenant de ces déplacements considérés independamment des causes qui peuvent les produire", Journal de Mathématiques Pures et Appliquées, vol. 5, 1st Series, 1840, pp. 380-440

2.4 Bisshop K.E., "Rodrigues' formula and the screw matrix" Journal of Engineering for Industry, Trans. ASME, Series B, vol. 91, February 1969.

2.5 Gupta V.K., "Kinematic analysis of plane and spatial mechanisms", Journal of Engineering for Industry, Trans. ASME, Series B, vol. 95, May 1973.

2.6 Fox E.A., Mechanics, Harper & Row, N. York, 1967, p. 11

2.7 Roth B., "The Kinematics of Motion Through Finitely Separated Positions", Journal of Applied Mechanics, Trans. ASME, Series E, September, 1967.

2.8 Truesdell C., A First Course in Rational Continuum Mechanics, vol. I. Academic Press, N. York, 1977. p. 38.

2.9 Kane T.R., Dynamics, Holt, Rinehart and Winston, Inc., N. York, 1968, p. 66.

2.10 Cowie A., Kinematics and Design of Mechanisms, International Texbook Company, Scranton, Penn, 1961, p.144.

2.11 Veldkamp G.R., "Acceleration axes and distribution in spatial motion" Journal of Engineering for Industry, Trans. ASME, Series B, vol. 91, Feb. 1969.

2.12 Coriolis G., "Mémoire sur les équations du mouvement relatif des systèmes des corps", J. Ecole Polytechnique, 15, cahier 24, 1835, pp. 142-154.

3. Generalities on Lower-Pair Kinematic Chains

3.1 INTRODUCTION

The coupling of rigid bodies constitutes a kinematic chain, each of its bodies thus being referred to as a link of the chain. If every link of a kinematic chain is coupled to at least two other links, the chain is said to be closed. If some links of the chain, however, are coupled to only one link, the chain is open. A given kinematic chain, though open, may contain a subset of links that constitute a closed kinematic subchain, which is referred to as a loop. Whether closed or open, kinematic chains can contain one or multiple loops, thus being referred to as single-loop or multiple-loop kinematic chains. Of course, an open kinematic chain may contain no loop.

Closed kinematic chains are referred to as linkages, whereas open ones are called manipulators. Linkages have been the main object of study within the modern Theory of Machines and Mechanisms for over a century [3.1], though it seems that linkages appeared in the 13th century. They are comprised within a wider class of mechanical devices known as mechanisms. A widely accepted definition of mechanism is [3.2]: "an assemblage of rigid or resistant bodies connected together for the purpose of transforming motion".

The interest for the study of mechanisms arose orginally in mechanical engineering. The underlying theory, however, has been found to embrace other areas such as biomechanics, and so, it finds wide applications in the study of some living entities like the locomotion systems of humans and animals. The wide variety of mechanisms defined previously, can be divided into two classes, namely, lower-pair and upper-pair mechanisms. These terms are discussed in the present chapter, together with other related items such as degree of freedom, kinematic pair and classification of kinematic chains.

3.2 KINEMATIC PAIRS

A kinematic pair is the coupling of two mechanical elements. If these two elements are rigid bodies, the pair can be either one of two kinds: i) lower pair or ii) upper pair. A lower pair exists when one element is coupled to the other via a wrapping action and contact takes place along a surface. If contact takes place along a line or a point, the resulting coupling is referred to as an upper pair.

3.3 DEGREE OF FREEDOM.

The degree of freedom of a kinematic chain is defined as an integer number corresponding to the minimum number of generalised [3.3] coordinates required to specify geometrically a configuration of the chain. If the degree of freedom of a chain is positive, it constitutes a mechanism; if it is zero, it corresponds to a statically determined structure, whereas, if negative, it corresponds to a statically underdetermined (hyperstatic) structure, the negative of its degree of freedom being referred to as its redundancy. The degree of freedom of a rigid body free to move in space is thus equal to six, namely, a translation along each of three noncoplanar directions, plus a rotation (e.g. Euler's angles) about each of three noncoplanar directions (not necessarily corresponding to the three previous directions). Constraints are imposed on rigid bodies to make their motion useful for a certain purpuse, thus diminishing their degree of freedom. Kinematic pairs are in fact constraints imposed on the involved bodies.

3.4 CLASSIFICATION OF LOWER PAIRS

All possible lower pairs can be put into one of six different types, namely, i) Revolute (R), ii) Prismatic (P), iii) Screw (H), iv) Cylindric (C), v) Spheric (S) and vi) Planar (E), physical models of which are shown in Fig 3.4.1, their kinematic models appearing in Fig 3.4.2.

The revolute pair only allows rotation about one axis, therefore imposing five constraints: prevention of translation along three directions and of rotation about two axes.

The prismatic pair only allows translation along one direction, also imposing five constraints: prevention of translation along two directions and of rotation about three axes.

The screw pair allows translation along one direction and rotation about the same direction, both being related. Thus, five constraints are also imposed by this pair, namely, prevention of tranlation along three (or, alternatively, two) directions and of rotation about two (or, alternatively, three) axes.

The cylindrical pair allows two independent motions, namely, translation about one axis and rotation about the same axis. This pair imposes clearly, four contraints.

The spherical pair allows rotation about three noncoplanar axes, thus imposing three constraints: prevention of translation along three noncoplanar directios.

The planar pair allows translation along two independent directions and rotation about one axis perpendicular to the plane of those directions. Thus, this pair imposes three constraints.

A link supplied with two kinematic pairs is referred to as a dyad. A dyad containing one revolute and one prismatic pair is referred to as an R-P dyad. Similarly are R-R, R-C, S-S, etc. dyads defined.

Hervé [3.4] has introduced the most rational classification of mechanisms by resorting to the theory of groups. In fact, each lower pair allows a set of relative motions of the two coupled bodies. These motions can be regarded as subsets of the group of rigid-body motions (See Sect.2.4). It can be readily shown that each of these subsets constitutes in turn a group. Thus, the set of relative motions allowed by a revolute pair constitutes the subgroup of rotations about a fixed axis passing through a given point A and parallel to a given unit vector $\underset{\sim}{e}$. All these rotations differ only by the angle of rotation. This group is in turn a subgroup of the group of rotations about point A, which is, in turn, a subgoup of the group of general rigid-body motions. The subgroup associated with the revolute pair can therefore be characterised by point A and vector $\underset{\sim}{e}$. Hence, it is represented by $g_R(A,\underset{\sim}{e})$.

On the other hand, the set of relative motions allowed by the prismatic pair is that of translations along a direction parallel to the unit vector $\underset{\sim}{e}$, which clearly constitutes a subgroup. This is in turn a subgroup of the group of pure translations, i.e. along three noncoplanar directions. The subgroup associated with the prismatic pair is thus characterised by the unit vector $\underset{\sim}{e}$ and is hence represented by $g_P(\underset{\sim}{e})$.

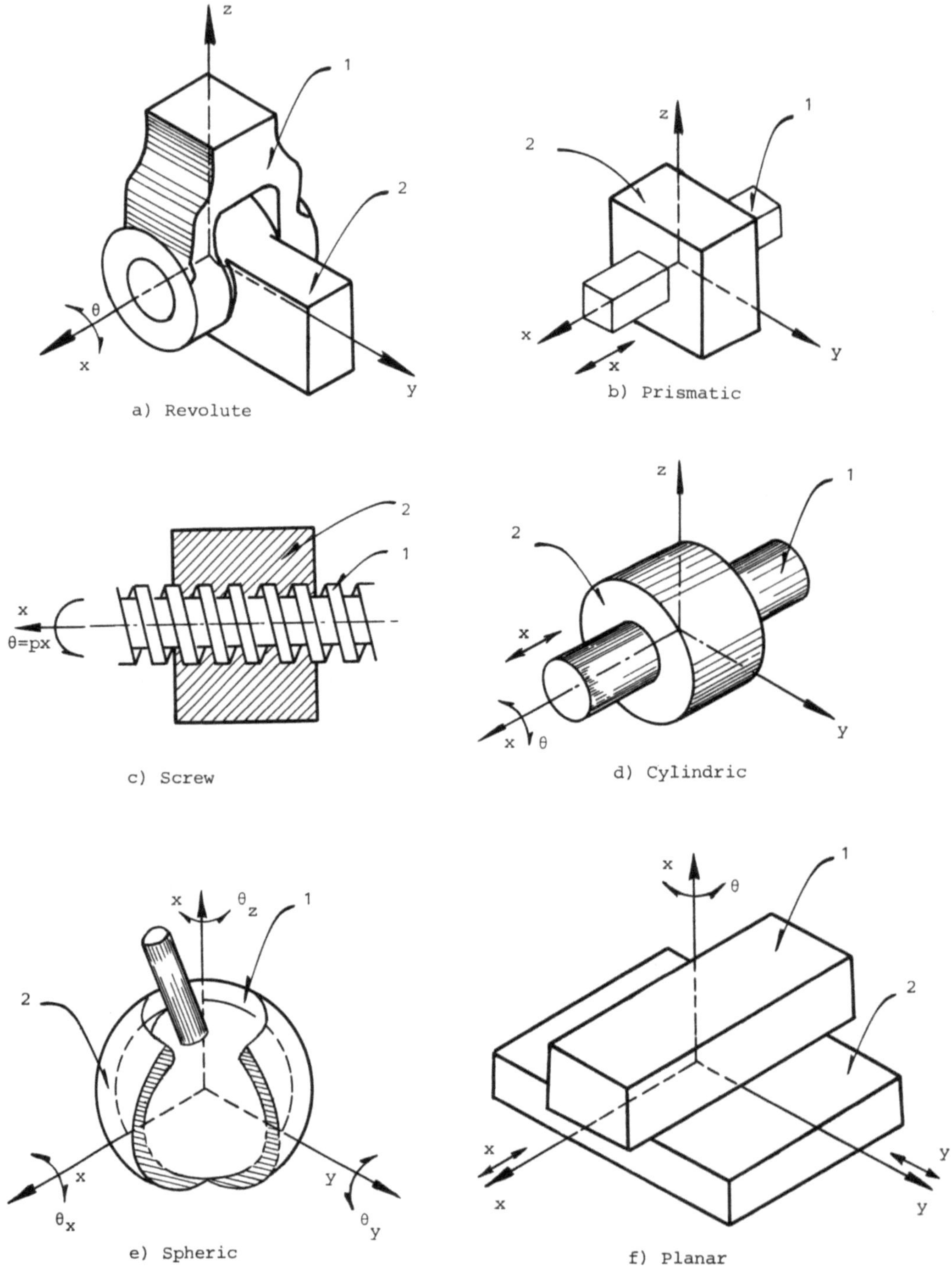

Fig. 3.4.1 Lower Pairs

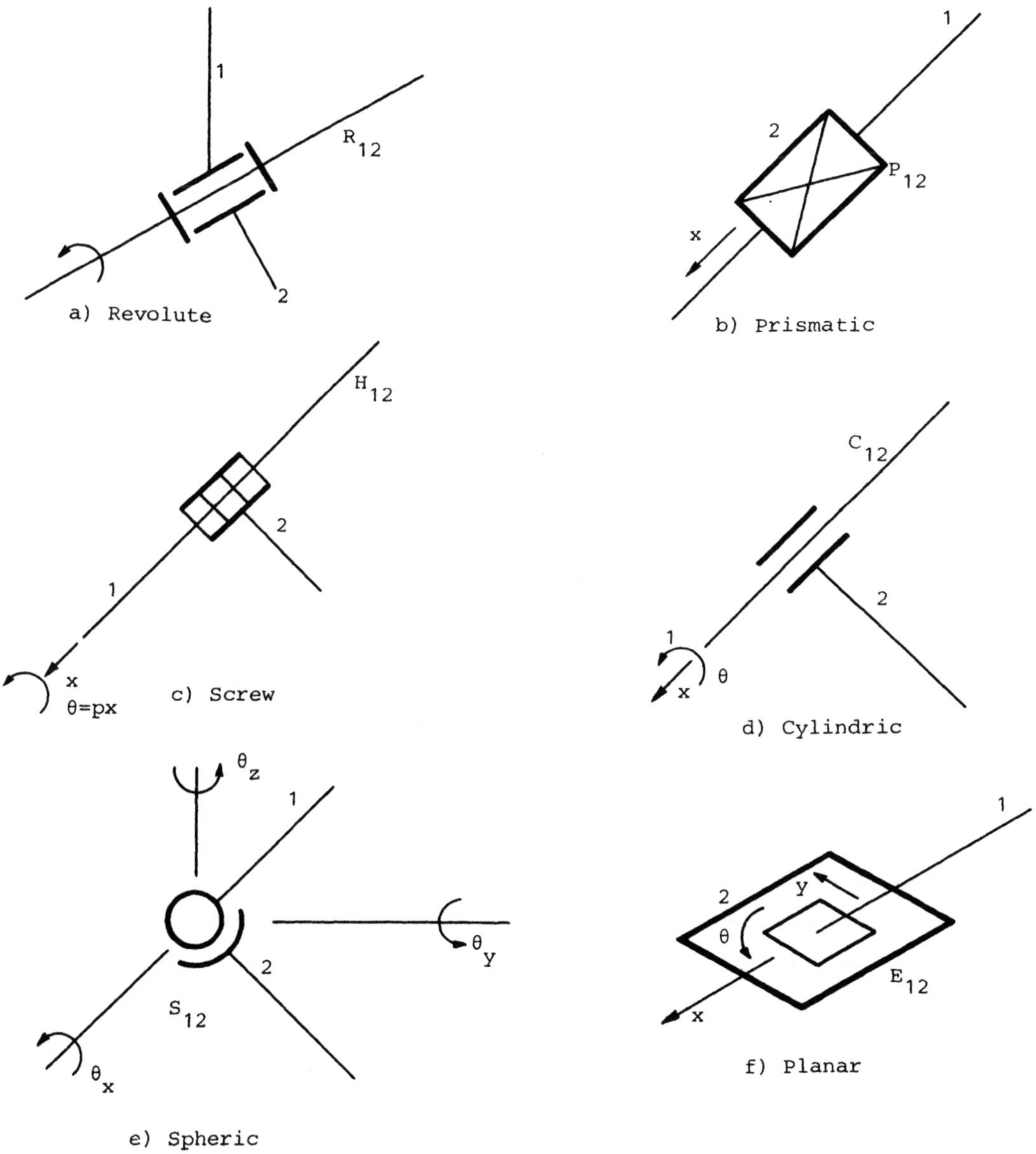

Fig. 3.4.2 Kinematic models
of lower pairs

The screw pair allows a relative motion of a rotation about an axis passing through point A and parallel to the unit vector $\underset{\sim}{e}$, plus a translation along a direction parallel to vector $\underset{\sim}{e}$. Both the rotation and the translation are not independent, however, for they are related by the pitch p of the screw. The corresponding subgroup, hence, is characterised by point A, the pitch p and vector $\underset{\sim}{e}$, its representation thus being $\mathcal{g}_H(A,p,\underset{\sim}{e})$.

The cylindrical pair defines in turn a subgroup being characterised by a rotation about an axis passing through point A and parallel to the unit vector $\underset{\sim}{e}$, plus a translation along a direction parallel to vector $\underset{\sim}{e}$. In this case, however, both the angle of rotation and the amount of translation, are independent. The subgroup associated with this pair is then characterised by point A and vector $\underset{\sim}{e}$, its representation being $\mathcal{g}_C(A,\underset{\sim}{e})$. Clearly,

$$\mathcal{g}_R(A,\underset{\sim}{e}),\mathcal{g}_P(\underset{\sim}{e}),\mathcal{g}_H(A,p,\underset{\sim}{e})\subset\mathcal{g}_C(A,\underset{\sim}{e})$$

The spherical pair defines a set of relative motions that constitutes the subgroup of rotations about a given point A. Hence, this subgroup is characterised by point A, its representation being $\mathcal{g}_S(A)$. Clearly

$$\mathcal{g}_R(A,\underset{\sim}{e})\subset\mathcal{g}_S(A)$$

The planar pair defines a set of relative motions that constitute the subgroup of motions lying in a plane perpendicular to the unit vector $\underset{\sim}{e}$. Hence, its representation is $\mathcal{g}_E(\underset{\sim}{e})$. Clearly

$$\mathcal{g}_R(A,\underset{\sim}{e})\subset\mathcal{g}_E(\underset{\sim}{e})$$

The reader can verify that all of the foregoing defined sets actually constitute subgroups, by recalling the definitions introduced in Section 2.4.

A word of caution is in order: if A and B are any two points and $\underset{\sim}{e}$ is a unit vector not parallel to line AB, then,

i) $\mathcal{g}_R(A,\underset{\sim}{e})$ and $\mathcal{g}_R(B,\underset{\sim}{e})$ are independent

ii) $g_H(A,p,\underset{\sim}{e})$ and $g_H(B,p,\underset{\sim}{e})$ are independent

iii) if $p \neq q$, $g_H(A,p,\underset{\sim}{e})$ and $g_H(B,q,\underset{\sim}{e})$ are independent. However, $g_H(A,p,e)$, $g_H(A,q,\underset{\sim}{e}) \sqsubset g_C(A,\underset{\sim}{e})$.

The validity of the foregoing statements is accepted almost intuitively. They will become clearer in the following discussion, which requires some previous definitions.

The number of independent variables that is required to define the relative position of two coupled links is referred to its degree of freedom, f. Thus, f_i is said to be the degree of freedom of the kinematic pair p_i. The concept of degree of freedom of a kinematic pair can be extended to any subgroup of the group g_r of rigid-body motions, the corresponding concept being that of *dimension*. Hence, the dimension of g_r is six, which is symbolically represented as

$$\dim g_r = 6$$

Subgroups of g_r that are not associated with one of the lower pairs already defined are the following:

The *identity transformation* i, leaving all points of space unchanged. One can think of two links coupled by a pair corresponding to i if the coupling is rigid.

The *plane translation* $g_T(\underset{\sim}{e}_1,\underset{\sim}{e}_2)$, corresponding to a relative translation along two independent directions defined by the nonidentical unit vectors $\underset{\sim}{e}_1$ and $\underset{\sim}{e}_2$.

The *spatial translation* $g_T(\underset{\sim}{e}_1,\underset{\sim}{e}_2,\underset{\sim}{e}_3)$, corresponding to a relative translation along three independent directions defined by the noncoplanar unit vectors $\underset{\sim}{e}_1$, $\underset{\sim}{e}_2$ and $\underset{\sim}{e}_3$.

The *translating-screw* transformation, $g_Y(p,\underset{\sim}{e}_1,\underset{\sim}{e}_2,\underset{\sim}{e}_3)$, corresponding to the motion of a screw of pitch p and axis parallel to the unit vector $\underset{\sim}{e}_3$, whose bolt can undergo a translation parallel to the plane defined by the nonidentical unit vectors $\underset{\sim}{e}_1$ and $\underset{\sim}{e}_2$. Vector $\underset{\sim}{e}_3$ is perpendicular to each $\underset{\sim}{e}_1$ and $\underset{\sim}{e}_2$.

The *translating gimbal* transformation, defining the motion of a rigid body mounted on a frame and free to rotate about one axis fixed on this frame and parallel to the unit vector $\underset{\sim}{e}$. Moreover, the frame can translate freely in the space. This transformation thus defines a subgroup represented by $g_X(\underset{\sim}{e})$.

The *general rigid-body motion* transformation g_r, allowing a rigid body to move freely with respect to a "fixed" frame. This can also be thought of a translating three-axis gimbal transformation, its three axis being noncoplanar.

The foregoing definitions are summarized in Table 3.4.1

Now, the concept of kinematic pair can be extended. In analysing a kinematic chain composed of bodies 1,2,...,i-1,i, one may be interested in studying the motion of body i relative to body 1. Whatever the couplings of bodies 1 and 2, 2 and 3,..., i-1 and i may be, the motion of body i with respect to body 1 will be a subset of the group of rigid-body motions. This subset may in turn be a subgroup itself or may not. Moreover, should this subset be a subgroup, it may or may not be associated to a kinematic lower pair. Thus, the coupling between bodies 1 and i is a generalization of the concept of lower pair and is henceforth referred to as the *liaison* L (1, i). In particular, if i = 2, i.e. if no intermediate links exist between links 1 and i, then L (1, i) coincides with the corresponding lower pair. One can think of L (1, i) as "the product" or "composition" of the liaisons L (1, 2), L (2, 3), ..., L (i-1, i), in this order, i.e. as

$$L(1,i) \equiv L(1,2,).L(2,3)^{\cdot}\ldots{}^{\cdot}L(i-1,i) \tag{3.4.1}$$

where each $L(k,k+1)$ is actually the pair connecting bodies k and k+1. Since to each liaison $L(k,k+1)$ there corresponds a kinematic lower pair, and each of these is associated with a subgroup, definition (3.4.1) must be actually interpreted as a product of subgroups, i.e. as

TABLE 3.4.1 Subgroups of the group of rigid-body motions

Group	dimension	associated kinematic lower pair
$\mathcal{i}$	0	rigid
$\mathcal{g}_P(\underset{\sim}{e})$	1	prismatic (P)
$\mathcal{g}_R(A,\underset{\sim}{e})$	1	revolute (R)
$\mathcal{g}_H(A,p,\underset{\sim}{e})$	1	screw (H)
$\mathcal{g}_T(\underset{\sim}{e}_1,\underset{\sim}{e}_2)$	2	none
$\mathcal{g}_C(A,\underset{\sim}{e})$	2	cylindric (C)
$\mathcal{g}_T(\underset{\sim}{e}_1,\underset{\sim}{e}_2,\underset{\sim}{e}_3)$	3	none
$\mathcal{g}_E(\underset{\sim}{e})$	3	planar (E)
$\mathcal{g}_Y(p,\underset{\sim}{e}_1,\underset{\sim}{e}_2,\underset{\sim}{e}_3)$	3	none
$\mathcal{g}_S(A)$	3	spheric (S)
$\mathcal{g}_X(\underset{\sim}{e})$	4	none
$\mathcal{g}_r$	6	none

$$\{L(1,i)\}=\{L(1,2)\}\cdot\{L(2,3)\}\ldots\{L(i-1,i)\}$$

The product of any two subgroups of Table 3.4.1 is a subset of $\mathcal{G}_r$ endowed with the properties of the two given subgroups. Thus, the product of $\mathcal{G}_P(\tilde{e})$ and $\mathcal{G}_R(A,\tilde{e})$ is $\mathcal{G}_C(A,\tilde{e})$, a subgroup of that table. In general, however, that product need not be in turn a subgroup.

Example 3.4.1. Assuming $L(1,2)$, $L(2,3)$ and $L(3,4)$ are identical to revolute pairs of parallel axes, then $\{L(1,2)\}.\{L(2,3)\}.\{L(3,4)\}$ equals $\mathcal{G}_E(\tilde{e})$, $\tilde{e}$ being parallel to the given revolute axes.

Example 3.4.2. The intersection of $\mathcal{G}_S(A)$ and $\mathcal{G}_R(A,\tilde{e})$ is $\mathcal{G}_R(A,\tilde{e})$. That of $\mathcal{G}_H(A,p,\tilde{e})$ and $\mathcal{G}_C(A,\tilde{e})$ is $\mathcal{G}_H(A,p,\tilde{e})$. However, that of $\mathcal{G}_T(\tilde{e}_1,\tilde{e}_2)$ and $\mathcal{G}_R(A,\tilde{e})$, $\tilde{e}$ being perpendicular to both $\tilde{e}_1$ and $\tilde{e}_2$, is $\mathcal{i}$.

The foundations needed to establish a classification of linkages and hence, formulae to compute their degree of freedom have now been laid down.

3.5 CLASSIFICATION OF KINEMATIC CHAINS

If links i,j and k of a kinematic chain are coupled so that i and j are coupled through $L(i,j)$ and j and k, through $L(j,k)$, the arising liaison $L(i,k)$ can be obtained, according to Section 6.4, as

$$\{L(i,k)\}\equiv\{L(i,j).L(j,k)\}=\{L(i,j)\}.\{L(j,k)\} \qquad (3.5.1)$$

The dimension of the composed liaison is related to that of the single ones as

$$\dim\{L(i,k)\}=\dim\{L(i,j)\}+\dim\{L(j,k)\}-\dim\{(Li,j)\}\cap\{L(j,k)\} \qquad (3.5.2)$$

If the groups $\{L(i,j)\}$ and $\{L(j,k)\}$ are such that their only common transformation is the identity, they are said to be independent. One then has

$$\{L(i,j)\}\cap\{L(j,k)\}=\mathcal{i}$$

the dimension of this intersection thus being 0.

Example 3.5.1. Let $L(i,j)$ be identical to $g_C(A,\underset{\sim}{e})$ and $L(j,k)$, to $g_H(A,p,\underset{\sim}{e})$. One has the kinematic chain shown in Fig 3.5.1

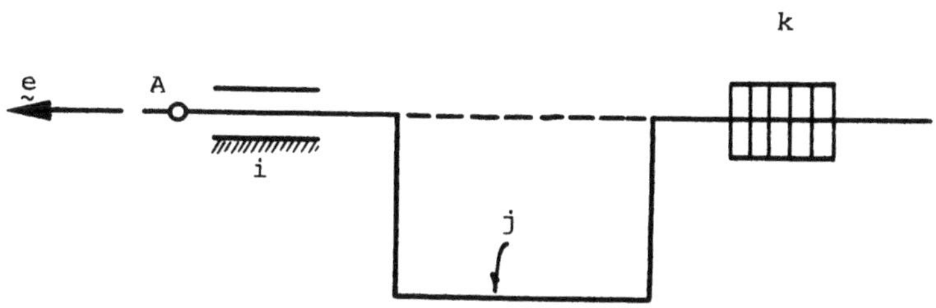

Fig 3.5.1 CR open chain

In that chain,

$$\dim \{L(i,j)\} = \dim g_C(A,\underset{\sim}{e}) = 2$$

$$\dim \{L(j,k)\} = \dim g_H(A,p,\underset{\sim}{e}) = 1$$

$$\dim \{L(i,j)\}\cdot\{L(j,k)\} = \dim g_H(A,p,\underset{\sim}{e}) = 1$$

Hence

$$\dim \{L(i,k)\} = 2+1-1 = 2$$

The foregoing example makes it clear that it may happen that the dimension of a kinematic chain composed of n links, $\dim \{L(1,n)\}$, does not equal the sum of the dimensions of its liaisons, $\sum_1^{n-1} \{L(i,i+1)\}$. This difference is, therefore, an integer m that can be different from 0. One then has

$$m \equiv \sum_1^{n-1} \dim \{L(i, i+1)\} - \dim \{L(1,n)\} \qquad (3.5.3)$$

In formula (3.5.3.), if m is different from 0, this indicates that there exist within the kinematic chain under study some degrees of freedom that are "idle" or "passive", i.e. degrees of freedom not affecting the transmission of motion from link 1 to link n. For example, the kinematic chain of Fig. 3.5.1 contains one idle degree of freedom, for the subgroup $g_C(A, \underset{\sim}{e})$ generated by its cylindrical pair can be regarded as the composition of the subgroup $g_R(A,\underset{\sim}{e})$ generated by a revolute pair and the subgroup $g_P(\underset{\sim}{e})$ generated by a prismatic pair. Its screw pair, in turn, generates a subgroup $g_H(A, p, \underset{\sim}{e})$ which can be regarded as the composition of the aforementioned subgroups $g_R(A, \underset{\sim}{e})$ and $g_P(\underset{\sim}{e})$, both of them related by the pitch p. The two pairs, thus, generate one identical subgroup, which can be thought of as being $g_H(A,p,\underset{\sim}{e})$. The dimension of this is 1, which is thus the said idle degree of freedom.

The computation of the degree of freedom of kinematic chains is next pursued. To this end, a classification due to Hervé [3.4] is now introduced. A kinematic chain is said to be *trivial* if any possible liaison between two of its links is a subset of <u>one</u> particular subgroup of g_r (see Table 3.4.1). It is said to be *exceptional* if any such possible liaison is a subset <u>not of one</u> particular subgroup of g_r, but of <u>several ones</u> of these. Finally, it is said to be *paradoxical* if its mobility cannot be analysed with the sole aid of the theory of groups. Experience shows that paradoxical chains possess certain axes of symmetry which produce their mobility.

Formula (3.5.3) can be applied to either open or closed kinematic chains. It can be applied to closed kinmematic chains if an *i* liaison between bodies 1 and n is introduced. This formula will thus help find the degree of freedom of a linkage, i.e. of a closed kinematic chain, provided that this is not paradoxical.

Now a formula providing the degree of freedom of trivial chains is introduced. Let g be the subgroup of the *least dimension* d, containing all possible liaisons between anty two links of a given trivial chain. Moreover, let f_i be the degree of freedom of the kinematic pair p_i, r_i being its restriction, defined as $d-f_i$. If the chain is composed of n links, its degree of freedom will be

$$F=d(n-1)-\sum_{1}^{n} r_i \qquad (3.5.4)$$

If the chain contains m idle degrees of freedom, as given by formula (3.5.3), these will be included in F. To compute the actual degree of freedom of the given trivial chain, F_a, then m should be subtracted from F, i.e.

$$F_a = F - m$$

where m, as given by eq. (3.5.3), is the number of passive degrees of freedom of the linkage. Formula (3.5.4) is referred to as a generalised Chebyshev or Kutzbach-Grübler formula.

The following examples are taken from Hervé [3.4].

Example 3.5.2. Consider the HHHRRH linkage of Fig 3.5.2, whose 4 screws have distinct pitches.

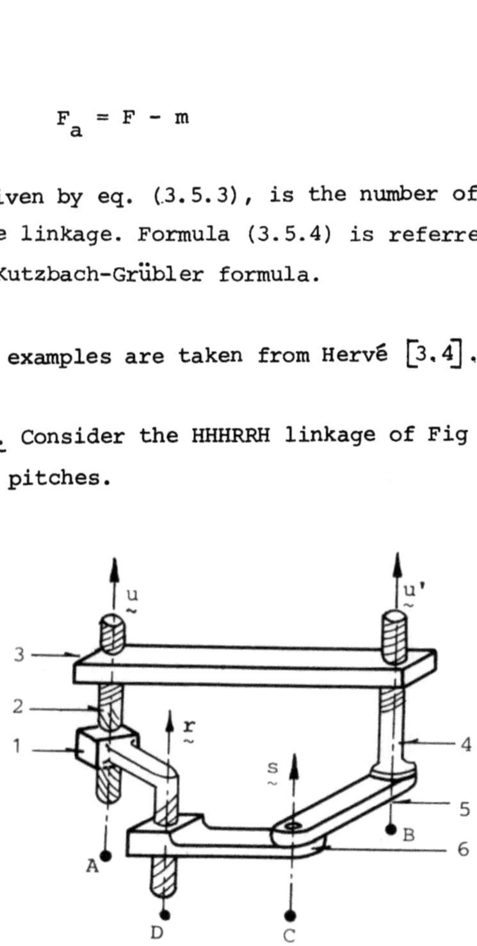

Fig 3.5.2 HHHRRH linkage

First of all, it will be decided whether the linkage contains passive or idle degrees of freedom. To this end, only the liaison $L(1,5)$ will be investigated, for it is clear that $L(5,6)$ and $L(6,1)$ are independent, i.e.

$$\{L(5,6)\}\ \{L6,1)\}=\mathcal{i}$$

One proceeds as follows

$$L(1,5)=L(1,3)\cdot L(3,5)$$

where

$$L(1,3)=L(1,2).L(2,3)$$

Now,

$$\{L(1,2)\}=g_H(A,p,\underset{\sim}{u})$$

and

$$\{L(2,3)\}=g_H(A,q,\underset{\sim}{u}),q\neq p$$

The composition of $g_H(A,p,\underset{\sim}{u})$ and $g_H(A,q,\underset{\sim}{u})$ is $g_C(A,\underset{\sim}{u})$ [3.4], i.e.

$$\{L(1,3)\}=g_C(A,\underset{\sim}{u})$$

Also,

$$L(3,5)=L(3,4).L(4,5)$$

where

$$\{L(3,4)\}=g_H(B,q',\underset{\sim}{u}')$$

and

$$\{L(4,5)\}=g_R(B,\underset{\sim}{u}')$$

The composition of $g_H(B,q'\underset{\sim}{u}')$ and $g_R(B,\underset{\sim}{u}')$ is $g_C(B,\underset{\sim}{u}')$ [3.4]. Thus,

$$\{L(3,5)\}=g_C(B,\underset{\sim}{u}')$$

Now,

$$\{L(1,5)\}=g_C(A,\underset{\sim}{u}).g_C(B,u')$$

which equals $g_C(A,\underset{\sim}{u}).g_R(B,\underset{\sim}{u}')$, both groups having a common translation along $\underset{\sim}{u}||\underset{\sim}{u}'$, which yields thus one passive degree of freedom.

Thus, the dimension of $\{L(1,5)\}$ equals that of $g_C(A,\underset{\sim}{u}).g_R(B,\underset{\sim}{u}')$, which is three, for one has

$$\dim[g_C(A,\underset{\sim}{u}).g_R(B,\underset{\sim}{u}')]=\dim g_C(A,\underset{\sim}{u})+\dim g_R(B,\underset{\sim}{u}')-\dim g_C(A,\underset{\sim}{u})\cap g_R(B,\underset{\sim}{u}')$$

From Table 3.4.1,

$$\dim\ g_C(A,\underset{\sim}{u})=2$$

$$\dim\ g_R(B-\underset{\sim}{u}')=1$$

and, since $g_C(A,\underset{\sim}{u})$ and $g_R(B,\underset{\sim}{u}')$ are independent

$$\dim\ g_C(A,\underset{\sim}{u})\cap g_R(B,\underset{\sim}{u}')=0$$

Thus,

$$\dim\ \{L(1,5)\}=3$$

However

$$\sum_{1}^{4}\dim\{L(i,i+1)\}=4$$

Thus

$$m = 4 - 3 = 1$$

thereby showing that the liaison $L(1,5)$ contains one passive degree of freedom. Moreover, the liaison $L(5,1)$ cannot contain extra passive degrees of freedom. Thus, the linkage contains one passive degree of freedom. In order to find the degree of freedom of the linkage, the group g of minimum dimension d, containing the subgroups of all pairs is now determined. This is done analysing the open chain $L(1,6)$.

$$\{L(1,6)\}=\{L(1,5)\}.\{L(5,6)\}=g_C(A,u).g_R(B,\underset{\sim}{u}').g_R(C,s)=g_X(\underset{\sim}{v})$$

where $\underset{\sim}{v}$ is parallel to $\underset{\sim}{u}$,$\underset{\sim}{u}'$ and $\underset{\sim}{s}$. Thus,

$$\mathcal{g}=\mathcal{g}_X(\underset{\sim}{v}),\ \dim \mathcal{g}=4$$

Furthermore, $f_i=1$ for each pair of the linkage. Hence, $r_i=4-1=3$, for each pair. Substituting these values into formula (3.5.4), one has

$$F=4(6-1)-6x3=2$$

Since m was found to be 1, one has

$$F_a = 2 - 1 =1$$

The passive degree of freedom of this linkage is the translation of link 3 parallel to vector $\underset{\sim}{u}$, which does not affect the motions of the other links.

Example 3.5.3 Consider now the linkage of Fig 3.5.3.

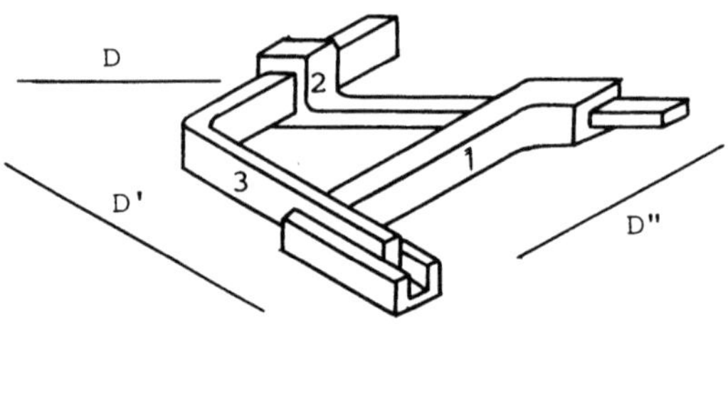

Fig 3.5.3 PPP linkage

In this linkage the directions D, D' and D" of the prismatic pairs were chosen coplanar. Hence, all links translate, without rotating, in the plane defined by these directions. Hence, $\mathcal{g}=\mathcal{g}_T(\underset{\sim}{e}_1,\underset{\sim}{e}_2)$,as given in Table 3.4.1. Here, vectors $\underset{\sim}{e}_1,\underset{\sim}{e}_2$ are parallel to the aforementioned plane. Thus,

$$\dim \mathcal{g} =2$$

Furthermore, $f_i=1$,for i=1,2,3 and hence, $r_i=1$, for i=1,2,3. The degree of freedom of the linkage is thus, from formula (3.5.4),

$$F=2(3-1)-3\times1=1$$

Moreover, since all pairs generate independent groups, the linkage contains no idle degrees of freedom and hence, F=1 is its actual degree of freedom.

Example 3.5.4. Consider now a modification of the previous linkage, depicted in Fig 3.5.4, where each prismatic pair was substituted by a cylindrical pair, vectors $\underset{\sim}{u}$, $\underset{\sim}{u}'$ and $\underset{\sim}{u}''$ being coplanar.

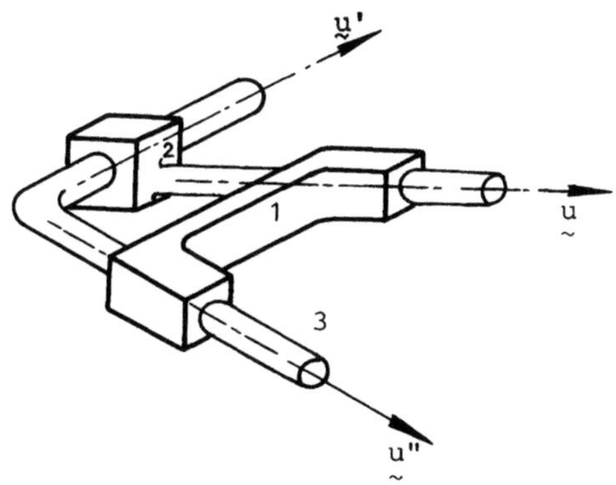

Fig 3.5.4. CCC linkage

This linkage is exceptional and thus, its degree of freedom cannot be found with the aid of formula (3.5.4). The said degree of freedom is still 1, for the rotational degrees of freedom introduced by the cylindrical pairs are not active. In fact, if A, B and C denote arbitrary points lying on the axes of the pairs associated with vectors $\underset{\sim}{u}$, $\underset{\sim}{u}'$ and $\underset{\sim}{u}''$, respectively, one has

$$\mathcal{G}_R(A,\underset{\sim}{u}) \cap \mathcal{G}_T(\underset{\sim}{u},\underset{\sim}{u}') = \mathcal{I}$$
$$\mathcal{G}_R(B,\underset{\sim}{u}') \cap \mathcal{G}_T(\underset{\sim}{u},\underset{\sim}{u}') = \mathcal{I}$$
$$\mathcal{G}_R(C,\underset{\sim}{u}'') \cap \mathcal{G}_T(\underset{\sim}{u},\underset{\sim}{u}') = \mathcal{I}$$

i.e. each rotation is independent of the plane translation generated by the translations along vectors $\underset{\sim}{u}$ and $\underset{\sim}{u}'$.

Examples of paradoxical chains are the Bennett mechanism [3.5], which is further discussed in section 5.3, and the 10-link planar linkage arising from the superposition of all three cognate linkages associated with the RRRR planar linkage, in accordance with the Roberts-Chebyshev Theorem [3.6]. Further examples of paradoxical chains are the mechanisms described by Bricard [3.7], all of which possess an axis of symmetry providing the otherwise nonexisting mobility.

Exercise 3.5.1. Determine the degree of freedom of each of the links appearing in Fig. 3.5.5.

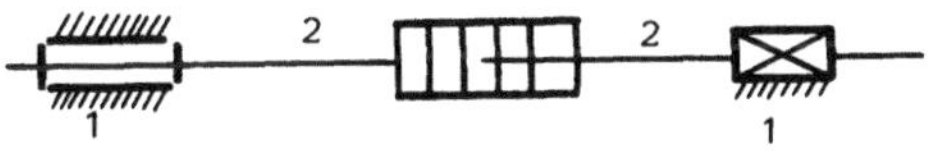

a) RHP coaxial linkage

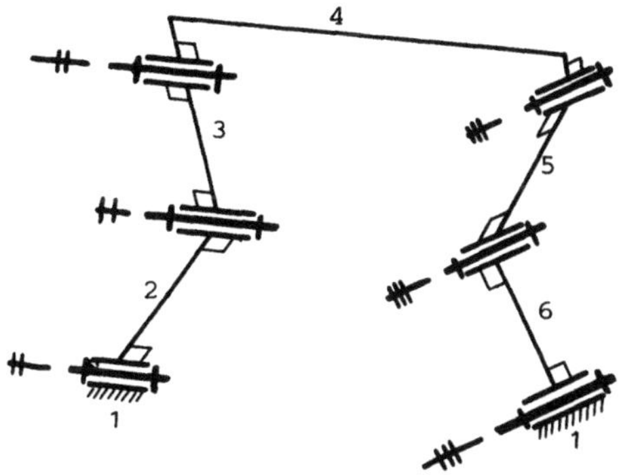

b) 6R linkage for which the axes of pairs R_{12}, R_{23} and R_{34} are parallel, as well as those of R_{45}, R_{56} and R_{61}. Link 4 undergoes pure translation.

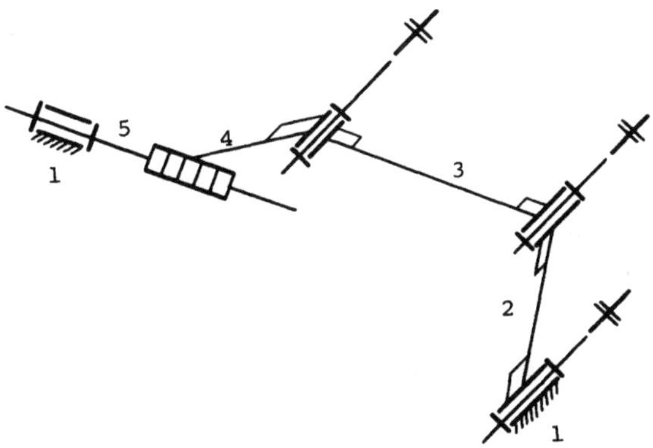

c) RHRRR linkage with the axes of pairs R_{12}, R_{23} and R_{34} being parallel, that of R_{51} being perpendicular to them.

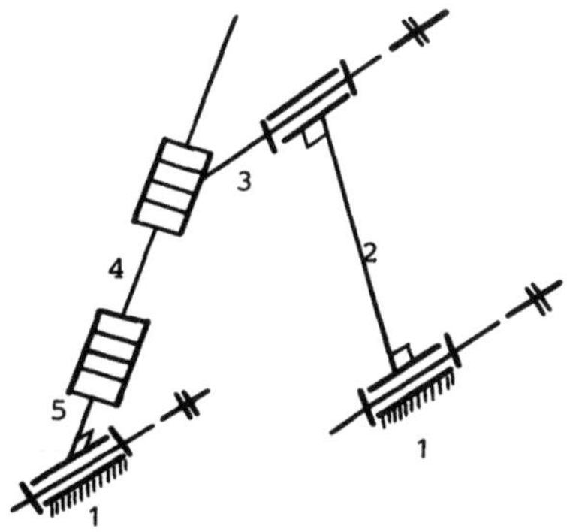

d) RRHHR linkage with the axes of R_{12}, R_{23} and R_{51} parallel.

Fig 3.5.5 Different types of linkages.

3.6 LINKAGE PROBLEMS IN THE THEORY OF MACHINES AND MECHANISMS.

Broadly speaking, there are two kinds of problems in The Theory of Machines and Mechanisms regarding lower-pair mechanisms or linkages, namely,

i) analysis

and

ii) synthesis

The analysis problem is concerned with obtaining the different variables of interest involved in the motion of a given linkage. What is meant by the latter term is that not only the topology of the mechanism (whether it is RRRR or RSCR, etc.) is given, but also its geometry, i.e. its relevant dimensions. The problem is said to be solved when all the variables of interest (output) of the linkage motion are obtained in terms of other arbitrarily assigned variables (input). If the linkage has a degree of freedom one, then the input of the linkage contains one single variable, sufficient to render the other variables determined. If the linkage has a double degree of freedom, then the input contains two variables, and so on. The variables of interest can be obtained via different means, namely, i) a system of algebraic* equations expressing explicity the output in terms of the input (This is very seldom obtained), ii) a discrete set of digital values of the involved** variables (most commonly being the case), iii) an analog readout (oscillogram), i.e. the cathode-ray-tube graph obtained in the oscilloscope of an analog computer, or iv) the actual plot obtained by mechanically recording the output of an existing linkage. In the first three cases the output does not exactly correspond to the mechanism itself, but to its mathematical model; the output is thus obtained via simulation.

---

* Algebraic equations as opposed to differential or integral equations, since no inertia is involved in a purely kinematic analysis.

** This set could be obtained in tabular form or, if a mechanical plotter is used, in graphical form.

In the fourth case, the output can correspond to the actual mechanism if it is accessible for measurements or, if it is not, then to its physical model. An example of the latter arises in the case of trying to measure joint motions in a living being, as is recorded in [3.8] where an experiment was made to measure rotations of the human subtalar and ankle-joint complex through the rotations of its physical model, an RRRR spherical linkage.

The synthesis problem is concerned with obtaining the relevant dimensions of a linkage of a given topology to perform a given operation. In this respect, the synthesis problem can also be thought of as one of system identification [3.9], since it is intended to obtain the parameters defining it starting with an input-output relationship. The synthesis problem can in turn be subdivided into two wide categories, namely, i) exact synthesis and ii) approximate synthesis. In the first case the arising system of equations is meant to be solved exactly, whereas in the second one, it is intended to obtain a "solution" satisfying the system with a minimum error. In either case, three basic synthesis problems can be defined, namely,

i) Function generation
ii) Rigid-body guidance
iii) Path generation

The function-generation synthesis problem is one of finding a linkage (of a given topology) such that its input and output links have given coordinate motions. The rigid-body guidance synthesis problem is concerned with finding a linkage (of a given topology) such that one of its links follows a prescribed set of configurations. Finally, in the path-generation problem, a mechanism of a given topology is sought, with the property that one of its links contains one point passing through a prescribed set of positions.

In any case, the term finite is associated with the synthesis problem if the data are finitely separated. Otherwise, the synthesis problem is said to be of an infinitesimal character, as would be the case when trying to satisfy conditions imposed on velocities or on acelerations. The analysis problem is discussed in Chapter 4, whereas the three synthesis problems, both exact and approximate, are discussed in Chapters 5 and 6, respectively.

References

3.1 Nolle H., "Linkage coupler curve synthesis : A historical review-I. Developments up to 1875", Mechanism and Machine Theory, Vol. 9, 1974, pp. 147-168.

3.2 Hartenberg R.S. and Denavit J., Kinematic Synthesis of Linkages, Mc Graw-Hill Book Co., New York, 1964, p. 28.

3.3. Synge J.L., "Classical Dynamics", in Flügge S., editor, Handbuch der Physik, Vol. III/1, Springer-Verlag, Berlin, 1960, pp. 38-43.

3.4 Hervé J.M., "Analyse structurelle des mécanismes par groupes de déplacements Mechanism and Machine Theory, Vol. 13, 1978, pp. 437-450.

3.5 Bennet G.T.,"A new mechanism", Engineering, Vol. 76, 1903, pp. 777-778.

3.6 Hartenberg R.S. and Denavit J., op. cit., pp. 169-176.

3.7 Bricard R., Leçons de cinématique, Vol. II, Gauthier-Villars et Cie, Paris, 1927, pp. 316-319.

3.8 Wright D.G., S.M. Desai and W.H. Henderson, "Action of the subtalar and ankle-joint complex during the stance phase of walking", The J. of Bone and Joint Surgery, Vol. 46-A, No 2, March 1964, pp. 361-382 and 464.

3.9 Eykhoff P., System Identification : Parameter and State Estimation, John Wiley and Sons Inc., London, 1974.

4. Analysis of Motions of Kinematic Chains

4.1 INTRODUCTION.

Two methods of analysis of linkage motions are presented in this chapter, both of them based on matrix computations. Methods based on Cartesian vector algebra or descriptive geometry can also be applied, as shown in (4.1). However, these are usually *ad hoc* methods and become cumbersome in many instances. The main aim of this chapter is to establish the methods required to obtain a (usually implicit) relationship between the input and the output variables of single-degree of freedom linkages.

4.2 THE METHOD OF DENAVIT AND HARTENBERG (4.5).

This method first appeared in (4.2) to (4.4) and is based on a closure relationship of successive affine transformations. An affine transformation is a change of coordinates involving a translation of the origin and a rotation of axes. Let X_1, Y_1, Z_1 and X_2, Y_2, Z_2 be two sets of coordinates related by an affine transformation, as appears in Fig 4.2.1

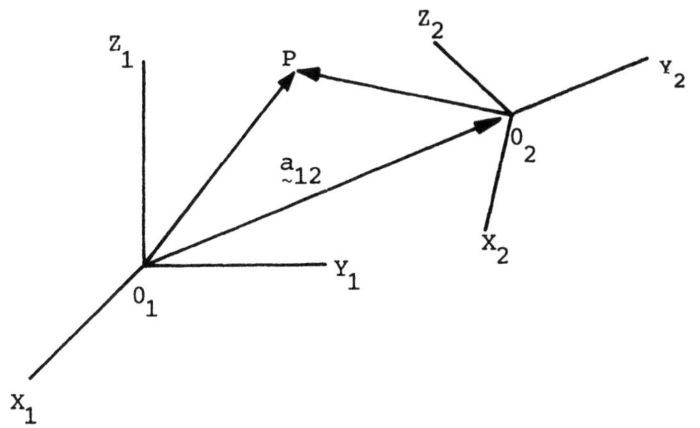

Fig 4.2.1 Translation and rotation of coordinate axes.

Thus, the position vector of any point P, referred to coordinates 1 and 2, can be expressed as

$$(\underset{\sim}{p})_1=(\underset{\sim}{a}_{12})_1+(\underset{\sim}{Q}_{12})_1(\underset{\sim}{p})_2 \tag{4.2.1}$$

where $\underset{\sim}{a}_{12}$ and $\underset{\sim}{Q}_{12}$ are the translation vector and the rotation matrix, from axes 1 to axes 2. Eq. (4.2.1) indicates the general form of an affine transformation. Symbolically, the transformation of (4.2.1) can be written as*

$$(\underset{\sim}{p})_1=(\underset{\sim}{T}_{12})_1(\underset{\sim}{p})_2 \tag{4.2.2}$$

Affine transformations contitute a group under the composition operation defined as

$$\underset{\sim}{T}_{13}=\underset{\sim}{T}_{23}\underset{\sim}{T}_{12} \tag{4.2.3}$$

$\underset{\sim}{T}_{23}$ is given through vector $\underset{\sim}{a}_{23}$ and matrix $\underset{\sim}{Q}_{23}$ as

$$(\underset{\sim}{p})_2=(\underset{\sim}{a}_{23})_2+(\underset{\sim}{Q}_{23})_2(\underset{\sim}{p})_3 \tag{4.2.4}$$

and $\underset{\sim}{T}_{13}$ is given through $\underset{\sim}{a}_{13}$ and $\underset{\sim}{Q}_{13}$, correspondingly, as

$$(\underset{\sim}{p})_1=(\underset{\sim}{a}_{13})_1+(\underset{\sim}{Q}_{13})_1(\underset{\sim}{p})_3 \tag{4.2.5}$$

Substitution of eq. (4.2.4) into eq. (4.2.1) yields

$$\begin{aligned}(\underset{\sim}{p})_1&=(\underset{\sim}{a}_{12})_1+(\underset{\sim}{Q}_{12})_1((\underset{\sim}{a}_{23})_2+(\underset{\sim}{Q}_{23})_2(\underset{\sim}{p})_3)=\\&=(\underset{\sim}{a}_{12})_1+(\underset{\sim}{Q}_{12})_1(\underset{\sim}{a}_{23})_2+(\underset{\sim}{Q}_{12})_1(\underset{\sim}{Q}_{23})_2(\underset{\sim}{p})_3\end{aligned} \tag{4.2.6}$$

Hence

$$(\underset{\sim}{a}_{13})_1=(\underset{\sim}{a}_{12})_1+(\underset{\sim}{Q}_{12})_1(\underset{\sim}{a}_{23})_2 \tag{4.2.7}$$

which, alternatively, can be written as

$$(\underset{\sim}{a}_{13})_1=(\underset{\sim}{a}_{12})_1+(\underset{\sim}{a}_{23})_1 \tag{4.2.7a}$$

---

* $(\underset{\sim}{T}_{ij})_i$ should not be mistaken as a matrix. It is in fact a nonlinear operator.

in agreement with the geometrical meaning, that is to say, the vector connecting the origin O_1 to the origin O_3 equals the sum of that connecting O_1 to O_2 plus that connecting O_2 to O_3. Also, from eq. (4.2.6),

$$(\underset{\sim}{Q}_{13})_1=(\underset{\sim}{Q}_{12})_1(\underset{\sim}{Q}_{23})_2 \tag{4.2.8}$$

thereby showing that the composition of two affine transformations is also affine. Let the <u>identity affine transformation</u> $\underset{\sim}{T}_{ii}$, be defined as

$$(\underset{\sim}{x})_i=(\underset{\sim}{T}_{ii})_i(\underset{\sim}{x})_i \tag{4.2.9}$$

i.e., $\underset{\sim}{T}_{ii}$ is the coordinate transformation from coordinates i into themselves. Clearly, its vector, $\underset{\sim}{a}_{ii}$, is the zero vector, and its matrix, $\underset{\sim}{Q}_{ii}$, is the identity matrix. To show that affine transformations in fact constitute a group under the composition previously noted, all that remains to be established is the existence of an <u>inverse transformation</u>, $\underset{\sim}{T}_{ij}^{-1}$ such that

$$(\underset{\sim}{T}_{ij}^{-1})_j(\underset{\sim}{T}_{ij})_i(\underset{\sim}{x})_j = (\underset{\sim}{T}_{ij})_i(\underset{\sim}{T}_{ij}^{-1})_j(\underset{\sim}{x})_j=(\underset{\sim}{x})_j \tag{4.2.10}$$

where

$$(\underset{\sim}{T}_{ij})_i(\underset{\sim}{x})_j=(\underset{\sim}{a}_{ij})_i+(\underset{\sim}{Q}_{ij})_i(\underset{\sim}{x})_j$$

and

$$(\underset{\sim}{T}_{ij}^{-1})_j(\underset{\sim}{x})_i=(\underset{\sim}{T}_{ji})_j(\underset{\sim}{x})_i=(\underset{\sim}{a}_{ji})_j+(\underset{\sim}{Q}_{ji})_j(\underset{\sim}{x})_i \tag{4.2.11}$$

Thus

$$(\underset{\sim}{T}_{ij}^{-1})_j(\underset{\sim}{T}_{ij})_i(\underset{\sim}{x})_j=(\underset{\sim}{a}_{ji})_j+(\underset{\sim}{Q}_{ji})_j((\underset{\sim}{a}_{ij})_i+(\underset{\sim}{Q}_{ij})_i(\underset{\sim}{x})_j)=$$

$$=(\underset{\sim}{a}_{ji})_j+(\underset{\sim}{Q}_{ji})_j(\underset{\sim}{a}_{ij})_i+(\underset{\sim}{Q}_{ji})_j(\underset{\sim}{Q}_{ij})_i(\underset{\sim}{x})_j \tag{4.2.12}$$

Substituting eq. (4.2.12) into eq. (4.2.10) one obtains

$$(\underset{\sim}{a}_{ji})_j+(\underset{\sim}{Q}_{ji})_j(\underset{\sim}{a}_{ij})_i= \underset{\sim}{0} \tag{4.2.13a}$$

and

$$(\underset{\sim}{Q}_{ji})_j(\underset{\sim}{Q}_{ij})_i= \underset{\approx}{I} \tag{4.2.13b}$$

Hence

$$(\underset{\sim}{Q}_{ji})_j = (\underset{\sim}{Q}_{ij})_i^T \tag{4.2.14a}$$

and

$$(\underset{\sim}{a}_{ji})_j = -(\underset{\sim}{Q}_{ij})_i^T (\underset{\sim}{a}_{ij})_i \tag{4.2.14b}$$

Next, a general composition law for n transformations is derived. Assuming that expressions similar to (4.2.7) and (4.2.8) hold for k transformations, it will be shown that they hold also for k+1, thereby obtaining general relationships by induction. Thus,

$$(\underset{\sim}{a}_{1k})_1 = (\underset{\sim}{a}_{12})_1 + (\underset{\sim}{a}_{23})_1 + \ldots + (\underset{\sim}{a}_{k-1,k})_1 \tag{4.2.15}$$

$$(\underset{\sim}{Q}_{1k})_1 = (\underset{\sim}{Q}_{12})_1 (\underset{\sim}{Q}_{23})_2 \ldots (\underset{\sim}{Q}_{k-1,k})_{k-1} \tag{4.2.16}$$

Then,

$$(\underset{\sim}{a}_{1,k+1})_1 = (\underset{\sim}{a}_{1k})_1 + (\underset{\sim}{a}_{k,k+1}) = (\underset{\sim}{a}_{12})_1 + (\underset{\sim}{a}_{23})_1 + \ldots + (\underset{\sim}{a}_{k-1,k})_1 + (\underset{\sim}{a}_{k,k+1})_1$$

and

$$(\underset{\sim}{Q}_{1,k+1})_1 = (\underset{\sim}{Q}_{k,k+1})_1 (\underset{\sim}{Q}_{1k})_1$$

Introducing a similary transformation to refer $\underset{\sim}{Q}_{k,k+1}$ to k-coordinates,

$$(\underset{\sim}{Q}_{1,k+1})_1 = (\underset{\sim}{Q}_{1k})_1 (\underset{\sim}{Q}_{k,k+1})_k (\underset{\sim}{Q}_{1k})_1^T (\underset{\sim}{Q}_{1k})_1 = (\underset{\sim}{Q}_{1k})_1 (\underset{\sim}{Q}_{k,k+1})_k =$$

$$= (\underset{\sim}{Q}_{12})_1 (\underset{\sim}{Q}_{23})_2 \ldots (\underset{\sim}{Q}_{k-1,k})_{k-1} (\underset{\sim}{Q}_{k,k+1})_k$$

Thus, in general

$$(\underset{\sim}{a}_{1n})_1 = \sum_1^{n-1} (\underset{\sim}{a}_{i,i+1})_1 \tag{4.2.17}$$

and

$$(\underset{\sim}{Q}_{1n})_1 = (\underset{\sim}{Q}_{12})_1 (\underset{\sim}{Q}_{23})_2 \ldots (\underset{\sim}{Q}_{n,n-1})_n \tag{4.2.18}$$

which are useful relationships because they enable the analyst, first, to compute the final rotation, from 1 to n, referred to coordinate system 1, in term of <u>successive</u> rotations, from i to i+1, <u>referred to coordinate system i</u>.

In general, however, $a_{i,i+1}$ will be more readily referred to coordinate system i, but the transformation to system 1 is easily performed as

$$\left(a_{i,i+1}\right)_1=\left(Q_{1i}\right)\left(a_{i,i+1}\right)_i$$

and $\left(Q_{1i}\right)_1$ can be obtained from (4.2.18) with n=i. The method of Denavit and Hartenberg (MDH) is based on the closure equation

$$T_n\cdots T_{23}T_{12}=T_{11} \quad (4.2.19)$$

or, equivalently,

$$\left(a_{12}\right)_1+\left(a_{23}\right)_1+\ldots+\left(a_{n-1,n}\right)_1+\left(a_{n1}\right)_1=0 \quad (4.2.20)$$

together with

$$\left(Q_{12}\right)_1\left(Q_{23}\right)_2\cdots\left(Q_{n-1,n}\right)_{n-1}\left(Q_{n1}\right)_n=I \quad (4.2.21)$$

where, according to the relationship (4.2.14a),

$$\left(Q_{n1}\right)_n=\left(Q_{1n}\right)_n^T \quad (4.2.22)$$

Next, the MDH is applied to a linkage composed of n links (rigid bodies) coupled by any of the six lower pairs (R,P,H,C,S or E) introduced in Chapter 3. Let the axis of the pair be defined as

i) the axis of rotation, if the pair is R,

ii) the direction of traslation, if the pair is P,

iii) the axis of rotation, which is identical to the direction of traslation, if the pair is H or C.

iv) the direction of the normal to the plane of contact if the pair is E. Only if the pair is spheric no single axis can be defined. In this case, there is freedom to choose the axis of the pair and so, the analyst can choose it to his best convenience.

To implement the MDH, number the links successively 1,2,...,n. Then

a) Let Z_1 be the axis of the pair coupling the first link (the fixed one) with the second one (the driving or input link), in such a way that the variable denoting the input be positive along Z_1.

b) In general, let Z_i be the axis of the pair connecting links i and i+1.

c) Let X_i be the common perpendicular to Z_{i-1} and Z_i, directed from Z_{i-1} to Z_i

d) Let a_i be the distance between Z_i and Z_{i+1}, always positive.

e) Let s_i be the coordinate of the intersection of axis X_{i-1} with axis Z_i, in frame X_i-Y_i-Z_i. Since it is a coordinate, its sign can be plus or minus, depending on the position of the said intersection. The absolute value of s_i is the distance between X_i and X_{i+1}.

f) Let α_i be the angle between Z_i and Z_{i+1}, measured along the positive direction of X_{i+1}

g) Let θ_i be the angle between X_i and X_{i+1}, measured along the positive direction of Z_i.

h) Construct the translation vectors $(\underset{\sim}{a}_{i,i+1})_i$ and the rotation matrices $(\underset{\sim}{Q}_{i,i+1})_i$, as is described next.

i) Apply the closure conditions (4.2.20) and (4.2.21) and from them obtain the sought input-output relationship.

In order to construct the translation vectors $(\underset{\sim}{a}_{i,i+1})_i$, it is necessary first to construct the rotation matrices $(\underset{\sim}{Q}_{i,i+1})_i$, which is done next. The relationship between coordinate systems i and i+1 is shown in Fig 4.2.1, in agreement with the notation of f) and g).

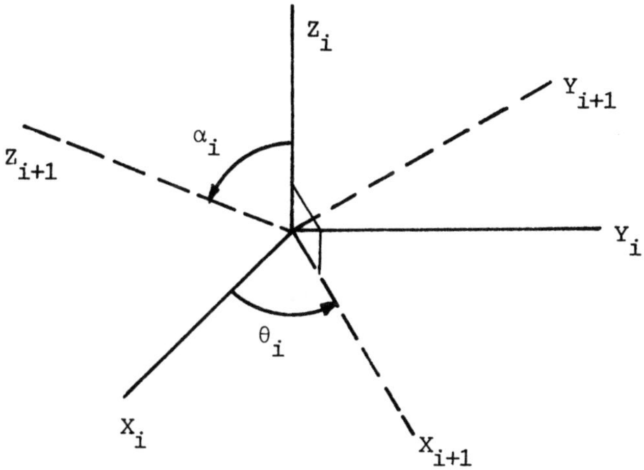

Fig 4.2.2. Relative position of coordinate systems i and i+1

Since X_{i+1} is perpendicular to Z_i (by definition), X_i and X_{i+1} lie in a plane perpendicular to Z_i. Thus, X_i can be made coincident with X_{i+1} by means of a rotation through an angle θ_i about Z_i, as shown in Fig 4.2.3, where X_i', Y_i', Z_i' are the original axes X_i, Y_i, Z_i after the said rotation, so that Z_i' coincides with Z_i.

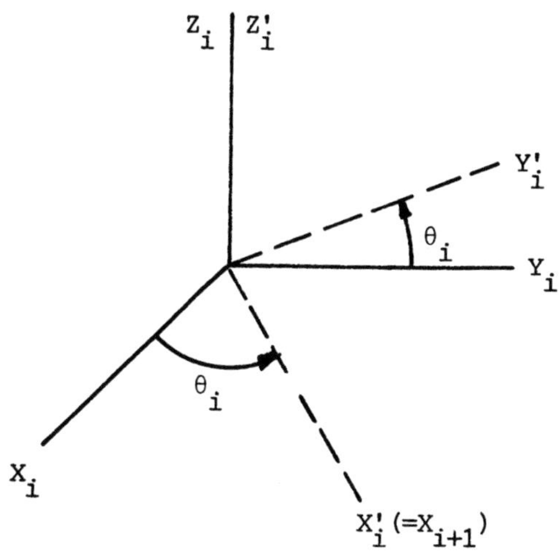

Fig 4.2.3 Rotation through an angle θ_i about axis Z_i.

Hence, from Section 2.3,

$$(\underset{\sim}{Q}_{ii'})_i = \begin{pmatrix} \cos\theta_i & -\sin\theta_i & 0 \\ \sin\theta_i & \cos\theta_i & 0 \\ 0 & 0 & 1 \end{pmatrix} \tag{4.2.23}$$

Next notice that X_{i+1} (i.e.X_i') is perpendicular to Y_i', Y_{i+1}', Z_i(i.e.Z_i') and Z_{i+1}. Hence, Y_i' and Z_i' can be made coincident with Y_{i+1} and Z_{i+1} by means of a rotation through an angle α_i about X_i'. The relative position of axes i' and i+1 is shown in Fig 4.2.4.

From Fig 4.2.4 and section 2.3,

$$(\underset{\sim}{Q}_{i',i+1})_{i'} = \begin{pmatrix} 1 & 0 & 0 \\ 0 & \cos\alpha_i & -\sin\alpha_i \\ 0 & \sin\alpha_i & \cos\alpha_i \end{pmatrix} \tag{4.2.24}$$

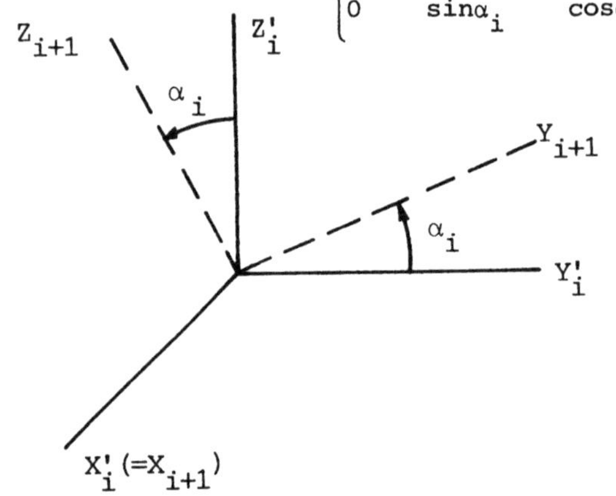

Fig 4.2.4 Relative position of axes i' and axes i+1

Finally, from eq. (4.2.18),

$$(\underset{\sim}{Q}_{i,i+1})_i = (\underset{\sim}{Q}_{i,i'})_i (\underset{\sim}{Q}_{i',i+1})_{i'} \tag{4.2.25}$$

the desired matrix is obtained as

$$(\underset{\sim}{Q}_{i,i+1})_i = \begin{pmatrix} \cos\theta_i & -\sin\theta_i\cos\alpha_i & \sin\theta_i\sin\alpha_i \\ \sin\theta_i & \cos\theta_i\cos\alpha_i & -\cos\theta_i\sin\alpha_i \\ 0 & \sin\alpha_i & \cos\alpha_i \end{pmatrix} \quad (4.2.26)$$

Now it is possible to construct vectors $(\underset{\sim}{a}_{i,i+1})_i$. The relative configuration of three successive links appears in Fig 4.2.5, where the notation of a) to g) has been followed.

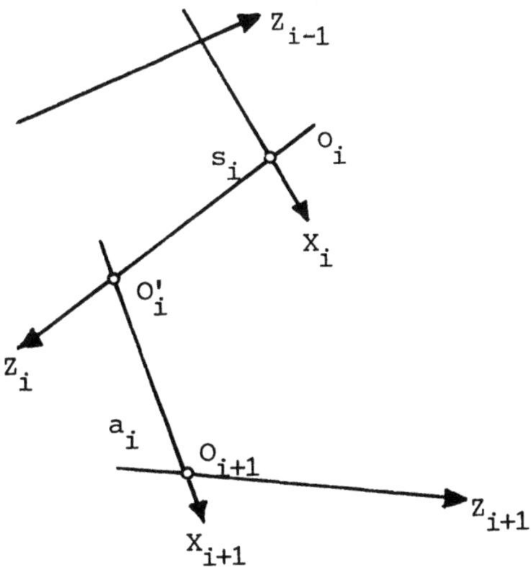

Fig 4.2.5 Three successive links of a linkage

From Fig 4.2.5,

$$(\underset{\sim}{a}_{i,i+1})_i = (\overrightarrow{O_iO_{i'}})_i + (\underset{\sim}{Q}_{i,i+1})_i(\overrightarrow{O_{i'}O_{i+1}})_{i+1} \quad (4.2.27)$$

with

$$(\overrightarrow{O_iO_{i'}})_i = (0,0,s_i)^T \quad (4.2.28)$$

and

$$(\overrightarrow{O'_iO_{i+1}})_{i+1} = (a_i,0,0)^T \quad (4.2.29)$$

$(\underset{\sim}{Q}_{i,i+1})_i$ being as given by eq. (4.2.26).

Substituting (4.2.26), (4.2.28) and (4.2.29) into (4.2.27),

$$(\underset{\sim}{a}_{i,i+1})_i = (a_i\cos\theta_i, a_i\sin\theta_i, s_i)^T \tag{4.2.30}$$

Expressions (4.2.26) and (4.2.30) enable the analyst to systematically construct the affine transformations required to establish the closure condition (4.2.19).

As is shown in Examples 4.2.1 and 4.2.2, it is not always necessary to apply both closure conditions (4.2.20) and (4.2.21) arising from (4.2.19), since only one suffices.

Example 4.2.1. Analysis of the universal joint. The layout of a universal (or Hooke's) joint appears in Fig 4.2.6, where the DH notation has been used. The universal joint is a special class of RRRR spherical linkages. Obtain an input-output relationship of the form $f(\theta_i, \theta_4) = 0$.

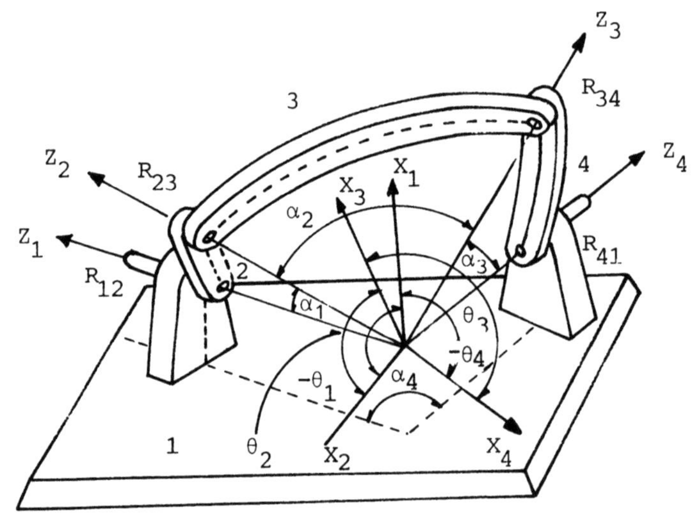

Fig 4.2.6 Universal Joint

Since all coordinate axes involved in this linkage have a common origin, the closure condition on the translation vectors is irrelevant and only that on the rotation matrices will be employed. The rotation matrices appearing in the present analysis are constructed from the fact that, in this case,

$$\alpha_1=\alpha_2=\alpha_3=90°$$

Constructing the rotation matrices according to eqs. (4.2.26) and (4.2.30), one obtains

$$(\underset{\sim}{Q}_{14})_1=(\underset{\sim}{Q}_{12})_1(\underset{\sim}{Q}_{23})_2(\underset{\sim}{Q}_{34})_3$$

Thus,

$$(\underset{\sim}{Q}_{12})_1(\underset{\sim}{Q}_{23})_2(\underset{\sim}{Q}_{34})_3=\begin{pmatrix} c\theta_1c\theta_2c\theta_3+s\theta_1s\theta_3 & c\theta_1s\theta_2 & c\theta_1c\theta_2s\theta_3-s\theta_1c\theta_3 \\ s\theta_1c\theta_2c\theta_3-c\theta_1s\theta_3 & s\theta_1s\theta_2 & s\theta_1c\theta_2s\theta_3+c\theta_1c\theta_3 \\ s\theta_2c\theta_3 & -c\theta_2 & s\theta_2s\theta_3 \end{pmatrix} \quad (4.2.31)$$

On the other hand, from (4.2.26) and noticing that, for i=n, i+1=1,

$$(\underset{\sim}{Q}_{41})_4=\begin{pmatrix} c\theta_4 & -s\theta_4c\alpha_4 & s\theta_4s\alpha_4 \\ s\theta_4 & c\theta_4c\alpha_4 & -c\theta_4s\alpha_4 \\ 0 & s\alpha_4 & c\alpha_4 \end{pmatrix}$$

But, from (4.2.14a),

$$(\underset{\sim}{Q}_{14})_1=(\underset{\sim}{Q}_{41})_4^T$$

Hence,

$$(\underset{\sim}{Q}_{14})_1=\begin{pmatrix} c\theta_4 & s\theta_4 & 0 \\ -s\theta_4c\alpha_4 & c\theta_4c\alpha_4 & s\alpha_4 \\ s\theta_4s\alpha_4 & -c\theta_4s\alpha_4 & c\alpha_4 \end{pmatrix} \quad (4.2.32)$$

Equating the (1,2)-and the (2,2)-entries of both forms of $(\underset{\sim}{Q}_{14})_1$ -eqs. (4.2.31) and (4.2.32) - one obtains

$$\cos\theta_1 \sin\theta_2 = \sin\theta_4 \tag{4.2.33a}$$

$$\sin\theta_1 \sin\theta_2 = \cos\theta_4 \cos\alpha_4 \tag{4.2.33b}$$

Eliminating θ_2 in the above expressions,

$$\tan\theta_1 = \cos\alpha_4 \cot\theta_4 \tag{4.2.34}$$

which is the input-output relationship meant to be obtained.

Example 4.2.2. Analysis of an RSRC linkage. A typical RSCR linkage is shown in Fig 4.2.7, where the DH notation has also been used. The parameters and variables have the values

$$\alpha_1 = \alpha_1 \quad \alpha_2 = \alpha_2 \quad \alpha_3 = -\frac{\pi}{2} \quad \alpha_4 = \frac{\pi}{2}$$

$$\theta_1 = -\phi \quad \theta_2 = \theta_2 \quad \theta_3 = \theta_3 \quad \theta_4 = \theta_4$$

$$a_1 = a \quad a_2 = b \quad a_3 = 0 \quad a_4 = 0$$

$$s_1 = c \quad s_2 = 0 \quad s_3 = 0 \quad s_4 = -s$$

For the analysis of this linkage, only the closure condition of the translation vectors will be needed. Since these vectors, as given by eq. (4.2.30), have to be expressed in one single coordinate system, the rotation matrices will also be constructed. Thus,

$$(\underset{\sim}{Q}_{12})_1 = \begin{pmatrix} c\phi & s\phi c\alpha_1 & -s\phi s\alpha_1 \\ -s\phi & c\phi c\alpha_1 & -c\phi s\alpha_1 \\ 0 & s\alpha_1 & c\alpha_1 \end{pmatrix} \tag{4.2.35a}$$

$$(\underset{\sim}{Q}_{23})_2 = \begin{pmatrix} c\theta_2 & -s\theta_2 c\alpha_2 & s\theta_2 s\alpha_2 \\ s\theta_2 & c\theta_2 c\alpha_2 & -c\theta_2 s\alpha_2 \\ 0 & s\alpha_2 & c\alpha_2 \end{pmatrix} \tag{4.2.35b}$$

$$(\underset{\sim}{Q}_{34})_3 = \begin{pmatrix} c\theta_3 & 0 & -s\theta_3 \\ s\theta_3 & 0 & c\theta_3 \\ 0 & -1 & 0 \end{pmatrix} \qquad (4.2.35c)$$

$$(\underset{\sim}{Q}_{41})_4 = \begin{pmatrix} c\theta_4 & 0 & s\theta_4 \\ s\theta_4 & 0 & -c\theta_4 \\ 0 & 1 & 0 \end{pmatrix} \qquad (4.2.35d)$$

and

$$(\underset{\sim}{a}_{12})_1 = (ac\phi, -as\phi, +c)^T \qquad (4.2.36a)$$

$$(\underset{\sim}{a}_{23})_2 = (bc\theta_2, bs\theta_2, 0)^T \qquad (4.2.36b)$$

$$(\underset{\sim}{a}_{34})_3 = (0,\ 0,\ 0)^T \qquad (4.2.36c)$$

$$(\underset{\sim}{a}_{41})_4 = (0,\ 0, -s)^T \qquad (4.2.36d)$$

The closure condition on the translation vectors requires that vectors (4.2.36) be all expressed in terms of the X_1, Y_1, Z_1 coordinate system, to yield

$$(\underset{\sim}{a}_{12})_1 + (\underset{\sim}{a}_{23})_1 + (\underset{\sim}{a}_{34})_1 + (\underset{\sim}{a}_{41})_1 = (\underset{\sim}{0})_1 \qquad (4.2.37)$$

Performing the corresponding trasformations,

$$(\underset{\sim}{a}_{23})_1 = (\underset{\sim}{Q}_{12})_1 (\underset{\sim}{a}_{23})_2 = \begin{pmatrix} b(c\phi c\theta_2 + s\phi s\theta_2 c\alpha_1) \\ b(-s\phi c\theta_2 + c\phi s\theta_2 c\alpha_1) \\ bs\theta_2 s\alpha_1 \end{pmatrix} \qquad (4.2.38a)$$

Since $(\underset{\sim}{a}_{34})_3$ vanishes,

$$(\underset{\sim}{a}_{34})_1 = (\underset{\sim}{0}) \qquad (4.2.38b)$$

$$(\underset{\sim}{a}_{41})_1 = (\underset{\sim}{Q}_{14})_1 (\underset{\sim}{a}_{41})_4 = \begin{pmatrix} 0 \\ s \\ 0 \end{pmatrix} \qquad (4.2.38c)$$

Substituting (4.2.36a) and (4.2.38a) - (4.2.38c) into eq. (4.2.37), the following scalar equations are obtained

$$a\cos\phi + b(\cos\phi\cos\theta_2 + \sin\phi\cos\alpha_1\sin\theta_2) = 0 \qquad (4.2.39a)$$

$$-a\sin\phi+b(-\sin\phi\cos\theta_2+\cos\phi\cos\alpha_1\sin\theta_2)-s=0 \quad (4.2.39b)$$

$$c+b\sin\alpha_1\sin\theta_2=0 \quad (4.2.39c)$$

which, for every value of the input angle $\phi(t)$, yields a nonlinear algebraic system for the corresponding three unknowns $\alpha_1(t), \theta_2(t)$ and $s(t)$.

One method to solve the foregoing system is via the Newton-Raphson method, as shown in Section 1.13, by application of the NRDAMP subroutine. However, in this particular case it is not necessary to spend so much computer time for, by introducing suitable trigonometric identities, the system (4.2.39) can be reduced to

$$s(t) = a\sin\phi + \sqrt{b^2 - c^2 - a^2\cos^2\phi} \quad (4.2.40)$$

which explicity provides the set of values of s for each value of ϕ.

A table of values for s(t) was obtained from a digital computer output, for the following values of mechanism parameters:

$a = 1.00$ m, $b = 3.00$ m, $c = 2.00$ m

for a value of $\dot{\phi}$ constantly equal to 1500 rpm. Similar tables for velocity and acceleration values were obtained via differentiation of s(t) by application of second central differences. Curves appearing in Fig 4.2.8 were obtained from the said tables.

Notice that eq. (4.2.40) could have also been obtained from the geometry of the linkage of Fig 4.2.7, due to the simplicity of the linkage.

In more general instances, however, the geometry is not so simple and the MDH becomes essential to perform the analysis.

In addition to the digital-computer method previously outlined, to obtain the output $s(t)$, $\dot{s}(t)$ and $\ddot{s}(t)$ out of eq. (4.2.40), analog computer methods can also be applied.

An analog computer is a (usually electrical) physical system whose behaviour is governed by the same mathematical model governing the system

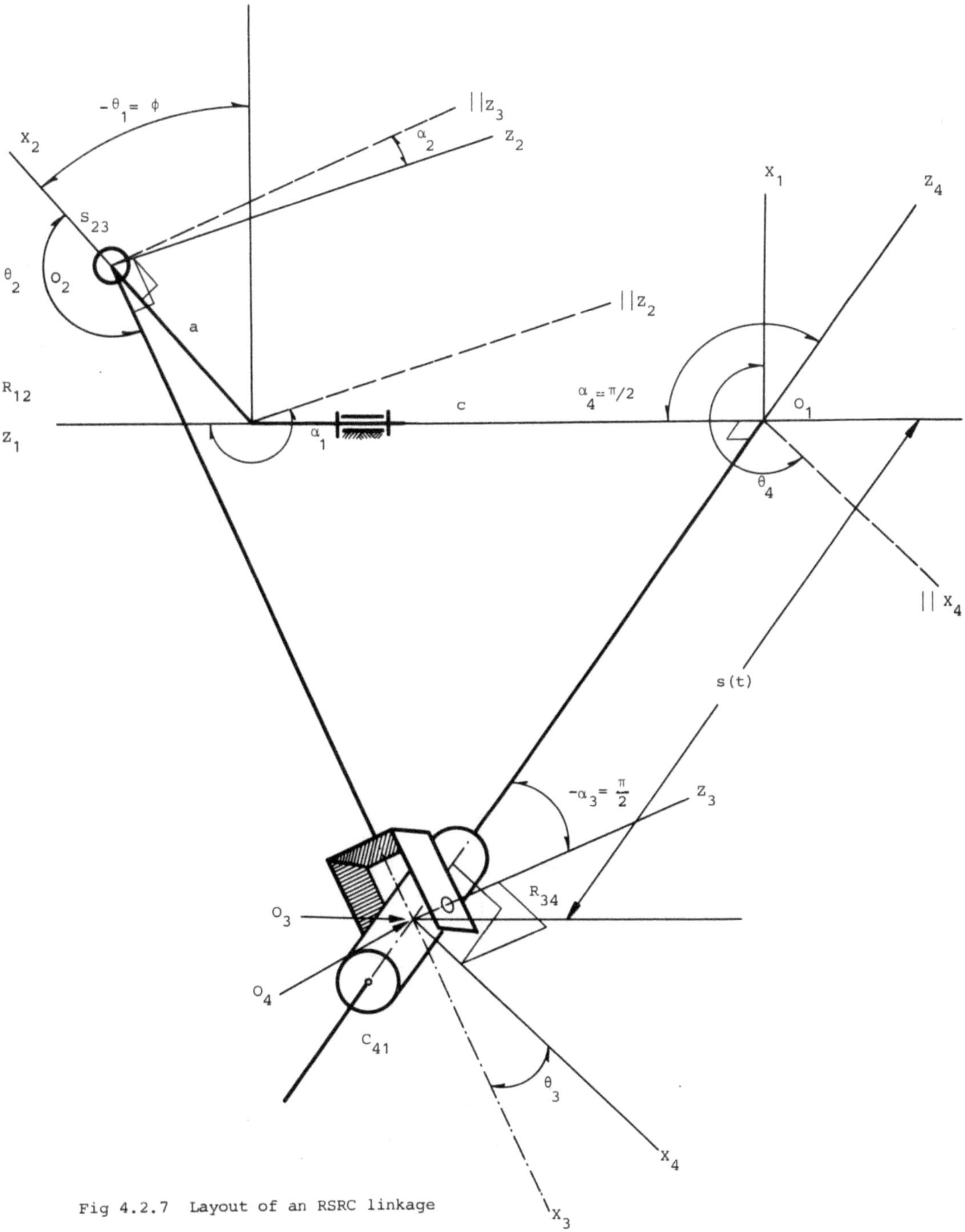

Fig 4.2.7 Layout of an RSRC linkage

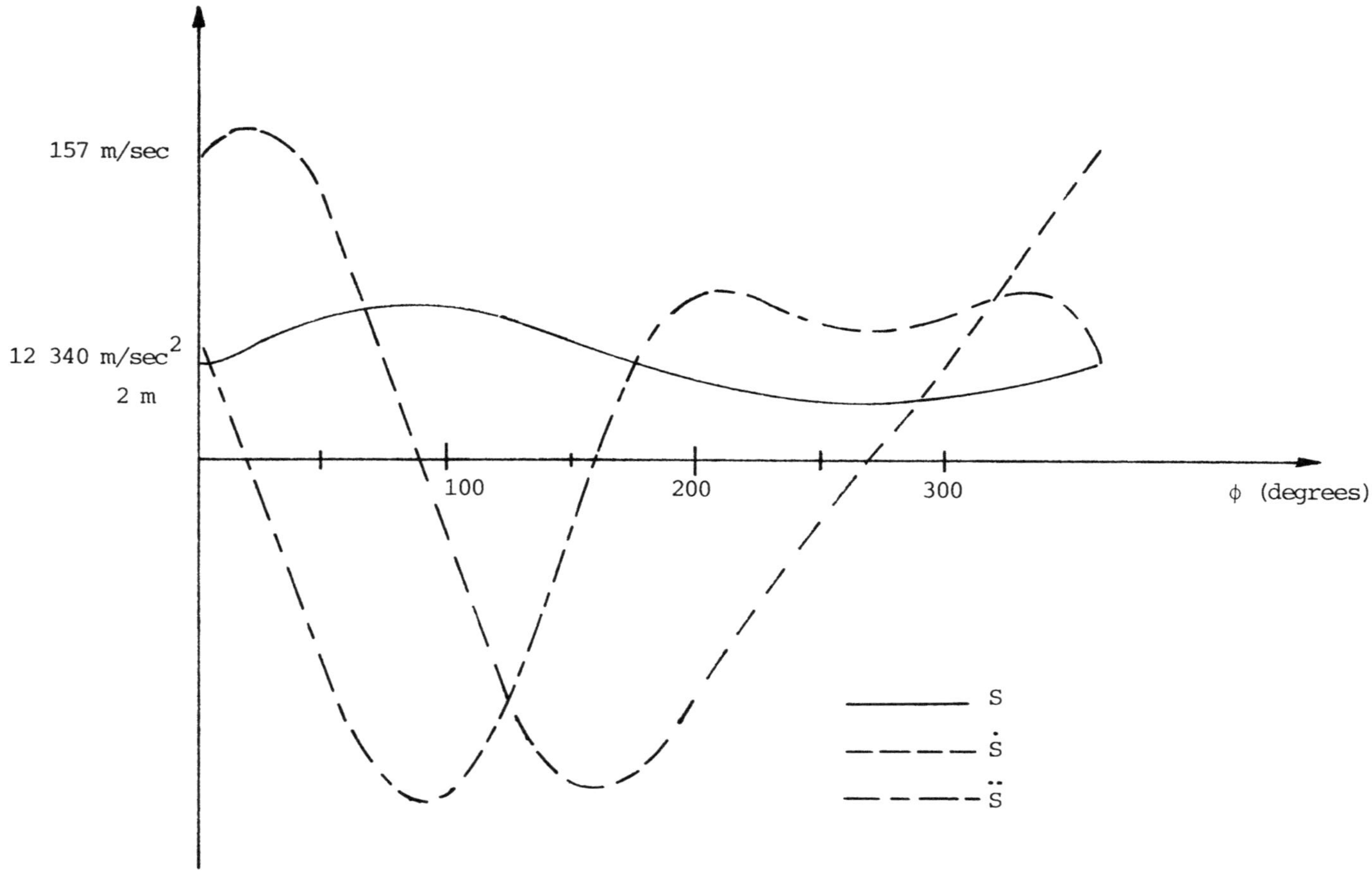

Fig 4.2.8 Displacement, velocity and acceleration curves of an RSRC linkage.

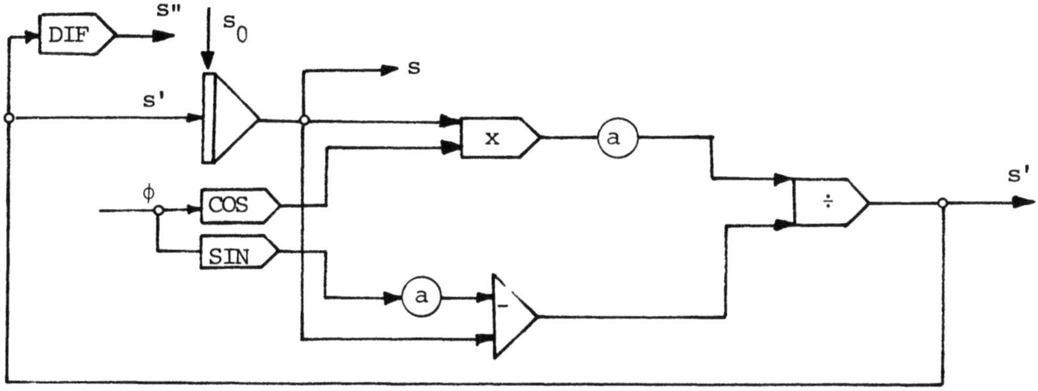

Fig 4.2.10 Analog realization of eq. (4.2.41)

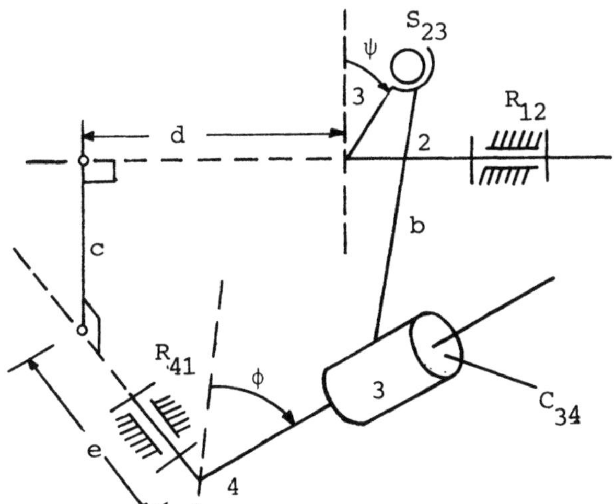

Fig 4.2.11 An RSCR linkage

under analysis. The elements constituting an analog computer perform the usual mathematical operations appearing in mathematical models, i.e., algebraic addition, multiplication, division, integration and differentiation. Besides these operations, the analog computer is also supplied with function generators. All these elements are symbolically represented as appearing in Fig 4.2.9. An analog computer representation of a mathematical model is usually called "a realisation" of the model, because via that representation the model is taken into physical reality.

Equation (4.2.40) could readily be realized in an analog computer, and two differentiations would have to be performed to obtain the acceleration $\ddot{s}(t)$. However, due to the noise present in every physical system, and the fact that a differentiator is a noise amplifier, it becomes undesirable to perform more than one differentiation. To avoid the second differentiation to obtain $\ddot{s}(t)$, then, eq. (4.2.40) is first differentiated to obtain

$$s'(\phi)=\frac{ds}{d\phi}=\frac{\dot{s}}{\dot{\phi}}=\frac{as\ \cos\phi}{s-a\ \sin\phi} \qquad (4.2.41)$$

which has to be integrated once, with initial conditions $s=s_0$ at $\phi=\phi_0$, and differentiated only once to obtain $\ddot{s}(t)$. The analog realization of eq. (4.2.41) appears in Fig 4.2.10

Details about analog simulation of linkages can be found in [4.6]. Concerning simulation of mechanical systemsin general, the reader can see [4.7].

A digital computer-oriented algorithm, based on an iterative procedure, to obtain the history of all variables of a linkage is presented in [4.8].

Exercise 4.2.10 Given the RSCR linkage of Fig 4.2.11, obtain an input-output relationship $f(\psi,\phi)$, an analog realization yielding $\phi, \dot{\phi}$ and $\ddot{\phi}$ and curves ϕ vs. t, $\dot{\phi}$ vs. t and $\ddot{\phi}$ vs. t, for the following values:

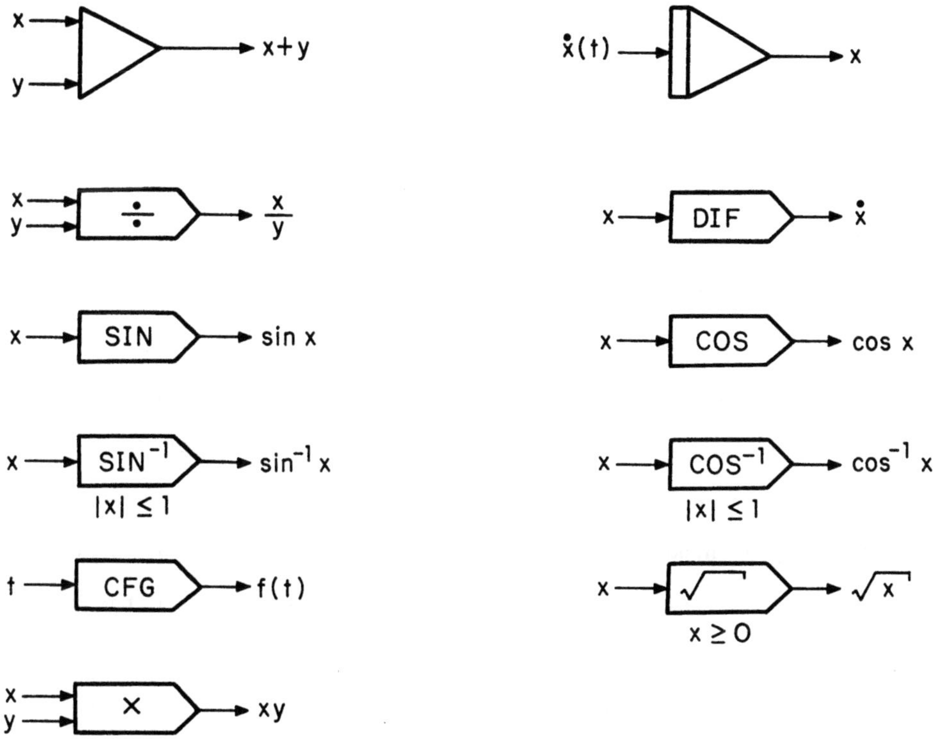

Fig 4.2.9. Elements for analog realisations

|a=0.5m, b=1.0m, c=0.25m, d=0.75m, e=0.5m, $\dot{\phi}$=1200rpm(const)

4.3 AN ALTERNATE METHOD OF ANALYSIS. Using the MDH one does not necessarily obtain one single relationship for the input and the output variables, but usually a system of nonlinear algebraic equations involving all the different linkage variables which appear strongly coupled, as occurs with eqs. (4.2.39). If the system is not very complicated, then it can happen that, after introducing appropriate trigonometric identities, one can obtain a single input-output relationship where the only variables appearing are the input and the output. If it is not obvious how to obtain the said single relationship, then one is forced to solve a system, instead of one single nonlinear equation. Gupta (4.10) has presented a method which does not require all the apparatus of the MDH and yields a single relationship between the input and the output. The method is based on an equation establishing the constancy of either a link length or a link angle, throughout the linkage motion. The linkage is assumed to have a single-loop which, for simplicity, is assumed to be composed of just four links*: the fixed link, the input link, the coupler link and the output link. Let A be a point on the input axis; B, a point of the pair connecting the input and the coupler links; C, a point of the pair connecting the coupler and the output links; and D, a point of the output axis. Furthermore, let $\underset{\sim}{a}_0$, $\underset{\sim}{b}_0$, $\underset{\sim}{c}_0$, $\underset{\sim}{d}_0$ and $\underset{\sim}{a}, \underset{\sim}{b}, \underset{\sim}{c}, \underset{\sim}{d}$, denote the position vectors of the corresponding points in both, a reference configuration and a current configuration, respectively.

Let also $\underset{\sim}{Q}$ and $\underset{\sim}{R}$ be the rotation matrices carrying the input and the output links from the reference configuration to the current configuration. Then, from the lenght constancy of the coupler link,

---

*This linkage could be RSSR, RSCR or its inversions. Spherical linkages are dealt with next.

$$||\underset{\sim}{b} - \underset{\sim}{c}||^2 = ||\underset{\sim}{b}_0 - \underset{\sim}{c}_0||^2 \qquad (4.3.1)$$

where, clearly

$$\underset{\sim}{b} = \underset{\sim}{a}_0 + \underset{\approx}{Q}(\underset{\sim}{b}_0 - \underset{\sim}{a}_0) \qquad (4.3.2a)$$

and

$$\underset{\sim}{c} = \underset{\sim}{d}_0 + \underset{\approx}{R}(\underset{\sim}{c}_0 - \underset{\sim}{d}_0) \qquad (4.3.2b)$$

Hence,

$$||\underset{\sim}{b}-\underset{\sim}{c}||^2 = ||\underset{\sim}{a}_0+\underset{\approx}{Q}(\underset{\sim}{b}_0-\underset{\sim}{a}_0) - \underset{\sim}{d}_0 - \underset{\sim}{R}(\underset{\sim}{c}_0 - \underset{\sim}{d}_0)||^2$$

which yiels the desired input-output scalar equation which was to be obtained, after substitution into eq. (4.3.1), namely

$$||\underset{\sim}{a}_0 + \underset{\approx}{Q}(\underset{\sim}{b}_0 - \underset{\sim}{a}_0) - \underset{\sim}{d}_0 - \underset{\sim}{R}(\underset{\sim}{c}_0 - \underset{\sim}{d}_0)||^2 = ||\underset{\sim}{b}_0 - \underset{\sim}{c}_0||^2 \qquad (4.3.3)$$

Equation (4.3.3) contains only two unknowns, the input and the output angles, thereby showing how one single scalar equation for these two variables can be obtained. The next example illustrates how to apply this method to spherical linkages.

<u>Example 4.3.1 Analysis of the general RRRR spherical linkage.</u>

The MDH can be applied, of course, to general spherical linkages, following the same procedure applied to the universal joint.

The fact that the MDH introduces other variables besides the input and the output, however, produces very cumbersome equations, algebraically difficult to handle. The alternate method proves, in this case, to be particulary helpful. Consider the RRRR spherical linkage appearing in Fig 4.3.1

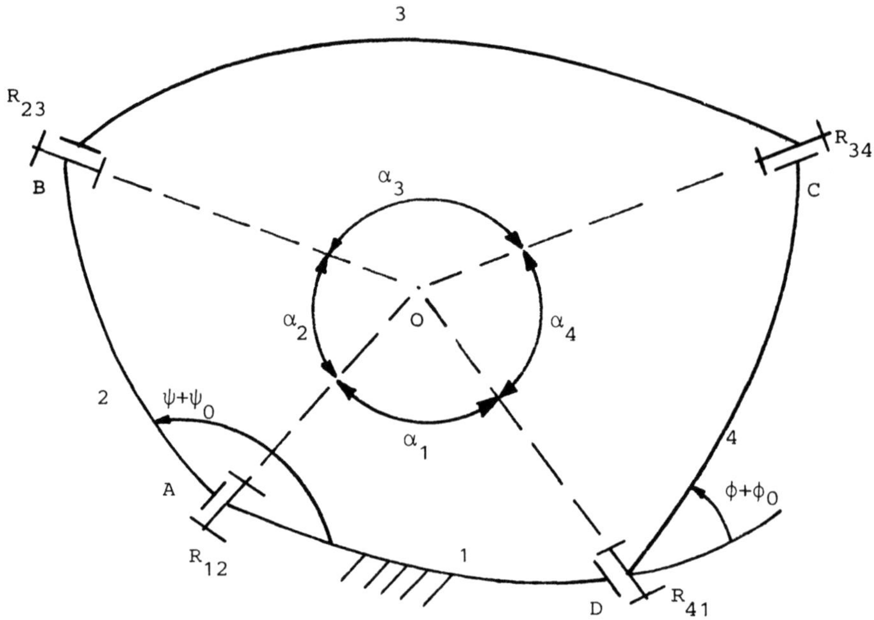

Fig 4.3.1 General RRRR spherical linkage

Let $\mathbf{a}_0$, $\mathbf{b}_0$, $\mathbf{c}_0$, $\mathbf{d}_0$ and $\mathbf{a},\mathbf{b},\mathbf{c},\mathbf{d}$ be the position vectors of points A, B, C and D in a reference configuration and in a time-varying configuration, respectively. Furthermore, let $\mathbf{Q}$ and $\mathbf{R}$ be the rotation matrices carruing links 2 and 4, respectively, from the reference to the current configurations. Thus,

$$\mathbf{b}=\mathbf{Q}\mathbf{b}_0 \tag{4.3.4}$$

and

$$\mathbf{c}=\mathbf{R}\mathbf{c}_0 \tag{4.3.5}$$

Clearly, $\mathbf{a}=\mathbf{a}_0$ and $\mathbf{d}=\mathbf{d}_0$. The cosine of α_3 in both the reference and the current configurations is

$$(\cos\alpha_3)_{ref} = \mathbf{b}_o^T\mathbf{c}_0 \tag{4.3.6a}$$

and

$$(\cos\alpha_3)_{cur} = \mathbf{b}^T\mathbf{c} \tag{4.3.6b}$$

where $||b_0||$ and $||c_0||$ are assumed unity, without loss of generality. Since link 3 is rigid, α_3 remains constant throughout the linkage motion and, since $||\underset{\sim}{b}|| = ||\underset{\sim}{b}_0||$ and $||\underset{\sim}{c}|| = ||\underset{\sim}{c}_0||$, one obtains

$$\underset{\sim}{b}^T\underset{\sim}{c} = \underset{\sim}{b}_0^T\underset{\sim}{c}_0 \qquad (4.3.7)$$

or, substituting the relations (4.3.4) and (4.3.5) in the above equation,

$$\underset{\sim}{b}_0^T\underset{\sim}{Q}^T\underset{\sim}{R}\underset{\sim}{c}_0 = \underset{\sim}{b}_0^T\underset{\sim}{c}_0 = \cos\alpha_3 \qquad (4.3.8)$$

which is the scalar input-output relationship meant to be obtained. The only variables appearing in eq. (4.3.8) are ψ, contained in $\underset{\sim}{Q}$, and ϕ, contained in $\underset{\sim}{R}$. Define the coordinate axes appearing in Fig 4.3.2, with X_i and X_o directed along the axes of R_{12} and R_{41}, respectively.

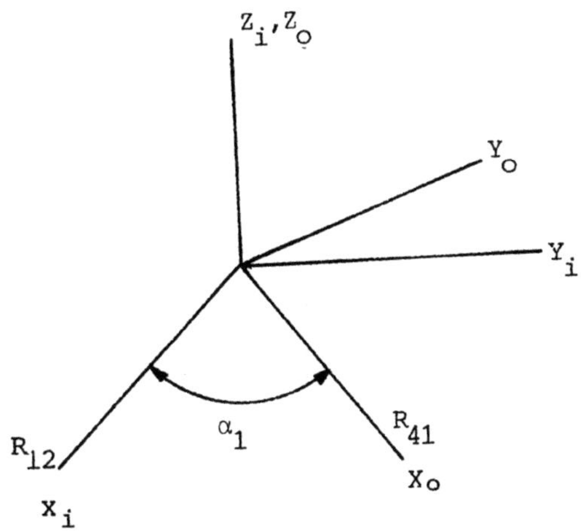

Fig 4.3.2 Fixed coordinate axes containing the axes of R_{12} and R_{41}

Matrices $\underset{\sim}{Q}$ and $\underset{\sim}{R}$, referred to i- and o- axes respectively, are given as

$$(\underset{\sim}{Q})_i = \begin{pmatrix} 1 & 0 & 0 \\ 0 & \cos\psi & -\sin\psi \\ 0 & \sin\psi & \cos\psi \end{pmatrix}, \quad (\underset{\sim}{R})_o = \begin{pmatrix} 1 & 0 & 0 \\ 0 & \cos\phi & -\sin\phi \\ 0 & \sin\phi & \cos\phi \end{pmatrix} \qquad (4.3.9)$$

Vectors $\underset{\sim}{b}_0$ and $\underset{\sim}{c}_0$ are shown in Figs 4.3.3a and 4.3.3b

Hence,

$$(\underset{\sim}{b}_0)_i = \begin{pmatrix} \cos\alpha_2 \\ \sin\alpha_2\cos\psi_0 \\ \sin\alpha_2\sin\psi_0 \end{pmatrix} , \quad (c_0)_o = \begin{pmatrix} \cos\alpha_4 \\ \sin\alpha_4\cos\phi_0 \\ \sin\alpha_4\sin\phi_0 \end{pmatrix} \tag{4.3.10}$$

In order to perform the products appearing in eq. (4.3.8) it is necessary to express all vectors and matrices in the same coordinate axes.

The transformation matrix carrying the i-axis into the o-axes, referred to the i-axes, is given as

$$(\underset{\sim}{S})_i = \begin{pmatrix} \cos\alpha_1 & -\sin\alpha_1 & 0 \\ \sin\alpha_1 & \cos\alpha_1 & 0 \\ 0 & 0 & 1 \end{pmatrix} \tag{4.3.11}$$

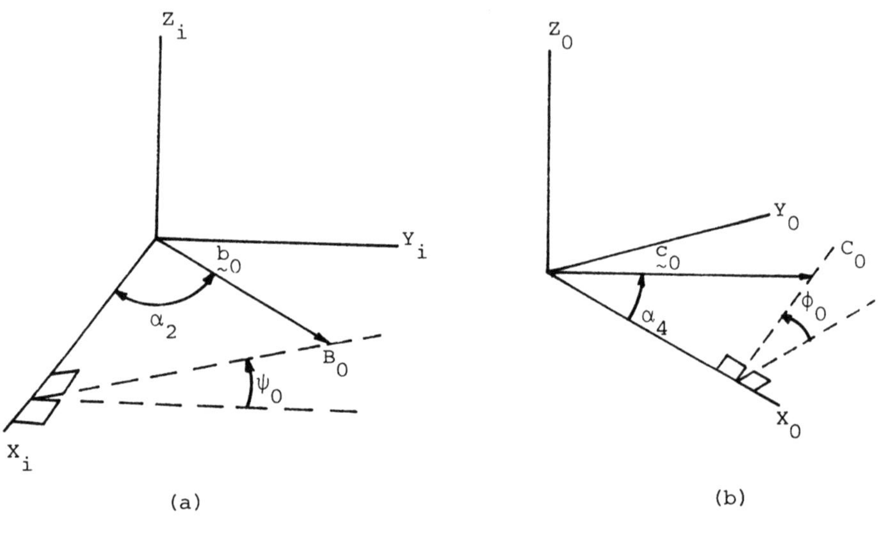

Fig 4.3.3 Reference configuration of points B and C.

The product $\underset{\sim}{b}_0^T \underset{\sim}{Q}^T \underset{\sim}{R}\, \underset{\sim}{c}_0$ needed in eq. (4.3.8) is next computed

$$\underset{\sim}{b}_0^T \underset{\sim}{Q}^T \underset{\sim}{R}\, \underset{\sim}{c}_0 = (\underset{\sim}{Q}\underset{\sim}{b}_0)_i^T \; (\underset{\sim}{R}\underset{\sim}{c}_0)_i = (\underset{\sim}{Q}\underset{\sim}{b}_0)_i^T \; (\underset{\sim}{S})_i (\underset{\sim}{R}\underset{\sim}{c}_0)_o \qquad (4.3.12)$$

which yiedls

$$\underset{\sim}{b}^T \underset{\sim}{Q}^T \underset{\sim}{R}\, \underset{\sim}{c}_0 = c\alpha_2\left(c\alpha_1 c\alpha_4 - s\alpha_1 s\alpha_4 c(\phi+\phi_0)\right) + s\alpha_2 c(\psi+\psi_0)\left(s\alpha_1 c\alpha_4 + c\alpha_1 s\alpha_4 c(\phi+\phi_0)\right) +$$

$$+ s\alpha_2 s\alpha_4 s(\psi+\psi_0)s(\phi+\phi_0) \qquad (4.3.13)$$

When eq. (4.3.13) is substituted into eq. (4.3.8) one obtaines the desired input-output equation

$$c\alpha_2\left(c\alpha_1 c\alpha_4 - s\alpha_1 s\alpha_4 c(\phi+\phi_0)\right) + s\alpha_2 c(\psi+\psi_0)\left(s\alpha_1 c\alpha_4 + c\alpha_1 s\alpha_4 c(\phi+\phi_0)\right) +$$

$$+ s\alpha_2 s\alpha_4 s(\psi+\psi_0)s(\phi+\phi_0) - c\alpha_3 = 0 \qquad (4.3.13a)$$

in which ψ and ϕ are measured from the reference values ψ_0 and ϕ_0, respectively, as defined previously. If the said angles are measured from the plane of the axes of R_{12} and R_{41}, instead, then the latter equation becomes

$$c\alpha_2(c\alpha_1 c\alpha_4 - s\alpha_1 s\alpha_4 c\phi) + s\alpha_2 c\psi(s\alpha_1 c\alpha_4 + c\alpha_1 s\alpha_4 c\phi) + s\alpha_2 s\alpha_4 s\psi s\phi - c\alpha_3 = 0 \qquad (4.3.14)$$

The input-output equation for the universal joint can now be obtained as a special case of eq. (4.3.14), letting $\alpha_2=\alpha_3=\alpha_4=90°$, thus obtaining

$$c\alpha_1 c\psi c\phi + s\psi s\phi = 0 \qquad (4.3.15)$$

or

$$\frac{c\alpha_1}{\tan\phi} + \tan\psi = 0$$

which is equivalent to eq. (4.2.34) previously obtained. In fact, α_4 of eq. (4.2.34) corresponds to α_1 in eq. (4.3.15) and

$$\psi = 180° - \theta_1, \phi = \theta_4$$

The method of Denavit and Hartenberg, as presented in Section 4.2, is a very well structured systematic procedure for the analysis of "single-loop linkages", i.e. Those whose links are all binary*; but problems arise when "multiple-loop linkages" are to be analyzed. Sheth and Uicker (4.9) have generalized the notation of Denavit and Hartenberg, however, to overcome the aforementioned situation and furthermore, to extend the application of the MDH to the analysis of higher-pair mechanisms.

* If a link is coupled to 2 other links, it is called binary; if it is coupled to 3, it is called ternary, and so on.

4.4 APPLICATIONS TO OPEN KINEMATIC CHAINS

The method of Denavit and Hartenberg, described in Section 4.2, can also be applied to open kinematic chains. An example is next described, showing the application of the method to the analysis of a 3-degree-of-freedom manipulator.

Fig 4.4.1 Shows a 3-degree-of-freedom manipulator whose function is to position point 0_4, of link number 4, in an arbitrary position given by the three cartesian coordinates x,y,z, referred to frame X_1-Y_1-Z_1. The manipulator is composed of 4 links and three revolute pairs. The axes of pairs R_{12} and R_{23} are perpendicular, whereas those of pairs R_{23} and R_{34} are parallel. Moreover, axis Z_4 is chosen parallel to these.

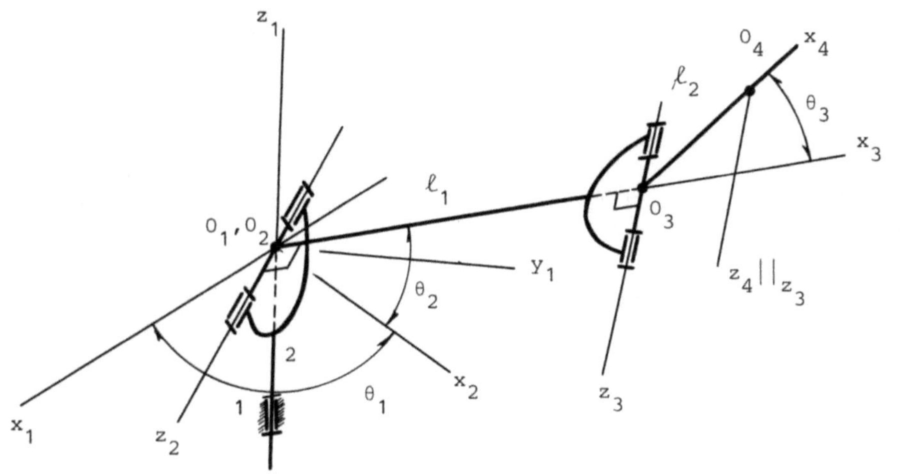

Fig 4.4.1 Three-degree-of-freedom manipulator.

Following Denavit and Hartenberg's notation, one has

$a_1=0,\ a_2=\overline{0_2 0_3}=\ell_1, a_3=\overline{0_3 0_4}=\ell_2$

$s_1=0, s_2=0, s_3=0$

$\alpha_1=90°, \alpha_2=0°, \alpha_3=0°$

The corresponding closure equation takes on the form

$$\underset{\sim}{a}_{12}+\underset{\sim}{a}_{23}+\underset{\sim}{a}_{34}=\underset{\sim}{r} \tag{4.4.1}$$

where $\underset{\sim}{r}$ is the positon vector of 0_4 and the foregoing relation is assumed to be referred to frame $X_1-Y_1-Z_1$. From eq. (4.2.30),

$$[\underset{\sim}{a}_{12}]_1=\underset{\sim}{0} \tag{4.4.2a}$$

$$[\underset{\sim}{a}_{23}]_2=\ell_1[c\theta_2, s\theta_2, 0]^T \tag{4.4.2b}$$

$$[\underset{\sim}{a}_{34}]_3=\ell_2[c\theta_3, s\theta_3\ 0]^T \tag{4.4.2c}$$

Since all vectors in eq. (4.4.1) should be described in the same reference frame,a change of coordinates should be introduced, namely

$$[\underset{\sim}{a}_{23}]_1=[\underset{\sim}{Q}_{12}]_1[\underset{\sim}{a}_{23}]_2 \tag{4.4.3a}$$

$$[\underset{\sim}{a}_{23}]_1=[\underset{\sim}{Q}_{13}]_1[\underset{\sim}{a}_{34}]_3=[\underset{\sim}{Q}_{12}]_1[\underset{\sim}{Q}_{23}]_2[\underset{\sim}{a}_{34}]_3 \tag{4.4.36}$$

From eq. (4.2.26) and the particular orientations of the axes of rotations,

$$[\underset{\sim}{Q}_{12}]_1=\begin{bmatrix} c\theta_1 & 0 & s\theta_1 \\ s\theta_1 & 0 & -c\theta_1 \\ 0 & 1 & 0 \end{bmatrix}, [\underset{\sim}{Q}_{23}]_2=\begin{bmatrix} c\theta_2 & -s\theta_2 & 0 \\ s\theta_2 & c\theta_2 & 0 \\ 0 & 0 & 1 \end{bmatrix} \tag{4.4.4}$$

Substituting expressions (4.4.4) into (4.4.3a and b),

$$[\underset{\sim}{a}_{23}]_1=\ell_1\begin{bmatrix} c\theta_1 c\theta_2 \\ s\theta_1 c\theta_2 \\ s\theta_2 \end{bmatrix}, [\underset{\sim}{a}_{34}]_3=\ell_2\begin{bmatrix} c\theta_1 c\theta_2 c\theta_3-c\theta_1 s\theta_2 s\theta_3 \\ s\theta_1 c\theta_2 c\theta_3-s\theta_1 s\theta_2 s\theta_3 \\ s\theta_2 c\theta_3 +c\theta_2 c\theta_3 \end{bmatrix} \tag{4.4.5}$$

Finally, substituting expressions (4.4.2a) and (4.4.5) into eq. (4.4.1) and recalling that $[\underset{\sim}{r}]_1=[x,y,z]^T$, one has

$$\ell_1 c\theta_1 c\theta_2 + \ell_2 (c\theta_1 c\theta_2 c\theta_3 - c\theta_1 s\theta_2 s\theta_3) = x \quad (4.4.6a)$$

$$\ell_1 s\theta_1 c\theta_2 + \ell_2 (s\theta_1 c\theta_2 c\theta_3 - s\theta_1 s\theta_2 s\theta_3) = y \quad (4.4.6b)$$

$$\ell_1 s\theta_2 + \ell_2 (s\theta_2 c\theta_3 + c\theta_2 c\theta_3) = z \quad (4.4.6c)$$

which are the sought relationships. These involve the angles of rotation of each revolute pair, θ_1, θ_2 and θ_3 and the cartesian coordinates x,y,z, with respect to frame $X_1-Y_1-Z_1$, of the point which the manipulator is to position. For a given manipulator, eqs. (4.4.6) allow to determine the set of values the three angles of rotation should be given in order to take 0_4 form its inital position, the one corresponding to $\theta_1=\theta_2=\theta_3=0$, i.e. $[\underset{\sim}{r}_0]_1=[\ell_1+\ell_2,0,0]^T$, into a prescribed position $[\underset{\sim}{r}]_1=[x,y,z]^T$.

On the other hand, if the problem at hand is one of synthesis, i.e. if the range of motions of point 0_4 is prescribed, the derived equations, (4.4.6), can be used to determine the lengths ℓ_1 and ℓ_2.

R E F E R E N C E S

4.1 Beggs J.S., Advanced Mechanism, The Macmillan Company, N. York, 1966, pp. 101-120 and 164-167.

4.2 Denavit J., "Displacement analysis of mechanisms based on (2x2) matrices and dual numbers", VDI Ber., Vol. 29, 1958.

4.3 Denavit J. and R.S. Hartenberg, "A Kinematic notation for lower-pair mechanisms based on matrices", ASME. J. Appl. Mech., Ser. E., vol. 22, June 1955

4.4 Denavit J. and R.S. Hartenberg "Approximate synthesis of spatial linkages", ASME. J. Appl. Mech., Ser. E., vol. 27, No. 1, March, 1960.

4.5 Denavit J. and R.S. Hartenberg, Kinematic Synthesis of Linkages, McGraw-Hill Book Co., N. York, 1964.

4.6 Timm R.F., "Analog simulation of rigid link mechanisms", ASME J. Eng. for Ind., Ser. B, vol. 89, No. 2, May 1967.

4.7 Shigley J. E., Simulation of Mechanical Systems. McGraw-Hill Book Co., N. York. 1967.

4.8 Uicker J. J., J. Denavit and R.S. Hartenberg, "An iterative method for the displacement analysis of spatial mechanism", ASME. J. Appl. Mech., Ser. E, vol. 31, June 1964.

4.9 Sheth P. N. and J.J. Uicker, "A generalized symbolic notation for mechanisms", ASME. J. Eng. for Ind., Ser. B, vol. 93, Feb. 1971.

4.10 Gupta V. K., "Kinematic analysis of plane and spatial mechanisms", ASME. J. Eng. for Ind., Ser. B, vol. 95, May 1973.

5. Synthesis of Linkages

5.1 INTRODUCTION. The problem of linkage synthesis of Applied Kinematics was outlined in Chapter 3. In the present chapter, the problem of exact synthesis is discussed and current methods of synthesis are presented. The three usual problems of synthesis are discussed, namely

i) funtion generation

ii) rigid body guidance

iii) path generation

and exact solutions to the resulting design equations are meant to be obtained, these solutions being exact up to round-off and/or measuring errors. Chapter 6 deals with the problem where no exact solution can be found, in which case the best approximation is sought, in the sense of rendering the minimum quadratic error in the approximation.

5.2 SYNTHESIS FOR FUNCTION GENERATION

Due to the fact that a linkage is a coupling of rigid bodies, a finite number of parameters (like those of the notation of Denavit and Hartenverg, Ch. 3) defines it. Hence, the set of design equations is of an algebraic character, i.e. no derivatives of the design parameters appear in them, and the number of these parameters is finite. Hence, no linkage can be obtained to produce an arbitrarily prescribed input/output function pointwise in the whole continuum of values of the input where the function is prescribed. The said function can only be produced at a finite set of input values, the number of this set being equal to the number of independent design parameters. Thus, the problem of linkage synthesis for function generation can be stated as:

"Given a function $f=f(x_i)$ $(i=1,2,\ldots,n)$, defined over a discrete set $\{x_i\}_1^n$, find the relevant dimensions of a linkage of a given topology* to produce an input-output relationship that coincides with the function f at the given discrete set $\{x_i\}_1^n$ of input values". The method to solve this problem consits of two stages, namely

i) Derivation of the input-output relationship for the prescribed topology, and

ii) Determination of the linkage parameters from the above relationship.

The first stage is now discussed. In ch. 4 it was shown that the MDH** can be applied to obtain an input-output relationship that, hopefully, does not contain other variables than the input and the output. It was also shown that an alternate method, less complex than that of Denavit and Hartenberg, guarantees that only the input and the output variables will appear in the input-output relationship. That method, however, is restricted to single-loop mechanisms, whereas that of DH can be extended to multiple-loop mechanisms. Hence, either method can be applied in the first stage of this problem, for single-loop linkages. The second stage is carried out by two different approaches, which are next discussed. It is assumed that an input-outout implicit function has been obtained, this function having the general form

$$f(x_i, y_i, \underset{\sim}{p})=0, i=1,2,\ldots,n \qquad (5.2.1)$$

where $\{(x_i, y_i)\}_1^n$ is a set of n pairs of values relating the <u>ith</u> value of the input x to the <u>ith</u> value of the output y, and $\underset{\sim}{p}$ is an n-dimensional

---

*See Section 3.6

**Method of Denavit and Hartenberg

vector containing the parameters of the linkage under consideration. Notice that (5.2.1) represents in fact a system of n synthesis equations in the n unknowns that constitute vector $\underset{\sim}{p}$. Hence, the synthesis problem can be solved through eq. (5.2.1), However, the system (5.2.1) is, in general, nonlinear, and no unique solution is guaranteed to exist; even more, the system might have no solution at all. If the system has one or more solutions, these can be obtained via the method of Newton-Raphson, as shown in Section 1.13. This is the first approach to the solution of the synthesis equations.

The second approach introduces a nonlinear transformation of the synthesis parameters which transforms the synthesis equations into a linear system. Let

$$\underset{\sim}{q}=\underset{\sim}{q}(\underset{\sim}{p}) \tag{5.2.2}$$

be this tranformation. When (5.2.2) is introduced into eq. (5.2.1), the following linear system is obtained

$$\underset{\sim}{A}\underset{\sim}{q}=\underset{\sim}{b} \tag{5.2.3}$$

System (5.2.3) can be solved very efficiently via the LU decomposition algorithm, as was shown in Section 1.12. Once the unique solution $\underset{\sim}{q}_u$ to (5.2.3) has been obtained, this is introduced into eq. (5.2.2), which is nonlinear if the original system of synthesis equations was nonlinear as well, and, if this tranformation -eq. (5.2.2)- is chosen in such a way that the synthesis parameters appear in it very weakly coupled, then the linkage parameter vector $\underset{\sim}{p}$ can be obtained without the need of a numerical method. The synthesis procedure is illustrated by means of the following example.

Example 5.2.1 Given the RSSR linkage appearing in Fig 5.2.1, determine its geometric parameters $a_1, a_2, a_3, a_4, a_1, s_4$ and α_4 that produce a given input-output relationship

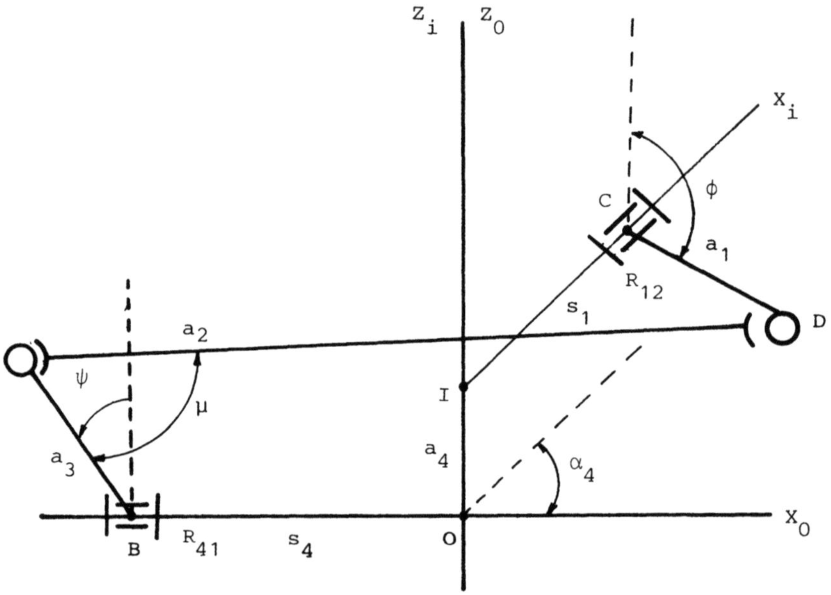

Fig 5.2.1 RSSR Function generating linkage

Using the method of Section 4.3, define sets of axes $X_i\ Y_i\ Z_i$ and $X_o\ Y_o\ Z_o$ whose X-axes coincide with the axes of R_{12} and R_{41} respectively, its Z-axes being parallel to the common normal OI, and define their positive directions from O to I. Finally, the Y-axes form with the previous ones right-hand rectangular sets of coordinate axes, as shown in Fig 5.2.2.

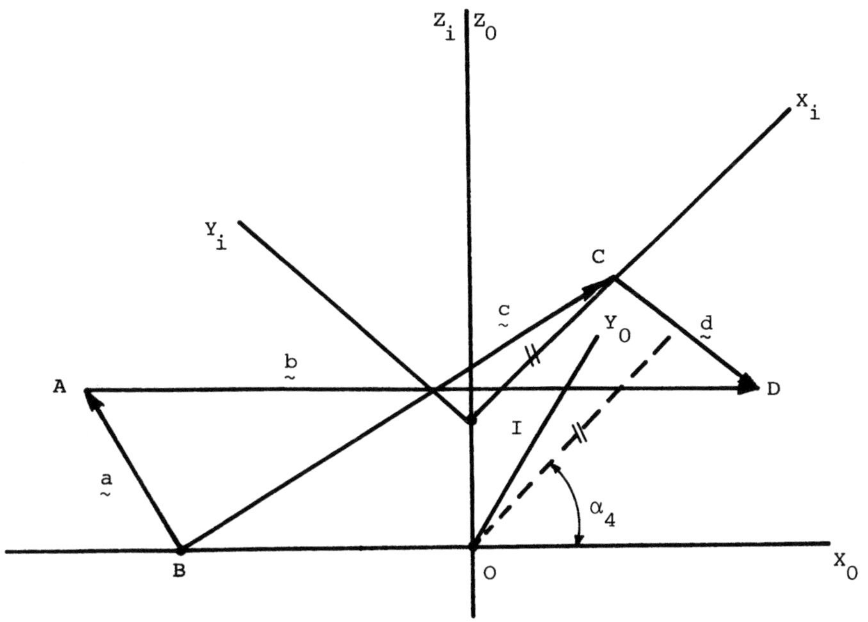

Fig 5.2.2 Coordinate axes fixed at input and output axes

Define:

$\mathbf{a}$ = vector directed from B to A

$\mathbf{b}$ = vector directed from A to D

$\mathbf{c}$ = vector directed from B to C

$\mathbf{d}$ = vector directed from C to D

From the geometry of the linkage, then

$$\mathbf{a}+\mathbf{b}=\mathbf{c}+\mathbf{d} \tag{5.2.4}$$

Hence,

$$\mathbf{b}=\mathbf{c}+\mathbf{d}-\mathbf{a} \tag{5.2.5}$$

and

$$\mathbf{b}^T=\mathbf{c}^T+\mathbf{d}^T-\mathbf{a}^T \tag{5.2.6}$$

Let $\mathbf{a}_j, \mathbf{b}_j$ and $\mathbf{d}_j$ $(j=0,1,2,\ldots,n)$ be the values attained by vectors $\mathbf{a}, \mathbf{b}$ and $\mathbf{d}$ respectively, at the successive configurations j.

Substituting the values of $\underset{\sim}{a}$, $\underset{\sim}{b}$ and $\underset{\sim}{d}$ at their current configuration j in eqs. (5.2.5a and b) and multiplying termwise the resulting equations leads to

$$||\underset{\sim}{b}_j||^2=||\underset{\sim}{c}||^2+||\underset{\sim}{d}_j||^2+||\underset{\sim}{a}_j||^2+2\underset{\sim}{c}^T\underset{\sim}{d}_j-2\underset{\sim}{c}^T\underset{\sim}{a}_j-2\underset{\sim}{d}_j^T\underset{\sim}{a}_j \qquad (5.2.7)$$

where, due to the rigid-body condition,

$$||\underset{\sim}{a}_j||^2=a_3^2 \qquad (5.2.8a)$$

$$||\underset{\sim}{b}_j||^2=a_2^2 \qquad (5.2.8b)$$

$$||\underset{\sim}{d}_j||^2=a_1^2 \qquad (5.2.8c)$$

From fig 5.2.2, vector $\underset{\sim}{c}$ is given as

$$\underset{\sim}{c} = \overrightarrow{BO} + \overrightarrow{OI} + \overrightarrow{IC} \qquad (5.2.9)$$

where, with reference to i-coordinates,

$$\overrightarrow{BO} = (s_4 c\alpha_4,\ -s_4 s\alpha_4,\ 0)^T \qquad (5.2.10a)$$

$$\overrightarrow{OI} = (0,\ 0,\ a_4)^T,\ [s_1,\ 0,\ 0]^T \qquad (5.2.10b)$$

where the notation cx, sx has been introduced to represent cosx and sinx, respectively, whenever the variable x has been previously defined as an angle. Thus, eq. (5.2.9) leads to

$$(\underset{\sim}{c})_i = (s_4 c\alpha_4+s_1, -s_4 s\alpha_4, a_4)^T \qquad (5.2.10c)$$

Furthermore,

$$(\underset{\sim}{d})_i = (0, -a_1\sin\phi, a_1\cos\phi)^T \qquad (5.2.11a)$$

$$(\underset{\sim}{a})_o = (0, -a_3\sin\psi, a_3\cos\psi)^T \qquad (5.2.11b)$$

Define $\underset{\sim}{S}$ as the rotation matrix carrying axes labelled i into those labelled o. Thus

$$(\underset{\sim}{S})_i = \begin{pmatrix} \cos\alpha_4 & \sin\alpha_4 & 0 \\ -\sin\alpha_4 & \cos\alpha_4 & 0 \\ 0 & 0 & 1 \end{pmatrix} \qquad (5.2.12)$$

and so

$$\underset{\sim}{(a)}_i = \underset{\sim}{(S)}_i \underset{\sim}{(a)}_o = a_3 \begin{pmatrix} -\sin\alpha_4 \sin\psi \\ -\cos\alpha_4 \sin\psi \\ \cos\psi \end{pmatrix} \qquad (5.2.13)$$

Other terms appearing in eq. (5.2.7), besides those of eqs. (5.2.8), can now be computed. These are

$$||\underset{\sim}{c}||^2 = s_1^2 + s_4^2 + 2s_1 s_4 \cos\alpha_4 + a_4^2 \qquad (5.2.14a)$$

$$\underset{\sim}{c}^T \underset{\sim}{d}_j = a_1 (s_4 \sin\alpha_4 \sin\phi_j + a_4 \cos\phi_j) \qquad (5.2.14b)$$

$$\underset{\sim}{c}^T \underset{\sim}{a}_j = a_3 (a_4 \cos\psi_j - s_1 \sin\alpha_4 \sin\psi_j) \qquad (5.2.14c)$$

$$\underset{\sim}{d}_j^T \underset{\sim}{a}_j = a_1 a_3 (\cos\alpha_4 \sin\psi_j \sin\phi_j + \cos\psi_j \cos\phi_j) \qquad (5.2.14d)$$

Substituting expressions (5.2.8) and (5.2.14) into eq. (5.2.7) one obtains

$$\begin{aligned} a_2^2 = {} & s_1^2 + s_4^2 + 2s_1 s_4 \cos\alpha_4 + a_4^2 + a_1^2 + a_3^2 + \\ & + 2a_1 (a_4 \cos\phi_j + s_4 \sin\alpha_4 \sin\phi_j) \\ & - 2a_3 (a_4 \cos\psi_j - s_1 \sin\alpha_4 \sin\psi_j) \\ & - 2a_1 a_3 (\cos\alpha_4 \sin\psi_j \sin\phi_j + \cos\psi_j \cos\phi_j) \end{aligned} \qquad (5.2.15)$$

which is the desired input-output relationship. Now, if angle ϕ is measured from certain reference ϕ_0 by letting

$$\phi_j = \phi_0 + p_j \qquad (5.2.16a)$$

$$k_1 = \frac{a_4 + s_4 \sin\alpha_4 \tan\phi_0}{a_3} \qquad (5.2.16b)$$

$$k_2 = \frac{s_4 \sin\alpha_4 - a_4 \tan\phi_0}{a_3} \qquad (5.2.16c)$$

$$k_3 = -\frac{a_4}{a_1 \cos\phi_0} \qquad (5.2.16d)$$

$$k_4 = \frac{s_1 \sin\alpha_4}{a_1 \cos\phi_0} \qquad (5.2.16e)$$

$$k_5 = \tan\phi_0 \qquad (5.2.16f)$$

$$k_6 = \frac{a_1^2 - a_2^2 + a_3^2 + a_4^2 + s_1^2 + s_4^2 + 2s_1 s_4 \cos\alpha_4}{2a_1 a_3 \cos\phi_0} \qquad (5.2.16g)$$

eq. (5.2.15) becomes

$$k_1\cos p_j+k_2\sin p_j+k_3\cos\psi_j+k_4\sin\psi_j+$$
$$+k_5(\cos\psi_j\sin p_j-\cos\alpha_4\sin\psi_j\cos p_j)+$$
$$+k_6=\cos\psi_j\cos p_j+\cos\alpha_4\sin\psi_j\sin p_j,\quad j=1,\ldots,6 \qquad (5.2.17)$$

Eqs. (5.2.17) constitute then a linear algebraic system of six equations in six unknowns $(k_1,\ldots,k_6)$. This system can be solved efficiently by application of subroutines DECOMP and SOLVE, of Sect. 1.12. In order to compute the linkage parameters a_1, a_2, a_3, a_4, s_1, s_4, $\cos\alpha_4$ and $\cos\phi_0$, however, the nonlinear system (5.2.16) has to be solved for the said parameters. The equations of this system, though nonlinear, are weakly coupled, for which reason its solution can be performed without having to resort to a numerical method. The aforementioned nonlinear system, nevertheless, contains a surplus of two unknowns. One of these unknowns can be eliminated through division by it, the surplus of unknowns thus reducing to one. In fact, the solution $a_2 = 0$ is ruled out, for this would lead to a topologically different layout, namely a conpling of three, instead of four, links. Scaling the linkage by a factor $1/a_2$, i.e. setting $a_2 = 1$ does not alter the input-output relationship, for this does not depend upon the absolute but upon the relative lengths of the different links. The remaining unknown in excess can be eliminated by assigning a value to it.

For the linkage appearing in Fig 5.2.1 let

i	1	2	3	4	5	6
ϕ_i	0	45°	60°	90°	60°	30°
ψ_i	0	30°	45°	60°	90°	180°

$a_2 = 1$, $\alpha_4 = 120°$

The arising linear system of equations was solved using subroutines DECOMP and SOLVE. The following results were obtained:

$k_1 = -0.840173$, $k_2 = 0.303087$, $k_3 = -1.350247$

$k_4 = -0.036659$, $k_5 = -1.140727$, $k_6 = 0.489926$

$a_1 = -0.175003$, $a_2 = 1.000000$, $a_3 = -0.724996$

$a_4 = 0.155768$, $s_1 = 0.007284$, $s_4 = -0.684495$

$\phi_0 = -0.851042$ rad

which is the solution to Example 5.2.1.

Mohan Rao et al. (5.2) have extended Denavit and Hartenberg's idea up to ten accuracy-point synthesis. For seven-point synthesis, they introduce the zero location of the input dial, ψ_0 , as one additional unknown in the above formulation, ending up with the following synthesis equations:

$$k_1 c\psi_j + k_2 s\psi_j + k_3 c\phi_j + k_4 s\phi_j - k_5 (c\alpha_4 c\phi_j s\psi_j - c\psi_j s\phi_j) - k_6 (s\phi_j c\psi_j c\alpha_4 - c\phi_j s\psi_j) +$$
$$+ k_7 - k_8 (c\alpha_4 c\phi_j c\psi_j + s\phi_j s\psi_j) = c\alpha_4 s\phi_j s\psi_j + c\phi_j c\psi_j , j=1,2,\ldots,7 \tag{5.2.20}$$

where

$$k_1 = \frac{s_1 s\alpha_4 \tan\psi_0 - a_4}{a_1 c\phi_0} \tag{5.2.21a}$$

$$k_2 = \frac{a_4 \tan\psi_0 + s_1 s\alpha_4}{a_1 c\phi_0} \tag{5.2.21b}$$

$$k_3 = \frac{a_4 + s_4 s\alpha_4 \tan\phi_0}{a_3 c\psi_0} \tag{5.2.21c}$$

$$k_4 = \frac{s_4 s\alpha_4 - a_4 \tan\phi_0}{a_3 c\psi_0} \tag{5.2.21d}$$

$$k_5 = \tan\phi_0 \tag{5.2.21e}$$

$$k_6 = \tan\psi_0 \tag{5.2.21f}$$

$$k_7 = \frac{a_1^2-1+a_3^2+a_4^2+s_1^2+s_2^2+2s_1s_4c\alpha_4}{2a_1a_3c\phi_0c\psi_0} \tag{5.2.21g}$$

$$k_8 = \tan\phi_0\tan\psi_0 \tag{5.2.21h}$$

Eqs. (5.2.20) constitute a linear system of seven equations in eight unknowns. However, the eight k_i are not independent, for they are related by

$$k_8=k_5k_6 \tag{5.2.22}$$

Thus, the synthesis equations comprise the seven equations (5.2.20) plus eq. (5.2.22), i.e. a system of eight equations in eight unknowns, out of which, seven are linear and one is nonlinear.

To solve this system, the aforementioned authors proposed a method based on the principle of superposition of linear systems*, i.e. the principle under which if $\underset{\sim}{x}_1$ and $\underset{\sim}{x}_2$ are solutions to $\underset{\sim}{A}\underset{\sim}{x}=\underset{\sim}{b}$ and $\underset{\sim}{A}\underset{\sim}{x}=\underset{\sim}{c}$, respectively, then $\beta\underset{\sim}{x}_1+\gamma\underset{\sim}{x}_2$ is the solution to $\underset{\sim}{A}\underset{\sim}{x}=\beta\underset{\sim}{b}+\gamma\underset{\sim}{c}$. The method is next outlined:

i) Write eqs. (5.2.20) in the form

$$k_1c\psi_j+ k_2s\psi_j+k_3c\phi_j+k_4s\phi_j-k_5(c\alpha_4c\phi_js\psi_j-c\psi_js\phi_j)-k_6(s\phi_jc\psi_jc\alpha_4-c\phi_js\psi_j)+$$

$$+k_7=(c\alpha_4s\phi_js\psi_j+c\phi_jc\psi_j)+k_8(c\alpha_4c\phi_jc\psi_j+s\phi_js\psi_j),j=1,2,\ldots,7 \tag{5.2.20a}$$

ii) Define the following vectors

$$\underset{\sim}{x}=(k_1,k_2,\ldots,k_7)^T$$
$$\underset{\sim}{b}=(b_1,b_2,\ldots,b_7)^T$$
$$\underset{\sim}{c}=(c_1,c_2,\ldots,c_7)^T$$

where

$$b_j=c\alpha_4s\phi_js\psi_j+c\phi_jc\psi_j,j=1,2,\ldots,7$$
$$c_j=c\alpha_4c\phi_jc\psi_j+s\phi_js\psi_j,j=1,2,\ldots,7$$

---

* See Section 1.11

iii) Rewrite system (5.2.20a) in the form

$$\underset{\sim}{A}\underset{\sim}{x}=\underset{\sim}{b}+k_8\underset{\sim}{c} \qquad (5.2.20b)$$

iv) Solve for $\underset{\sim}{x}_1$ and $\underset{\sim}{x}_2$ from

$$\underset{\sim}{A}\underset{\sim}{x}_1=\underset{\sim}{b},\underset{\sim}{A}\underset{\sim}{x}_2=\underset{\sim}{c}$$

v) Write the solution to system (5.2.20b) as

$$\underset{\sim}{x}=\underset{\sim}{x}_1+k_8\underset{\sim}{x}_2 \qquad (5.2.23)$$

where k_8 is not known as yet

vi) Letting λ_1 and μ_1 be the 5th and 6th components of $\underset{\sim}{x}_1$, respectively, and defining λ_2 and μ_2 analogously, one obtains from (5.2.23),

$$k_5=\lambda_1+k_8\lambda_2 \qquad (5.2.24)$$

$$k_6=\mu_1+k_8\mu_2 \qquad (5.2.25)$$

vii) Substitute (5.2.24) and (5.2.25) into (5.2.22), thus obtaining

$$\lambda_2\mu_2k_8^2+(\lambda_1\mu_2+\lambda_2\mu_1-1)k_8+\lambda_1\mu_1=0 \qquad (5.2.26)$$

from which two values for k_8 can be obtained

viii) With one real value of k_8, obtain the values of $k_1,k_2,\ldots,k_7$ from eq. (5.2.23)

ix) The linkage parameters can now be obtained from eqs. (5.2.21) thus completing the proposed synthesis.

By leaving α_4 unspecified in the above formulation, the same authors (5.2) state the 8-point synthesis problem as

$$k_1c\psi_j+k_2s\psi_j+k_3c\phi_j+k_4s\phi_j-k_5c\phi_js\psi_j-k_6c\psi_js\phi_j-k_7s\phi_js\psi_j+k_8=c\psi_jc\phi_j,$$

$$j=1,2,\ldots,8 \qquad (5.2.27)$$

thus obtaining a <u>linear system</u> of eight equations in eight unknowns, which can readily be solved via the LU algorithm. In the above system

$$k_1=\frac{s_1\tan\psi_0 s\alpha_4-a_4}{a_1c\phi_0(1+\tan\phi_0\tan\psi_0 c\alpha_4)} \qquad (5.2.28a)$$

$$k_2 = \frac{a_4 \tan\psi_0 + s_1 a\alpha_4}{a_1 c\phi_0 (1+\tan\phi_0 \tan\psi_0 c\alpha_4}$$ (5.2.28b)

$$k_3 = \frac{a_4 + s_4 \tan\phi_0 s\alpha_4}{a_3 c\psi_0 (1+\tan\phi_0 \tan\psi_0 c\alpha_4)}$$ (5.2.28c)

$$k_4 = \frac{s_4 s\alpha_4 - a_4 \tan\phi_0}{a_3 c\psi_0 (1+\tan\phi_0 \tan\psi_0 c\alpha_4)}$$ (5.2.28d)

$$k_5 = \frac{\tan\phi_0 c\alpha_4 - \tan\psi_0}{1+\tan\phi_0 \tan\psi_0 c\alpha_4}$$ (5.2.28e)

$$k_6 = \frac{\tan\psi_0 c\alpha_4 - \tan\phi_0}{1+\tan\phi_0 \tan\psi_0 c\alpha_4}$$ (5.2.28f)

$$k_7 = \frac{c\alpha_4 + \tan\phi_0 \tan\psi_0}{1+\tan\phi_0 \tan\psi_0 c\alpha_4}$$ (5.2.28g)

$$k_8 = \frac{a_1^2 + a_2^2 + a_3^2 + a_4^2 + s_1^2 + s_4^2 2 s_1 s_4 c\alpha_4}{2 a_1 a_3 c\phi_0 c\psi_0 (1+\tan\phi_0 \tan\psi_0 c\alpha_4)}$$ (5.2.28h)

The obtention of the linkage parameters from eqs. (5.2.28) is not a simple matter, for these appear strongly coupled in those equations. A method to solve for the said parameters is also presented in [5.2]. In the same paper, the authors propose that the RSSR linkage synthesis can be extended up to 10 points if scale parameters for the function intended to be generated are introduced.

Since the RSSR linkage lends itself very suitably to be used as a function generator, it has received much attention. Luck presents in [5.3] a method of synthesis of this linkage that allows for optimization by the introduction of a free parameter. Referring to Fig 5.2.3, this author writes the input-output function of the RSSR linkage in the form

$$f_1(\phi,\psi) = A_0 c\psi + B_0 s\phi + C_0 = 0$$ (5.2.29)

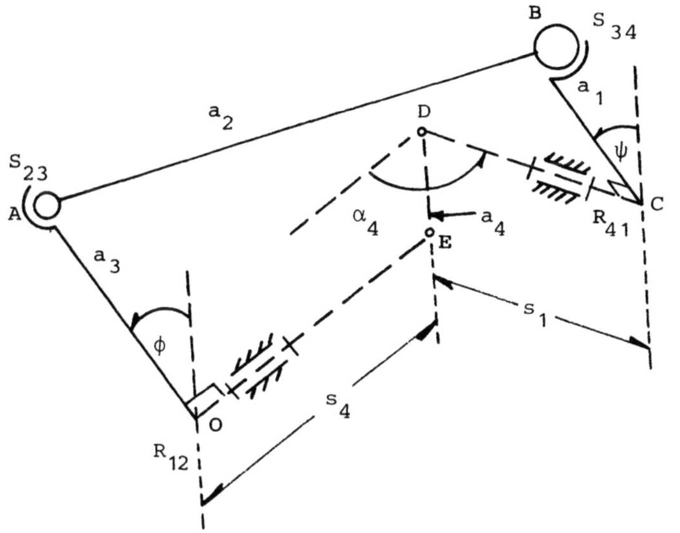

Fig. 5.2.3. RSSR function generating linkage

where

$$A_0=2a_1(s_4s\alpha_4-a_3c\alpha_4c\phi) \tag{5.2.30a}$$

$$B_0=2a_1(a_4-a_3s\phi) \tag{5.2.30b}$$

$$C_0=s_1^2+s_4^2+a_1^2-a_2^2+a_3^2a_4^2-2(s_1s_4c\alpha_4+s_1a_3s\alpha_3c\phi+a_3a_4s\phi) \tag{5.2.30c}$$

Normalizing the linkage lengths with respect to s_4, the following variables are defined

$$x_1=\frac{a_3}{s_4},\ x_2=\frac{a_2}{s_4},\ x_3=\frac{a_1}{s_4},\ \lambda=\frac{s_1}{s_4},\ e=\frac{a_4}{s_4},\ s_4=1 \tag{5.2.31}$$

The input-output function is then transformed into

$$f\ (\phi,\psi)=x_1^2-x_2^2+x_3^2+1+\lambda^2-2\lambda c\alpha_4+e\ -2x_1(\lambda s\alpha_4c\phi+es\phi)+2x_3(es\psi+s\alpha_4c\psi)$$
$$-2x_1x_3(s\psi s\phi+c\alpha_4c\psi c\phi)=0 \tag{5.2.32}$$

Moreover, define

$$\phi_j=\alpha+\Delta\phi_j,\psi_j=\beta+\Delta\psi_j,\ j=1,2,\ldots,n \tag{5.2.33}$$

with

$$\Delta\psi_0=\Delta\phi_0\equiv 0$$

Subtracting $f(\phi_0,\psi_0)=0$ from $f(\phi_j,\psi_j)=0$ $(j=1,2,\ldots,n)$ leads to the following linear homogenous system

$$f(\Delta\phi_j,\Delta\psi_j)\equiv f(\phi_j,\psi_j)-f(\phi_0,\psi_0)=A_ju_1+B_ju_2+C_ju_3+D_ju_4+E_ju_5+F_ju_6=0 \tag{5.2.34}$$

where

$$u_1=\frac{e}{x_1},\ u_2=\frac{s\alpha_4}{x_1},\ u_3=-\frac{e}{x_3} \tag{5.2.35a}$$

$$u_4=-\frac{\lambda s\alpha_4}{x_3},\ u_5=-c\alpha_4,\ u_6=1 \tag{5.2.35b}$$

$$A_j=s(\beta+\Delta\psi_j)-s\beta \tag{5.2.36a}$$

$$B_j=c(\beta+\Delta\psi_j)-c\beta \tag{5.2.36b}$$

$$C_j=s(\alpha\Delta\phi_j)-s\alpha \tag{5.2.36c}$$

$$D_j=c(\alpha+\Delta\phi_j)-c\alpha \tag{5.2.36d}$$

$$E_j=c(\alpha+\Delta\phi_j)c(\beta+\Delta\psi_j)-c\alpha c\beta \tag{5.2.36e}$$

$$F_j=-s(\alpha+\Delta\phi_j)s(\beta+\Delta\psi_j)+s\alpha s\beta \tag{5.2.36f}$$

System (5.2.34) has non-trivial solutions if and only if its determinant vanishes. In the problem of synthesis for function generation, the values of $\Delta\psi_j$ and $\Delta\phi_j$ $(j=1,2,\ldots,n)$ are given. Hence, the determinant δ is a function of α and β only. This determinant has, in fact, the following form

$$\begin{vmatrix} A_2 & B_2 & \cdots & F_2 \\ A_3 & B_3 & \cdots & F_3 \\ \cdot & \cdot & & \cdot \\ \cdot & \cdot & & \cdot \\ \cdot & \cdot & & \cdot \\ A_7 & B_7 & \cdots & F_7 \end{vmatrix} = \delta(\alpha,\beta) \tag{5.2.37}$$

The function $\delta(\alpha,\beta)$ vanishes along the curve $\beta=\beta(\alpha)$ defined over the plane $\alpha-\beta$, as shown in Fig 5.2.4

Luck proposes in (5.3) a numerical method to find pairs of values (α_i,β_i) along which δ vanishes, this method being based on the "regula-falsi" algorithm. A different method is proposed here.

For a given value of β, say β_i, δ can be regarded as a function of one real variable, namely, α, i.e.

$$\delta_i=\delta_i(\alpha)$$

To find the roots of δ_i, the method of Newton-Raphson is applied, as follows

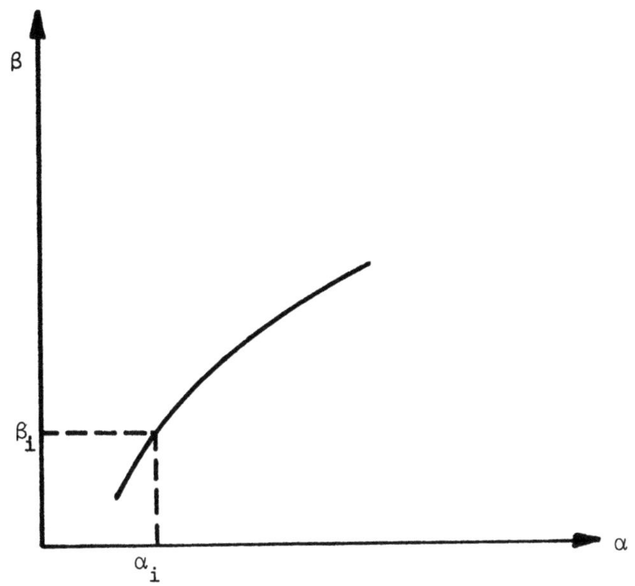

Fig 5.2.4 Set of values of α and β along which δ vanishes.

1.i) Estimate (or guess) a starting value of α, say α^o, to begin the iterative procedure

1.ii) From the current value α^k of the sought root, α_i, of δ_i, compute an improved value of α_i, say α^{k+1} as

$$\alpha^{k+1}=\alpha^k-\frac{\delta_i(\alpha^k)}{\delta_i'(\alpha^k)} \tag{5.2.38}$$

1.iii) If the correction value $\delta_i(\alpha^k)/\delta_i'(\alpha^k)$ is greater than a given tolerence ε, return to Step 1.ii. Otherwise, verity if $|\delta_i(\alpha^k)|\leq\varepsilon$. If so, stop the procedure and accept the current value α^k as the value of the sough root, α_i. If not, repeat the procedure starting at Step 1.i, with a different value of α^o.

Expression (5.2.38) greaty simplifies when the formula for the derivative of a determinant is introduced, namely

$$\delta_i'(\alpha)=\delta_i(\alpha)\mathrm{Tr}\left[\underset{\sim}{M}^{-1}(\alpha)\underset{\sim}{M}'(\alpha)\right] \tag{5.2.39}$$

Where $\underset{\sim}{M}$ is the matrix whose determinant is δ and $\underset{\sim}{M}'(\alpha)$ is the matrix whose entries are the derivatives, with respect to α, of the corresponding entries of $\underset{\sim}{M}$. Since formula (5.2.39) is not very popular, its derivation is presented in Appendix 2. Substitution of eq. (5.2.39) into eq. (5.2.38) leads to

$$\alpha^{k+1}=\alpha^k-\frac{1}{\mathrm{Tr}\left(\underset{\sim}{M}^{-1}(\alpha^k)M'(\alpha^k)\right)} \tag{5.2.40}$$

to compute $\underset{\sim}{M}^{-1}(\alpha^k)\underset{\sim}{M}'(\alpha^k)$, let

$$\underset{\sim}{M}^{-1}(\alpha^k)\underset{\sim}{M}'(\alpha^k)=\underset{\sim}{N}$$

i.e., after dropping the argument for shortness,

$$\underset{\sim}{M}\underset{\sim}{N}=\underset{\sim}{M}' \tag{5.2.41}$$

In eq. (5.2.41), let $\underset{\sim}{n}_p$ and $\underset{\sim}{m}_p'$ denote vectors identical to the pth columns of matrices $\underset{\sim}{N}$ and $\underset{\sim}{M}'$, respectively. Hence, matrix $\underset{\sim}{N}$ can be computed by solving the 6 linear systems

$$\underset{\sim}{M}\underset{\sim}{n}_p = \underset{\sim}{m}'_p \ , \quad p=1\ 2,\ldots,6 \qquad (5.2.42)$$

which can be done via the LU decomposition algorithm. At this stage there are two simplifications, namely,

2.i) Subroutine DECOMP need only once be applied at the kth iteration for, once the LU decomposition of $\underset{\sim}{M}$ is obtained, this can be used to solve the 6 systems appearing in (5.2.42)

2.ii) Since A_j and B_j do not contain α explicitly, the first two columns of $\underset{\sim}{M}'$ vanish. Therefore, $\underset{\sim}{n}_1=\underset{\sim}{n}_2=\underset{\sim}{0}$ and subroutine SOLVE need be applied only four times. Funthermore,

2.iii) If at any of the iterations DECOMP detects matrix $\underset{\sim}{M}$ to be singular, the system of (5.2.42) cannot be solved, but this is not necessary any more, for precisely what one is seeking is that value of α, for a given value of β_i, that renders matrix $\underset{\sim}{M}$ singular, i.e. that makes δ zero.

Once the value α_i that, for a given β_i makes the determinant vanish, has been found, a new value of β_i, say β_{i+1}, is introduced and the process starting in 1.i. is repeated, except that a new starting value α^o need not be guessed, for a good estimate for α_{i+1} is, of course, α_i, provided β_i and β_{i+1} are sufficiently close to each other. This way, a set of discrete values $\{\alpha_i,\beta_i\}$, that zero δ, can be given in tabular form; however, if a continuous function $\beta=\beta(\alpha)$ is necessary, this can be approximated by interpolation. The most efficient way of interpolating such a function is by means of spline functions (5.4).

SUBROUTINE ZERDET, appearing in Fig 5.2.5 implements the algorithm outlined in 1.i-1.iii. This subroutine was used to find the graph β vs. α arising from the following problem for seven-point accuracy synthesis (Example 1 of (5.3)).

$$\Delta\phi_2=30°, \quad \Delta\psi_2=-16.1°$$

$$\Delta\phi_3=75°, \quad \Delta\psi_3=-11.5°$$

$$\Delta\phi_4=135°, \quad \Delta\psi_4=22.5°$$

$$\Delta\phi_5=195°, \quad \Delta\psi_5=53.5°$$

$$\Delta\phi_6=240°, \quad \Delta\psi_6=57.9°$$

$$\Delta\phi_7=300°, \quad \Delta\psi_7=41.3°$$

This graph appears in Fig 5.2.6

Given one pair of values of the aforementioned set $\{\alpha_i,\beta_i\}$ and recalling that $u_6=1$, the following linear inhomogeneous system is obtained

$$\underset{\sim}{G}\underset{\sim}{u}=-\underset{\sim}{f} \qquad (5.2.43)$$

where $\underset{\sim}{G}$ is the submatrix of $\underset{\sim}{M}$ containing the first five rows and columns of it, $\underset{\sim}{u}=(u_1,u_2,\ldots,u_5)^T$ and $\underset{\sim}{f}=(F_2,F_3,\ldots,F_6)^T$. Vector $\underset{\sim}{u}$ can be obtained

```
      SUBROUTINE ZERDET(N,MAX,P,X,A,DERA,EPSIL)
      DIMENSION P(1),A(N,N),DERA(N,N),IP(100),C(100)
C
C
C     THIS SUBROUTINE FINDS THE VALUE OF X THAT ZEROES THE DETERMINANT
C     OF MATRIX A(X), A BEING OF NXN AND X, A SCALAR.
C     X = THE UNKNOWN.
C     P = AN ARRAY WHOSE COMPONENTS ARE PARAMETERS APPEARING IN THE
C         DETERMINANT. THE USER DEFINES IT ACCORDINGLY.
C     IP= A VECTOR DEFINED IN SUBROUTINE 'DECOMP'.
C     C = AUXILIARY VECTOR USED TO COMPUTE THE TRACE.
C     THE FOLLOWING SUBROUTINES ARE USED WITHIN THIS SUBROTINE :
C     DECOMP AND SOLVE___FIND THE SOLUTION OF A NXN NONSINGULAR LINEAR
C                        ALGEBRAIC SYSTEM. (MOLER C.B.,ALGORITHM 423
C                        LINEAR EQUATION SOLVER(F4), COMMUNICATIONS OF THE
C                        ACM, VOL. 15, NO. 4, P. 274).
C     MATRIX_____________COMPUTES THE ENTRIES OF MATRIX A. THE USER
C                        PROVIDES IT.
C     DERMAT_____________COMPUTES THE ENTRIES OF THE DERIVATIVE OF A
C                        WHIT RESPECT TO X. THE USER PROVIDES IT.
C     IMPRIM_____________PRINTS RESULTS.
C     IMPRIM PRINTS THE MESSAGE 'NO CONVERGENCE----------ERROR GROWS'
C     AND RETURNS TO MAIN PROGRAM, WHEN THE ABSOLUTE VALUE OF THE
C     CORRECTION TO THE UNKNOWN GROWS FROM ONE ITERATION TO THE NEXT ONE,
C     THUS INDICATING THE POSSIBILITY OF DIVERGENCE. IN THIS CASE, THE
C     FIRST PARAMETER OF IMPRIM IS SET EQUAL TO 1. IT PRINTS THE MESSAGE
C     'NO CONVERGENCE' ALONE WHEN, EVEN IF THE VALUE OF THE DETERMINANT
C     DECREASES, THIS VALUE DOES NOT MEET THE PRESCRIBED TOLERANCE, FOR
C     THE ALLOWED NUMBER OF ITERATIONS. IN THE LATTER CASE THAT PARAMETER
C     OF IMPRIM IS SET EQUAL TO 2 AND ZERDET RETURNS TO MAIN PROGRAM. IF
C     CONVERGENCE IS REACHED, THE AFOREMENTIONED PARAMETER IS SET EQUAL
C     TO 3.
C
C
C     NEWTON-RAPHSON METHOD STARTS
C
      ITER=0
C
C     THE ENTRIES OF THE MATRIX ARE COMPUTED
 1    CALL MATRIX(N,X,P,A)
      CALL DECOMP(N,N,A,IP)
      DET=1.0
      DO 2 LM=1,N
 2    DET=DET*A(LM,LM)
      DET=DET*IP(N)
      IF(IP(N).EQ.0) GO TO 8
C   COMPUTING THE TRACE OF (A**(-1))*DERA, WHERE DERA IS THE DERIVATIVE
C   OF A WITH RESPECT TO X
*
```

Fig 5.2.5 Listing of SUBROUTINE ZERDET

```
          CALL DERMAT(N,X,P,DERA)
          DO 4 I=1,N
              DO 3 J=1,N
  3           C(J)=DERA(J,I)
              CALL SOLVE(N,N,A,C,IP)
  4       TRACE=TRACE+C(I)
C
C    DETERMINES THE EVOLUTION OF THIS TRACE
          TRAINV=1.0/TRACE
          IF(ITER.EQ.0) GO TO 5
          IF(ABS(XV-X).LT.EPSIL) GO TO 8
          IF(ABS(TRAINV).GT.TRAVI) GO TO 6
  5       XV=X
          X=X-TRAINV
          ITER=ITER+1
          IF(ITER.GE.MAX) GO TO 7
          TRAVI=ABS(TRAINV)
          GO TO 1
  6       CALL IMPRIM(1,ITER,X,DET,P)
          RETURN
  7       CALL IMPRIM(2,ITER,X,DET,P)
          RETURN
  8       CALL IMPRIM(3,ITER,X,DET,P)
          RETURN
          END
#
```

Fig. 5.2.5 Listing of SUBROUTINE ZERDET (Continued)

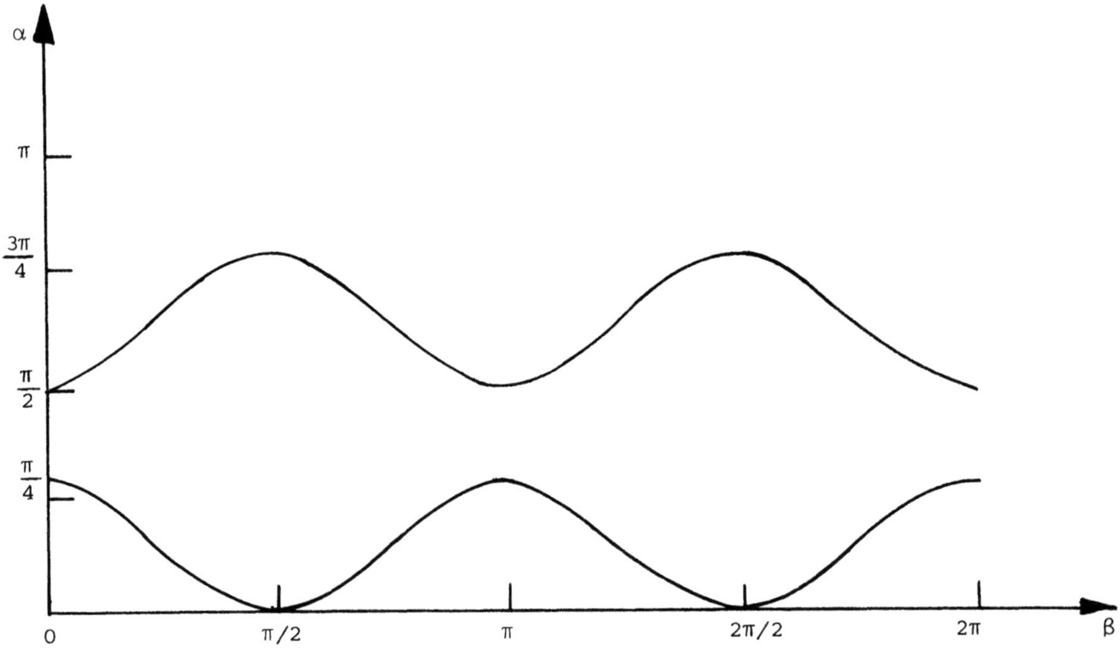

Fig 5.2.6 Free-parameter relationship for the synthesis of a function generator linkage with seven specified points.

from eq. (5.2.43) via the LU decomposition algotithm. With $\underset{\sim}{u}$ known, the normalized lengths x_1, x_3, λ, e and the angle α_4 can be obtained by substitution into eqs. (5.2.35). The remaining length, x_2, is readily obtained from eq. (5.2.32) when, for convenience, the values of ψ_1 and ϕ_1 are introduced in that equation. Thus, the synthesis problem is totally solved.

Two observations are in order: i) This method allows to synthesize a function generator for up to seven precision points and ii) since infinitely many combinations of α and β (that turn the determinant δ zero) exist, that combination rendering the best transmission angle can be used, thus allowing for optimization.

So far no velocity nor acceleration nor higher-derivative conditions have been considered. The introduction of such conditions, rendering a synthesis problem for infinitesimally-separated positions is, however, straightforward. In fact, all that need be done is to differentiate the input-output function with respect to time as many times as necessary, substitute in the resulting equations the prescribed values of the input, the output and their derivatives and form a system of synthesis equations similar to that appearing in eqs. (5.2.17). This is next illustrated with an example.

Example 5.2.2 Synthesis of the RSSR linkage for function generation with prescribed dead-points.

It is required to determine the dimensions of the linkage shown in Fig 5.2.1, for a given value of angle α_4, to produce an oscillation of link BC of 90° in such a way that the return motion (under no load) be performed twice as faster as the first half (under full load). The input link should be a crank, i.e. it should rotate through 360°.

Solution:

In the synthesis equation 5.2.17, let

$$k_1=\frac{s_1}{a_1},\ k_2=\frac{a_4}{a_1},\ k_3=\frac{a_4}{a_3},\ k_4=\frac{s_4}{a_3} \tag{5.2.44a}$$

$$k_5=-\frac{a_1^2-a_2^2+a_3^2+a_4^2+s_1^2+s_4^2+2s_1s_4c\alpha_4}{2a_1a_3} \tag{5.2.44b}$$

thus obtaining the suitable synthesis equation in the form

$$k_1s\alpha_4s\phi-k_2c\phi+k_3c\psi+k_4s\alpha_4s\psi+k_5-(c\phi c\psi+c\alpha_4s\phi s\psi) = 0 \tag{5.2.45}$$

To meet the problem conditions, assume that the no-load motion is performed during a 120° rotation of the input crank, the load motion being executed during the remaining 240°rotation of the said crank. Thus, the following conditions can be imposed

$$\phi_1=0,\psi_1=0,\dot{\psi}_1=0 \tag{5.2.46a}$$

$$\phi_2=120°,\psi_2=90°,\dot{\psi}_2=0 \tag{5.2.46b}$$

These conditions do not suffice for the present problem for, even if they are met, the output linkage could rotate through an angle of 270° and not one of 90°, as required. To ensure the proper motion to be performed by the linkage, the additional following condition is imposed

$$\phi_3=180°,\psi_3=67.5° \tag{5.2.46c}$$

which arises from the assumption that, when the input link has rotated through an angle of 60°(= ϕ_3 - ϕ_2) of the load motion (1/4 of this motion), the output link has rotated through an angle of 22.5°(= ψ_2 - ψ_3) of the same motion (1/4 of this motion, also).

To specify the velocity conditions, the synthesis equation (5.2.45) is differentiated with respect to time, thus obtaining

$$k_1\dot{\phi}s\alpha_4c\phi+k_2\dot{\phi}s\phi-k_3\dot{\psi}s\psi+k_4\dot{\psi}s\alpha_4c\psi-(-\dot{\phi}s\phi c\psi-\dot{\psi}c\phi s\psi$$
$$+\dot{\phi}c\alpha_4c\phi s\psi+\dot{\psi}c\alpha_4s\phi c\psi)=0 \tag{5.2.47}$$

It is then noticed that five synthesis equations can be obtained to produce five unknowns, thereby justifying the use of a synthesis equation of the form of eq. (5.2.45). The said five synthesis equations are:

For $\phi_1=0$ and $\psi_1=0$, eq. (5.2.45) leads to

$$-k_2+k_3+k_5-1=0 \tag{5.2.48a}$$

For $\phi_1=0$, $\psi_1=0$ and $\dot{\psi}_1=0$, with $\dot{\phi}_1\neq 0$, which can then be dropped, eq. (5.2.47) leads to

$$k_1 s\alpha_4=0 \tag{5.2.48b}$$

For $\phi_2=120°$ and $\psi_2=90°$, eq. (5.2.45) leads to

$$\frac{\sqrt{3}}{2}s\alpha_4 k_1+\frac{1}{2}k_2+s\alpha_4 k_4+k_5-\frac{\sqrt{3}}{2}\,c\alpha_4=0 \tag{5.2.48c}$$

For $\phi_2=120°$, $\psi_2=90°$ and $\dot{\psi}_2=0$, with $\dot{\phi}_2\neq 0$ eq. (5.2.47) leads to

$$-\frac{1}{2}s\alpha_4 k_1+\frac{\sqrt{3}}{2}k_2+\frac{1}{2}c\alpha_4=0$$

Finally, from condition (5.2.46c), eq. (5.2.45) leads to

$$k_2+c67.5°k_3+s67.5°s\alpha_4 k_4+k_5+c67.5°=0 \tag{5.2.48d}$$

Solving for the five unknowns in the foregoing system (5.2.48), one obtains

$$k_1 = 0$$

$$k_2 = -\frac{\sqrt{3}}{3}c\alpha_4$$

$$k_3 = \frac{1}{1-c}\left(1 + c + \frac{2\sqrt{3}}{3}\,c\alpha_4(s-1)\right)$$

$$k_4 = \frac{2\sqrt{3}}{3}\cot\alpha_4$$

$$k_5 = 1 - \frac{\sqrt{3}}{3}c\alpha_4 - \frac{1}{1-c}\left(1 + c + \frac{2\sqrt{3}}{3}c\alpha_4(s-1)\right)$$

where $c\equiv\cos 67.5°$, $s\equiv\sin 67.5°$

Putting the above expressions into a computer and computing them for different values of α_4 yields numerous linkages, out of which the best (in a given sense, e.g., the one with the best transmission angle) can be selected.

The transmission angle, μ, defined as angle BAD of the linkage appearing in Fig 5.2.1, is a measure of the force transmission of the linkage. From Fig 5.2.1, the definition of vectors $\underset{\sim}{a}, \underset{\sim}{b}, \underset{\sim}{c}$ and $\underset{\sim}{d}$ following it, and eqs. (5.2.5), (5.2.11), (5.2.14) and (5.2.16c), it follows that

$$\cos\mu = -\frac{\underset{\sim}{b}^T \underset{\sim}{a}}{a_2 a_3} = -\frac{(\underset{\sim}{c}+\underset{\sim}{d}-\underset{\sim}{a})^T \underset{\sim}{a}}{a_2 a_3}$$

from which

$$\cos\mu = -\frac{a_4 c\psi - s_1 s\alpha_4 s\psi + a_1 c\alpha_4 s\psi s\phi + a_1 c\psi c\phi - a_3}{a_2} \qquad (5.2.49)$$

The reason why the transmission angle is a measure of the force transmission of the linkage is the following: disregarding the inertia forces of the links, the input torque, M_ϕ, is transmitted from link CD to link AB (Fig 5.2.1) through the coupler link AD by means of a force collinear with line AD. If this force, F, is resolved into two components, one parallel to line AB and the other perpendicular to it, the working component is that perpendicular to AB and its value is $F\sin\mu$, whereas the nonworking component is that parallel to AB and its value is $F\cos\mu$. Thus, for a "good" force transmission, the parallel component of F should be as small as possible, which is attained if $\cos\mu$ lies close to zero, i.e. if μ lies close to either 90° or 270°

The mechanical advantage of a linkage is another feature used to evaluate its operation. It is defined as the ratio of output torque, M_ψ, to the input torque, M_ϕ. An expression for the mechanical advantage is next obtained.

Disregarding the effect of friction forces for simplicity, the output power is equated to the input power, i.e.

$$M_\phi \dot{\phi} = M_\psi \dot{\psi} \qquad (5.2.50)$$

from which, the mechanical advantage, m, is obtained as

$$m = \frac{M_\psi}{M_\phi} = \frac{\dot{\phi}}{\dot{\psi}} \qquad (5.2.51)$$

From eq. (5.4.47),

$$m = \frac{k_3 s\psi - k_4 s\alpha_4 c\psi - s\psi c\phi + c\alpha_4 c\phi s\psi}{k_1 s\alpha_4 c\phi + k_2 s\phi + c\psi s\phi - c\alpha_4 c\phi s\psi} \qquad (5.2.52)$$

An analogous procedure can be followed to determine the mechanical advantage of other types linkages.

Exercise 5.2.1 Determine the mechanical advantage of the following linkages: Spherical RRRR, RSRP, RSRC

Exercise 5.2.2. Using the results of Example 5.2.2, find the dimensions of the linkage whose transmission angle lies between 40° and 140° (or between 220° and 1320°) throughout the performance of the linkage.

Mobility analysis

In synthesising function- generating linkages one is interested in producing a link with certain mobility conditions. These conditions refer to the range of motion of either the input- or the output link or of both. In this respect, one of these links is to be either a crank (possibility of rotation through 360°) or a rocker (possibility of rotation through a fraction of a complete turn). The conditions under which these mobility specifications are met are next discussed.

Referring to Fig 5.2.1 the following analysis is performed to establish the conditions under which the input link, CD, is a crank.

For a given position of the input link, i.e. for a given value of ϕ, the relative position of the pairs D and A is depicted in Fig 5.2.7. In that Figure, the locus of A is the circle centered at B with a radius a_3, $\underset{\sim}{n}$ being the unit normal to the plane of the circle, i.e. the vector paralel to the axis of the the revolute pair R_{41}.

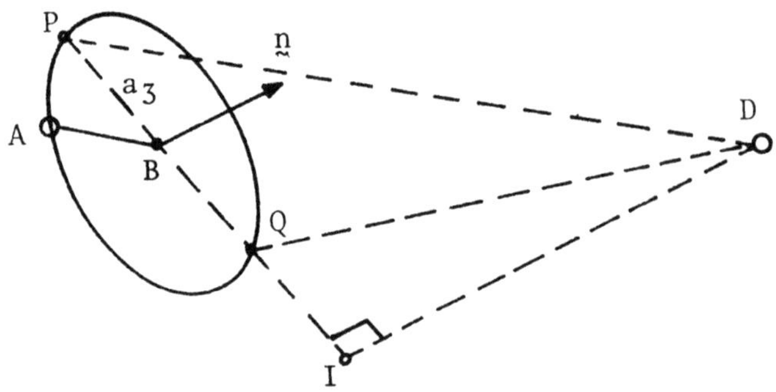

Fig 5.2.7 Relative position of pairs A and D of Fig 5.2.1

In what follows the following Theorem will be resorted to

THEOREM 5.2.1 *Given a circle and a point not lying on its circumference and not necessarily in the plane of the circle, the points on the circumference lying the closest and the farthest from the point, Q and P, respectively, have the property that lines QD and PD are perpendicular to the tangent to the circle passing through Q and through P.*

Exercise 5.2.3 Using the method of the Lagrange multipliers, prove Theorem 5.2.1

From Theorem 5.2.1 it follows that P and Q are determined as the intersections of the circle with line IB. Line IB is in turn determined by the center of the circle, B, and the intersection of the plane of the circle with its perpendicular from D. Let $\ell(\phi)$ and $L(\phi)$ be the lenght of segments DQ and DP, respectively. Hence, the condition for the input link to be a crank is that a_2, the lenght of the coupler link, lie within the maximum value of ℓ and the minimum value L, i.e.

$$\ell_{max}<a_2<L_{min}$$

<u>Exercise 5.2.4</u> Derive the following expressions.

$$\ell(\phi)=\{\left(\sqrt{a_1^2+s_1^2+a_4^2+2a_4a_1\cos\phi - (s_1\cos\alpha_4+a_1\operatorname{sen}\alpha_4\operatorname{sen}\phi)^2} -a_3\right)^2$$

$$+ (s_4+s_1\cos\alpha_4+a_1\operatorname{sen}\alpha_4\operatorname{sen}\phi)^2\}^{1/2}$$

$$L(\phi)=\{\left(\sqrt{a_1^2+s_1^2+a_4^2+2a_4a_1\cos\phi - (s_1\cos\alpha_4+a_1\operatorname{sen}\alpha_4\operatorname{sen}\phi)^2}+a_3\right)^2$$

$$+ (s_4+s_1\cos\alpha_4+a_1\operatorname{sen}\alpha_4\operatorname{sen}\phi)^2\}^{1/2}$$

5.3 MECHANISM SYNTHESIS FOR RIGID-BODY GUIDANCE.

A complete account of the theory and applications of this subject appears in (5.5-5.12). Different approaches to this problem are presented, all of them regarding the calculation of geometric parameters of one dyad* at a time, but the most unified treatment is that of Tsai and Roth (5.11). The method introduced by these authors is based on formulae relating the different screw angles and displacements of composed motions (carrying the rigid body intended to be guided from its reference configuration to its n specified successive configurations) to the directions and positions of the screw axes involved, as is shown next.

Let a rigid body B occupy configuration B_0 initially (henceforth called "the reference configuration") and assume it is intended to conduct this rigid body through n successive configurations B_j $(j=1,2,...,n)$, all of them being finitely separated, i.e. the screw motions relating one of these configurations to each other and to the reference one contain parameters (angle and displacement) with only finite values (Fig 5.3.1)

The motion carrying B from B_0 to B_j can be regarded as the composition of two motions: one, given by a screw $\underset{\sim}{M}_j$** carrying B from B_0 to B_j', followed by a second one, given bv a screw $\underset{\sim}{F}_j$, carriying B from B_j' to B_j. The axes of both screws $\underset{\sim}{M}_j$ and $\underset{\sim}{F}_j$ are lines M_j and F_j, shown in Fig 5.3.1, where F_j is fixed in space, i.e. its position relative to B_0 does not change after the motion is completed, whereas line M_j is a moving one. Let u_j and w_j be the scalar displacements of $\underset{\sim}{M}_j$ and $\underset{\sim}{F}_j$, respectively, and α_j and γ_j their

---

*See Section 3.2.

**$\underset{\sim}{M}_j$ denotes a screw, whereas M_j, its axis.

respective angles. The screw $\underset{\sim}{S}_j$ resulting from the composition of $\underset{\sim}{M}_j$ and $\underset{\sim}{F}_j$

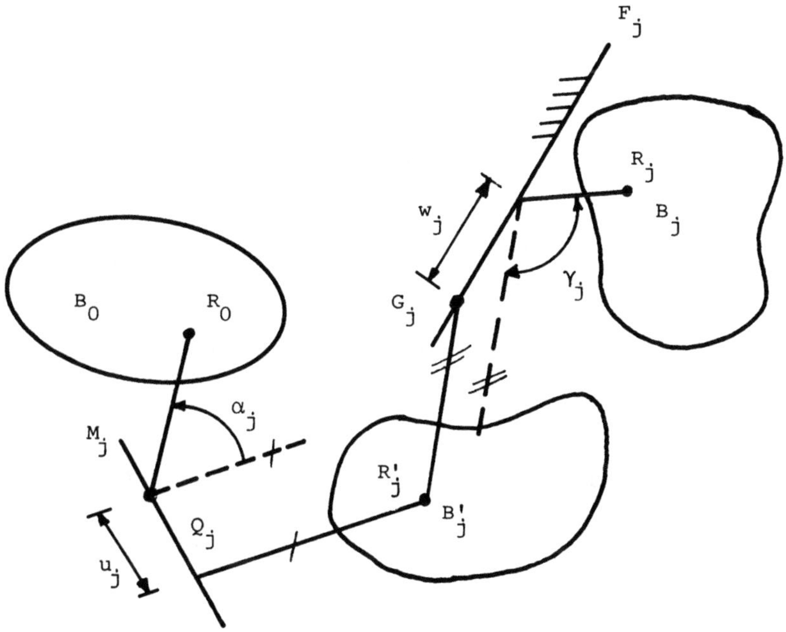

Fig 5.3.1 Successive configurations of a rigid body

has an axis S_j passing through point A_j, its displacement being denoted by t_j and its angle by θ_j. All three screws form what Roth calls a "screw triangle", Next, relationships among the screw displacements, angle and axes are obtained.

In what follows, let $\underset{\sim}{m}_j$, $\underset{\sim}{f}_j$ and $\underset{\sim}{s}_j$ be unit vectors parallel to the axes of $\underset{\sim}{M}_j$, $\underset{\sim}{F}_j$ and $\underset{\sim}{S}_j$. Moreover, let $\underset{\sim}{q}_j$, $\underset{\sim}{g}_j$, and $\underset{\sim}{a}_j$ be the position vectors of points Q_j, G_j and A_j, located on M_j, F_j and S_j, respectively. For short, the indices are dropped from the screw parameters in the following derivations.

Starting with Rodrigues' formula (2.5.3) the following relation is readily obtained (5.13,5.14)

$$\tan\frac{\theta}{2}\mathbf{s} = \frac{\tan\frac{\alpha}{2}\mathbf{m}+\tan\frac{\gamma}{2}\mathbf{f}+\tan\frac{\alpha}{2}\tan\frac{\gamma}{2}\mathbf{f}\times\mathbf{m}}{1-\tan\frac{\alpha}{2}\tan\frac{\gamma}{2}\mathbf{m}\cdot\mathbf{f}} \tag{5.3.1}$$

Multiplying both sides of the above equations times $\times\mathbf{f}$ yields

$$\tan\frac{\theta}{2}\mathbf{s}\times\mathbf{f} = \frac{\tan\frac{\alpha}{2}\mathbf{m}\times\mathbf{f}+\tan\frac{\alpha}{2}\tan\frac{\gamma}{2}(\mathbf{f}\times\mathbf{m})\times\mathbf{f}}{1-\tan\frac{\alpha}{2}\tan\frac{\gamma}{2}\mathbf{m}\cdot\mathbf{f}} \tag{5.3.2}$$

Multiplying both sides of the above equation times $\cdot\mathbf{s}$, the left hand side clearly vanishes thus leading to

$$\tan\frac{\alpha}{2}\left(\mathbf{m}\times\mathbf{f}\cdot\mathbf{s}+\tan\frac{\gamma}{2}(\mathbf{f}\times\mathbf{m})\times\mathbf{f}\cdot\mathbf{s}\right)=0 \tag{5.3.3}$$

which, for non-zero values of α, vanishes only if the term in brackets does, i.e. if

$$\tan\frac{\gamma}{2}= \frac{\mathbf{m}\times\mathbf{f}\cdot\mathbf{s}}{(\mathbf{m}\times\mathbf{f})\cdot(\mathbf{f}\times\mathbf{s})} \tag{5.3.4}$$

which is one of the relations sought.

To stablish a second relation, multiply both sides of eq. (5.3.1) times $\times\mathbf{m}$, thus obtaining

$$\tan\frac{\theta}{2}\mathbf{s}\times\mathbf{m}= \frac{\tan\frac{\gamma}{2}\mathbf{f}\times\mathbf{m}+\tan\frac{\alpha}{2}\tan\frac{\gamma}{2}(\mathbf{f}\times\mathbf{m})\times\mathbf{f}}{1-\tan\frac{\alpha}{2}\tan\frac{\gamma}{2}\mathbf{m}\cdot\mathbf{f}} \tag{5.3.5}$$

Multiplying both sides of the latter equation times $\cdot\mathbf{s}$ renders its left hand side zero, thus leading to

$$\tan\frac{\gamma}{2}\left(\mathbf{f}\times\mathbf{m}\cdot\mathbf{s}+\tan\frac{\alpha}{2}(\mathbf{f}\times\mathbf{m})\times\mathbf{m}\cdot\mathbf{s}\right)=0 \tag{5.3.6}$$

which, for non-zero values of γ, vanishes only if the term in brackets does, i.e. if

$$\tan\frac{\alpha}{2} = \frac{\underset{\sim}{f}.\underset{\sim}{s}x\underset{\sim}{m}}{(\underset{\sim}{s}x\underset{\sim}{m}).(\underset{\sim}{m}x\underset{\sim}{f})} \qquad (5.3.7)$$

thereby obtaining one second relationship

To obtain a third relationship, one connecting θ to $\underset{\sim}{m}$,$\underset{\sim}{f}$ and $\underset{\sim}{s}$, proceed as follows:

Eq. (5.3.1) was obtained by first rotating B through an angle α about an axis parallel to $\underset{\sim}{m}$, and then through an angle γ about an axis parallel to $\underset{\sim}{f}$. The resulting rotation is equivalent to a single one through an angle θ about an axis parallel to $\underset{\sim}{s}$. If now configuration B_j' is regarded as the reference one, B_0 can be reached from B_j' through B_j.B_j can be reached from B_j' in exactly the same way as described previously, i.e. by means of a rotation through an angle γ about an axis parallel to $\underset{\sim}{f}$; but now B_0 can be reached from B_j by means of the inverse of the rotation carrying B_0 to B_j, i.e., via a rotation through an angle -θ abount an axis parallel to $\underset{\sim}{s}$.

Writing the equation equivalent to (5.3.1) for this composition of rotations

$$\tan(-\frac{\alpha}{2})\underset{\sim}{m} = \frac{\tan\frac{\gamma}{2}\underset{\sim}{f}-\tan\frac{\gamma}{2}\underset{\sim}{s}-\tan\frac{\theta}{2}\tan\frac{\gamma}{2}\underset{\sim}{s}x\underset{\sim}{f}}{1+\tan\frac{\gamma}{2}\tan\frac{\theta}{2}\underset{\sim}{f}.\underset{\sim}{s}} \qquad (5.3.8)$$

Proceeding in a fashion analogous to the one used to obtain (5.3.4) and (5.3.7), i.e. multiplying both sides of eq. (5.3.8) times x$\underset{\sim}{s}$.$\underset{\sim}{m}$, one finally obtains

$$\tan\frac{\theta}{2} = -\frac{\underset{\sim}{f}.\underset{\sim}{s}x\underset{\sim}{m}}{(\underset{\sim}{f}x\underset{\sim}{s}).(\underset{\sim}{s}x\underset{\sim}{m})} \qquad (5.3.9)$$

as the third relationship sought.

In the following, expressions for u, w and t are obtained. The composition of screws $\underset{\sim}{M}$ and $\underset{\sim}{F}$ is shown in Fig. 5.3.2. In This figure, $\underset{\sim}{r}$ is the original position vector of a point P of a rigid body undergoing first a screw motion of scalar displacement u, parallel to vector $\underset{\sim}{m}$, and rotation α about an axis passing through Q and parallel to $\underset{\sim}{m}$. After this screw motion the position vector of P is $\underset{\sim}{r}'$. Next, the rigid body undergoes one second screw motion of scalar displacement w parallel to $\underset{\sim}{f}$, and rotation γ about an axis passing through G and parallel to $\underset{\sim}{f}$. The final position vector of point P is $\underset{\sim}{r}''$.

The composition of both screws is equivalent to one single screw of scalar displacement t parallel to $\underset{\sim}{s}$, and rotation θ about an axis passing through A and parallel to $\underset{\sim}{s}$. This screw is shown in Fig 5.3.3.

Let $\underset{\sim}{Q}_1$ and $\underset{\sim}{Q}_2$ be the rotations of screws $\underset{\sim}{M}$ and F, respectively, $\underset{\sim}{Q}_3$ being that of the equivalent screw $\underset{\sim}{S}$. Thus, applying eqs. (2.6.18) and (2.6.19),

$$\underset{\sim}{r}''=\underset{\sim}{r}'+w\underset{\sim}{f}+(\underset{\sim}{Q}_2-\underset{\sim}{I})(\underset{\sim}{r}'-\underset{\sim}{g}) \qquad (5.3.10)$$

where, applying the same result again,

$$\underset{\sim}{r}'=\underset{\sim}{r}+u\underset{\sim}{m}+(\underset{\sim}{Q}_1-\underset{\sim}{I})(\underset{\sim}{r}-\underset{\sim}{q}) \qquad (5.3.11)$$

Substitution of (5.3.11) into (5.3.10) yields, after cancellations and rearrangement of terms,

$$\underset{\sim}{r}''=\underset{\sim}{r}+w\underset{\sim}{f}+u\underset{\sim}{Q}_2\underset{\sim}{m}+(\underset{\sim}{Q}_3-\underset{\sim}{I})\underset{\sim}{r}-(\underset{\sim}{Q}_2-\underset{\sim}{I})\underset{\sim}{g}-\underset{\sim}{Q}_3\underset{\sim}{q}+\underset{\sim}{Q}_2\underset{\sim}{q} \qquad (5.3.12)$$

in which $\underset{\sim}{Q}_3=\underset{\sim}{Q}_2\underset{\sim}{Q}_1$, has been substituted.

On the other hand, using the equivalent screw to compute $\underset{\sim}{r}''$,

$$\underset{\sim}{r}''=\underset{\sim}{r}+t\underset{\sim}{s}+(\underset{\sim}{Q}_3-\underset{\sim}{I})(\underset{\sim}{r}-\underset{\sim}{a}) \qquad (5.3.13)$$

Equating the right hand sides of (5.3.12) and (5.3.13), one obtains

$$w\underset{\sim}{f}+u\underset{\sim}{Q}_2\underset{\sim}{m}-t\underset{\sim}{s}+(\underset{\sim}{Q}_3-\underset{\sim}{I})(\underset{\sim}{a}-\underset{\sim}{q})+(\underset{\sim}{Q}_2-\underset{\sim}{I})(\underset{\sim}{q}-\underset{\sim}{g})=\underset{\sim}{0} \qquad (5.3.14)$$

from (5.13,p.5),

$$\underset{\sim}{Q}_2\underset{\sim}{m}=\underset{\sim}{m}+(1-c\gamma)\underset{\sim}{f}\times(\underset{\sim}{f}\times\underset{\sim}{m})+s\gamma\underset{\sim}{f}\times\underset{\sim}{m} \qquad (5.3.15)^*$$

---

*$c\gamma\equiv\cos\gamma$, $s\gamma\equiv\sin\gamma$

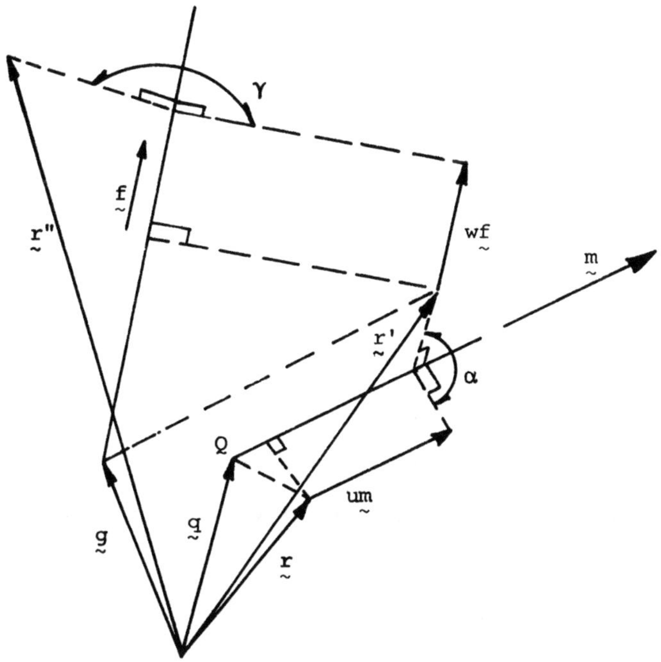

Fig 5.3.2 Composition of screws $\underset{\sim}{M}$ and $\underset{\sim}{F}$

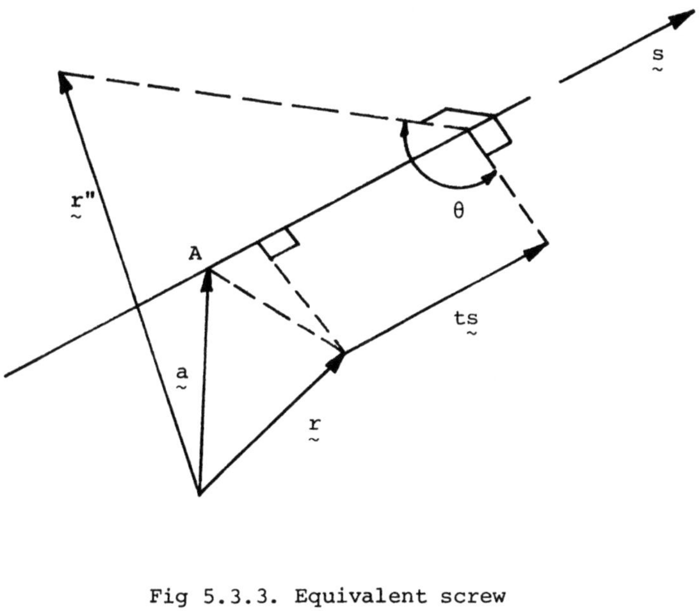

Fig 5.3.3. Equivalent screw

$$(\underset{\sim}{Q}_2-\underset{\sim}{I})(\underset{\sim}{q}-\underset{\sim}{g})=(1-c\gamma)\underset{\sim}{f}x\left(\underset{\sim}{f}x(\underset{\sim}{q}-\underset{\sim}{g})\right)+s\gamma\underset{\sim}{f}x(\underset{\sim}{q}-\underset{\sim}{g}) \qquad (5.3.16)$$

$$(\underset{\sim}{Q}_2-\underset{\sim}{I})(\underset{\sim}{a}-\underset{\sim}{q})=(1-c\theta)\underset{\sim}{s}x\left(\underset{\sim}{s}x(\underset{\sim}{a}-\underset{\sim}{q})\right)+s\theta\underset{\sim}{s}x(\underset{\sim}{a}-\underset{\sim}{q}) \qquad (5.3.17)$$

Substitution of (5.3.15)-(5.3.17) into (5.3.14) yields

$$w\underset{\sim}{f}-t\underset{\sim}{s}+u\underset{\sim}{m}+u(1-c\gamma)\underset{\sim}{f}x(\underset{\sim}{f}x\underset{\sim}{m})+us\gamma\underset{\sim}{f}x\underset{\sim}{m}+(1-c\theta)\underset{\sim}{s}x\left(\underset{\sim}{s}x(\underset{\sim}{a}-\underset{\sim}{q})\right)+s\theta\underset{\sim}{s}x(\underset{\sim}{a}-\underset{\sim}{q})$$
$$+(1-c\gamma)\underset{\sim}{f}x\left(\underset{\sim}{f}x(\underset{\sim}{q}-\underset{\sim}{g})\right)+s\gamma\underset{\sim}{f}x(\underset{\sim}{q}-\underset{\sim}{g})=\underset{\sim}{0} \qquad (5.3.18)$$

Multiplying both sides of eq. (5.3.18) times $x\underset{\sim}{f}.\underset{\sim}{s}$ yields

$$u\underset{\sim}{m}x\underset{\sim}{f}.\underset{\sim}{s}+u(1-c\gamma)\left(\underset{\sim}{f}x(\underset{\sim}{f}x\underset{\sim}{m})\right)x\underset{\sim}{f}.\underset{\sim}{s}+us\gamma(\underset{\sim}{f}x\underset{\sim}{m})x\underset{\sim}{f}.\underset{\sim}{s}+$$
$$+(1-c\theta)\{\underset{\sim}{s}x\left(\underset{\sim}{s}x(\underset{\sim}{a}-\underset{\sim}{q})\right)\}x\underset{\sim}{f}.\underset{\sim}{s}+s\theta\left(\underset{\sim}{s}x(\underset{\sim}{a}-\underset{\sim}{q})\right)x\underset{\sim}{f}.\underset{\sim}{s}+$$
$$+(1-c\gamma)\{\underset{\sim}{f}x\left(\underset{\sim}{f}x(\underset{\sim}{q}-\underset{\sim}{g})\right)\}x\underset{\sim}{f}.\underset{\sim}{s}+s\gamma\left(\underset{\sim}{f}x(\underset{\sim}{q}-\underset{\sim}{g})\right)x\underset{\sim}{f}.\underset{\sim}{s}=0 \qquad (5.3.19)$$

Define

$$p_1=\underset{\sim}{m}x\underset{\sim}{f}.\underset{\sim}{s};\ p_2=(\underset{\sim}{m}x\underset{\sim}{f}).(\underset{\sim}{f}x\underset{\sim}{s});\ p_3=(\underset{\sim}{f}x\underset{\sim}{s}).(\underset{\sim}{s}x\underset{\sim}{m}) \qquad (5.3.20)$$

From eq. (5.3.4), the corresponding trigonometric identities and definitions (5.3.20), one obtains

$$c\gamma=\frac{p_2^2-p_1^2}{p_1^2+p_2^2},\quad 1-c\gamma=\frac{2p_1^2}{p_1^2+p_2^2},\quad s\gamma=\frac{2p_1p_2}{p_1^2+p_2^2} \qquad (5.3.21)$$

$$c\theta=\frac{p_3^2-p_1^2}{p_1^2+p_3^2},\quad 1-c\theta=\frac{2p_1^2}{p_1^2+p_3^2},\quad s\theta=\frac{2p_1p_3}{p_1^2+p_3^2} \qquad (5.3.22)$$

Substituting eqs. (5.3.20)-(5.3.22) into eq. (5.3.19),

$$u\{p_1+\frac{2p_1^2}{p_1^2+p_2^2}\left(\underset{\sim}{f}x(\underset{\sim}{f}x\underset{\sim}{m})\right)x\underset{\sim}{f}.\underset{\sim}{s}+\frac{2p_1p_2}{p_1^2+p_2^2}(\underset{\sim}{f}x\underset{\sim}{m})x\underset{\sim}{f}.\underset{\sim}{s}+$$
$$+\frac{2p_1^2}{p_1^2+p_2^2}\{\underset{\sim}{f}x\left(fx(\underset{\sim}{q}-\underset{\sim}{g})\right)\}x\underset{\sim}{f}.\underset{\sim}{s}+\frac{2p_1p_2}{p_1^2+p_2^2}\left(\underset{\sim}{f}x(\underset{\sim}{q}-\underset{\sim}{g})\right)x\underset{\sim}{f}.\underset{\sim}{s}+$$
$$+\frac{2p_1^2}{p_1^2+p_3^2}\{\underset{\sim}{s}x\left(\underset{\sim}{s}x(\underset{\sim}{a}-\underset{\sim}{q})\right)\}x\underset{\sim}{f}.\underset{\sim}{s}-\frac{2p_1p_3}{p_1^2+p_3^2}\left(\underset{\sim}{s}x(\underset{\sim}{a}-\underset{\sim}{q})\right)x\underset{\sim}{f}.\underset{\sim}{s}=0 \qquad (5.3.23)$$

where

$$\left(\underset{\sim}{f}\times(\underset{\sim}{f}\times\underset{\sim}{m})\right)\times\underset{\sim}{f}\cdot\underset{\sim}{s}=-\underset{\sim}{m}\times\underset{\sim}{f}\cdot\underset{\sim}{s}=-p_1 \qquad (5.3.24)$$

and

$$(\underset{\sim}{f}\times\underset{\sim}{m})\times\underset{\sim}{f}\cdot\underset{\sim}{s}=(\underset{\sim}{f}\times\underset{\sim}{m})\cdot(\underset{\sim}{f}\times\underset{\sim}{s})=-p_2 \qquad (5.3.25)$$

Let A be the coefficient of u. Then,

$$A=p_1-\frac{2p_1^3}{p_1^2+p_2^2}-\frac{2p_1p_2^2}{p_1^2+p_2^2}=p_1\left(1-\frac{2p_1^2}{p_1^2+p_2^2}-\frac{2p_2^2}{p_1^2+p_2^2}\right)=$$

$$=p_1\,\frac{p_1^2+p_2^2-2(p_1^2+p_2^2)}{p_1^2+p_2^2}=-p_1 \qquad (5.3.26)$$

Let B be the sum of the next two terms in (5.3.23). Then

$$B=\frac{2p_1}{p_1^2+p_2^2}\left(p_1\underset{\sim}{s}\times\underset{\sim}{f}+p_2(\underset{\sim}{f}\times\underset{\sim}{s})\times\underset{\sim}{f}\right)\cdot(\underset{\sim}{q}-\underset{\sim}{g}) \qquad (5.3.27)$$

Let C be sum of the last two terms in (5.3.23),

$$C=-\frac{2p_1}{p_1^2+p_3^2}\left(p_1\underset{\sim}{f}\times\underset{\sim}{s}+p_3(\underset{\sim}{f}\times\underset{\sim}{s})\times\underset{\sim}{s}\right)\cdot\;(\underset{\sim}{a}-\underset{\sim}{g}) \qquad (5.3.28)$$

Substituting (5.3.26)-(5.3.28) into eq. (5.3.23), dropping the common factor p_1 in it and solving for u/2, one obtains

$$\frac{u}{2}=\underset{\sim}{c}_1\cdot(\underset{\sim}{q}-\underset{\sim}{g})-\underset{\sim}{c}_2\cdot(\underset{\sim}{q}-\underset{\sim}{q}) \qquad (5.3.29)$$

where

$$\underset{\sim}{c}_1=\frac{1}{p_1^2+p_2^2}\left(p_1\underset{\sim}{s}\times\underset{\sim}{f}+p_2(\underset{\sim}{f}\times\underset{\sim}{s})\times\underset{\sim}{f}\right),\;\underset{\sim}{c}_2=\frac{1}{p_1^2+p_3^2}\left(p_1\underset{\sim}{f}\times\underset{\sim}{s}+p_3(\underset{\sim}{f}\times\underset{\sim}{s})\times\underset{\sim}{s}\right) \qquad (5.3.30)$$

To simplify expression (5.3.29), expand the terms in the denominators and those in the brackets of expressions (5.3.30).

$$p_1^2+p_2^2=(\underset{\sim}{m}\times\underset{\sim}{f}\cdot\underset{\sim}{s})^2+\left((\underset{\sim}{m}\times\underset{\sim}{f})\cdot(\underset{\sim}{f}\times\underset{\sim}{s})\right)^2=(\underset{\sim}{m}\times\underset{\sim}{f}\cdot\underset{\sim}{s})^2+||\underset{\sim}{m}\times\underset{\sim}{f}||^2||\underset{\sim}{f}\times\underset{\sim}{s}||^2\cos^2(\underset{\sim}{m}\times\underset{\sim}{f},\underset{\sim}{f}\times\underset{\sim}{s}) \qquad (5.3.31)$$

But

$$\cos^2(\mathbf{m}\times\mathbf{f},\mathbf{f}\times\mathbf{s})=1-\sin^2(\mathbf{m}\times\mathbf{f},\mathbf{f}\times\mathbf{s})=1-\frac{||(\mathbf{m}\times\mathbf{f})\times(\mathbf{x}\mathbf{s})||^2}{||\mathbf{m}\times\mathbf{f}||^2||\mathbf{f}\times\mathbf{s}||^2} \tag{5.3.32}$$

Substituting expression (5.3.32) into (5.3.31), one obtains

$$p_1^2+p_2^2=(\mathbf{m}\times\mathbf{f}.\mathbf{s})^2+||\mathbf{m}\times\mathbf{f}||^2||\mathbf{f}\times\mathbf{s}||^2-||(\mathbf{m}\times\mathbf{f})\times(\mathbf{f}\times\mathbf{s})||^2 \tag{5.3.33}$$

where

$$||\mathbf{m}\times\mathbf{f}||^2=(\mathbf{m}\times\mathbf{f}).(\mathbf{m}\times\mathbf{f})=(\mathbf{m}\times\mathbf{f})\times\mathbf{m}.\mathbf{f}=(\mathbf{f}-\mathbf{f}.\mathbf{m}\mathbf{m}).\mathbf{f}=1-(\mathbf{m}.\mathbf{f})^2 \tag{5.3.34}$$

and, similarly,

$$||\mathbf{f}\times\mathbf{s}||^2=1-(\mathbf{s}.\mathbf{f})^2 \tag{5.3.35}$$

Furthermore,

$$\begin{aligned} ||(\mathbf{m}\times\mathbf{f})\times(\mathbf{f}\times\mathbf{s})||^2 &= ||(\mathbf{m}\times\mathbf{f}).\mathbf{s}\mathbf{f}-(\mathbf{m}\times\mathbf{f}).\mathbf{f}\mathbf{s}||^2 \\ &= ||(\mathbf{m}\times\mathbf{f}.\mathbf{s})\mathbf{f}|| \\ &= (\mathbf{m}\times\mathbf{f}.\mathbf{s})^2||\mathbf{f}||^2=(\mathbf{m}\times\mathbf{f}.\mathbf{s})^2 \end{aligned} \tag{5.3.36}$$

Substitution of eqs. (5.3.34)-(5.3.36) into eq. (5.3.33) leads to

$$p_1^2+p_2^2=\left(1-(\mathbf{m}.\mathbf{f})^2\right)\left(1-(\mathbf{s}.\mathbf{f})^2\right) \tag{5.3.37}$$

and, similarly

$$p_1^2+p_3^2=\left(1-(\mathbf{s}.\mathbf{f})^2\right)\left(1-(\mathbf{s}.\mathbf{m})^2\right) \tag{5.3.38}$$

Moreover,

$$p_1\mathbf{s}\times\mathbf{f}+p_2(\mathbf{f}\times\mathbf{s})\times\mathbf{f}=p_1\mathbf{s}\times\mathbf{f}+p_2(\mathbf{s}-\mathbf{s}.\mathbf{f}\mathbf{f})=\mathbf{m}\times\mathbf{f}.\mathbf{s}\mathbf{s}\times\mathbf{f}+(\mathbf{m}\times\mathbf{f}).(\mathbf{f}\times\mathbf{s})\left(\mathbf{s}-\mathbf{s}.\mathbf{f}\mathbf{f}\right)$$

$$=\mathbf{m}.\mathbf{f}\times\mathbf{s}\mathbf{s}\times\mathbf{f}+\mathbf{m}.\mathbf{f}\times(\mathbf{f}\times\mathbf{s})\mathbf{s}.(\mathbf{1}-\mathbf{f}\mathbf{f})$$

$$=\mathbf{m}.\left(\mathbf{f}\times\mathbf{s}\mathbf{s}\times\mathbf{f}+(\mathbf{f}\mathbf{f}-\mathbf{1}).\mathbf{s}\mathbf{s}.(\mathbf{1}-\mathbf{f}\mathbf{f})\right) \qquad (5.3.39)^{*}$$

Each of the two terms in the brackets of expression (5.3.39) are next expanded

Let

$$\mathbf{f}\times\mathbf{s}=\mathbf{a},\ \mathbf{s}\times\mathbf{f}=\mathbf{b} \qquad (5.3.40)$$

Thus, the dyadic $\mathbf{f}\times\mathbf{s}\mathbf{s}\times\mathbf{f}$ can be written as**

$$\mathbf{f}\times\mathbf{s}\mathbf{s}\times\mathbf{f}=\mathbf{a}\mathbf{b}$$

---

* $\mathbf{1}$ is the identity dyadic, i.e. a dyadic that is isomorphic to the 3x3 identity matrix. Thus, in matrix notation, if the components of $\mathbf{f}$ are $f_1 f_2$ and f_3, then

$$\left(\mathbf{1}-\mathbf{f}\mathbf{f}\right)=\begin{pmatrix} 1-f_1^2 & -f_1f_2 & -f_1f_3 \\ -f_1f_2 & 1-f_2^2 & -f_1f_3 \\ -f_1^2f_3 & -f_2f_3 & 1-f_3^2 \end{pmatrix}$$

** For a short account on the algebra of dyadics see Appendix 1.

Introducing the usual index notation (5.15),

$$(\underset{\sim}{f}x\underset{\sim}{s}\underset{\sim}{s}x\underset{\sim}{f})_{ij}=a_i b_j \tag{5.3.41}$$

where, from definitions (5.3.40),

$$a_i=\varepsilon_{ik\ell}f_k s_\ell \ , \ b_j=\varepsilon_{jmn}s_m f_n \tag{5.3.42}$$

Hence

$$(\underset{\sim}{f}x\underset{\sim}{s}\underset{\sim}{s}x\underset{\sim}{f})_{ij}=\varepsilon_{ik\ell}\varepsilon_{jmm}f_k s_\ell s_m f_n \tag{5.3.43}$$

where, as is shown in Appendix 3,

$$\varepsilon_{ik\ell}\varepsilon_{jmn}=\delta_{ij}(\delta_{km}\delta_{\ell n}-\delta_{kn}\delta_{\ell m})+\delta_{jk}(\delta_{m\ell}\delta_{in}-\delta_{im}\delta_{\ell n})$$

$$+\delta_{j\ell}(\delta_{in}\delta_{kn}-\delta_{km}\delta_{in}) \tag{5.3.44}$$

Substituting (5.3.44) into (5.3.43), one obtains finally

$$\underset{\sim}{f}x\underset{\sim}{s}\underset{\sim}{s}x\underset{\sim}{f}=\underset{\sim}{1}\left((\underset{\sim}{f}.\underset{\sim}{s})^2-1\right)+\underset{\sim}{f}\underset{\sim}{f}+\underset{\sim}{s}\underset{\sim}{s}-\underset{\sim}{s}.\underset{\sim}{f}(\underset{\sim}{s}\underset{\sim}{f}+\underset{\sim}{f}\underset{\sim}{s}) \tag{5.3.45}$$

Substitution of the latter expression into (5.3.39) and simplifying the resulting expression leads to

$$p_1\underset{\sim}{s}x\underset{\sim}{f}+p_2(\underset{\sim}{f}x\underset{\sim}{s})x\underset{\sim}{f}=\underset{\sim}{m}.\left(1-(\underset{\sim}{f}.\underset{\sim}{s})^2\right)(\underset{\sim}{f}\underset{\sim}{f}-\underset{\sim}{1}) \tag{5.3.46}$$

Similarly,

$$p_1\underset{\sim}{f}x\underset{\sim}{s}+p_3(\underset{\sim}{f}x\underset{\sim}{s})x\underset{\sim}{s}=\underset{\sim}{m}.\left(1-(\underset{\sim}{f}.\underset{\sim}{s})^2\right)(\underset{\sim}{1}-\underset{\sim}{s}\underset{\sim}{s}) \tag{5.3.47}$$

Substituting (5.3.37), (5.3.38), (5.3.46) and (5.3.47) into (5.3.30) and the corresponding expressions for $\underset{\sim}{c}_1$ and $\underset{\sim}{c}_2$ into (5.3.29), one obtains.

$$\frac{u}{2}=\frac{\underset{\sim}{m}.(\underset{\sim}{1}-\underset{\sim}{f}\underset{\sim}{f})}{1-(\underset{\sim}{m}.\underset{\sim}{f})^2}\ .\ (\underset{\sim}{g}-\underset{\sim}{q})+\frac{\underset{\sim}{m}.(\underset{\sim}{1}-\underset{\sim}{s}\underset{\sim}{s})}{1-(\underset{\sim}{s}.\underset{\sim}{m})^2}\ .\ (\underset{\sim}{a}-\underset{\sim}{q}) \tag{5.3.48}$$

which is identical to the corresponding expression obtained by Roth and Tsai (5.11).

Multiplying eq. (5.3.14) times $\underset{\sim}{Q}_2^T$ one obtains

$$w\tilde{f}+u\tilde{m}-t\tilde{Q}_2^T\tilde{s}+(\tilde{Q}_1-\tilde{Q}_2^T)(\tilde{a}-\tilde{q})-(\tilde{Q}_2^T-\tilde{I})(\tilde{q}-\tilde{g})=\tilde{0} \qquad (5.3.49)$$

Multiplying the latter equation times $x\tilde{f}.\tilde{m}$ and proceeding in a manner similar to that leading to eq. (5.3.48) starting from eq. (5.3.15), one obtains

$$\frac{t}{2}=-\frac{\tilde{s}-(\tilde{s}.\tilde{m})\tilde{m}}{1-(\tilde{s}.\tilde{m})^2}.(\tilde{q}-\tilde{a})+\frac{\tilde{s}-(\tilde{s}.\tilde{f})\tilde{f}}{1-(\tilde{s}.\tilde{f})^2}.(\tilde{g}-\tilde{a}) \qquad (5.3.50)$$

Now multiplying eq. (5.3.14) times $\tilde{Q}_3^T$ leads to

$$w\tilde{Q}_3^T\tilde{f}+u\tilde{m}-t\tilde{s}-(\tilde{Q}_3^T-\tilde{I})(\tilde{a}-\tilde{q})+(\tilde{Q}_1^T-\tilde{Q}_3^T)(\tilde{q}-\tilde{g})=\tilde{0} \qquad (5.3.51)$$

Finally, multiplying eq. (5.3.51) times $x\tilde{m}.\tilde{s}$ and proceeding as before, one obtains

$$\frac{w}{2}=\frac{\tilde{f}-(\tilde{f}.\tilde{s})\tilde{s}}{1-(\tilde{f}.\tilde{s})^2}\cdot(\tilde{a}-\tilde{g})-\frac{\tilde{f}-(\tilde{f}.\tilde{m})\tilde{m}}{1-(\tilde{f}.\tilde{m})^2}\cdot(\tilde{q}-\tilde{g}) \qquad (5.3.52)$$

Equations (5.3.4), (5.3.7), (5.3.9), (5.3.48), (5.3.50) and (5.3.52) are the <u>synthesis equations</u> which were meant to be obtained.

<u>Remarks about the synthesis equations</u>:

i) They are useful to synthesise spatial linkages, but not plane ones, because in the latter case, $\tilde{f}$, $\tilde{m}$ and $\tilde{s}$ are parallel. Hence angles α, γ and θ are undetermined in eqs. (5.3.4), (5.3.7) and (5.3.9) and displacements u, t and w are undertermined in eqs. (5.3.48), (5.3.50) and (5.3.52). Thus, to synthesise plane linkages, other equations should be used, like those developed by Suh [5.16] or by Angeles [5.17]. Equations appearing in [5.17] are derived in the Appendix.

ii) They enable the designer to synthesise dyads of any combination of the six kinematic lower pairs introduced in Ch. 3, except for the planar one. Notice, however, that since the axis of one of the two pairs of the dyad is fixed and the other one is moving, the dyad is not symmetric, for which reason the R-S dyad, for instance, is different from the S-R dyad. Hence, the total number of dyads that can be designed with the foregoing equations is $5^2=25$. The syntheses of all these dyads, except for the P-P one, are discussed in (5.11). As Roth and Tsai point out in (5.11), unless the guided rigid body undergoes pure translation, in general a P-P dyad does not exist for an arbitrary rigid body motion.

iii) If the different configurations of a rigid body meant to be guided are specified, not via their screws (referred to a common original configuration), but via the successive positions of a set of three non-collinear points of the rigid body, then the corresponding screws must first be computed. This can be done with the computer subroutine SCREW, whose listing appears in Fig 2.6.5.

iv) The path-generation problem of synthesis, discussed in Section 5.5, can also be treated using the equations under consideration.

The outlined synthesis procedure is illustrated by means of an example regarding the design of an R-R dyad for rigid-body guidance. The synthesis of this dyad has been studied extensively. It was first shown that the maximum number of specified configurations of the rigid body is three (See Section 5.4), if exact solutions are to be obtained, these solutions being exact up to round-off and/or measuring errors. This statement can be readily proved by the reader; besides, it is reported in several papers, e.g. in (5.11,p.94). Roth (5.18) showed that the aforementioned exact synthesis problem has no more than 24 real solutions, whereas Suh (5.19) showed that these solutions always come in pairs, each pair forming a Bennett mechanism (5.20), i.e. an RRRR spatial linkage with degree of freedom 1*. Finally, Roth and Tsai (5.12) showed that this problem has only one pair of real solutions, which constitute a Bennett mechanism.

Example 5.3.1 Synthesise an RRRR spatial linkage to guide a rigid body whose points A,B and C attain the following successive positions:

$A_0(0,0,1)$ $A_1(0,1,1)$ $A_2(-1,1,1)$

$B_0(0,1,0)$ $B_1(-1,1,0)$ $B_2(-1,2,0)$

$C_0(1,0,0)$ $C_1(0,0,0)$ $C_2(-2,1,0)$

The two involved screws, $\underset{\sim}{S}_1$ and $\underset{\sim}{S}_2$, transporting the rigid body from configuration 0 to configurations 1 and 2, respectively were obtained with the aid of SUBROUTINE SCREW, the resulting parameters being those obtained from the computer printout of the program that was written for this purpose. These are the following:

$$\underset{\sim}{s}_1=\begin{Bmatrix} 0.707 \\ -0.707 \\ 0.000 \end{Bmatrix},\ \underset{\sim}{a}_1=\begin{Bmatrix} 0.250 \\ 0.250 \\ 0.500 \end{Bmatrix}, t_1=-0.707, \theta_1=-70.529°$$

$$\underset{\sim}{s}_2=\begin{Bmatrix} 0.000 \\ -0.707 \\ 0.707 \end{Bmatrix}\ \ \underset{\sim}{a}_2=\begin{Bmatrix} -0.250 \\ 1.000 \\ 1.000 \end{Bmatrix}, t_2=-0.707, \theta_2=-109.471°$$

In this problem, θ_j, $\underset{\sim}{s}_j$, $\underset{\sim}{a}_j$, u_j and $w_j (j=1,2)$ are known. In fact, $u_j=w_j=0 (j=1,2)$, for the corresponding motions are produced by revolute pairs, thereby not allowing for any sliding; the remaining aforementioned parameters are those obtained via SUBROUTINE SCREW. The synthesis equations that are applied are eqs. (5.3.9), (5.3.48), (5.3.50) and (5.3.52), all of

* If the Grübler-Kutzbach (3.5) formula is applied to this linkage, it is found that its degree of freedom is -2.

them taken twice, once for each value of j.

Additionally, the unity-magnitude (i.e. normality) condition on vectors $\underset{\sim}{f}$ and $\underset{\sim}{m}$ yields two more equations. Finally, specifying the location of points G and Q, on the axes of the $\underset{\sim}{F}$ and $\underset{\sim}{M}$ screws, respectively, in such a way as to render GQ perpendicular to both axes, one obtains two additional equations. Summarizing, the problem leads to a system of 12 nonlinear algebraic equations in 12 unknowns. Roth and Tsai (5.12) introduced an algorithm that reduces the problem to finding the unique positive real root of a cubic equation. In the course of the description of their algorithm, they show that the system admits exactly two real different solutions which, as Suh (5.19) proved before, are bound to constitute a Bennett mechanism. Roth and Tsai also show that spurious solutions to the said system also appear, which contain, as solutions to $\underset{\sim}{f}$ or $\underset{\sim}{m}$, either $\underset{\sim}{s}_1$ or $\underset{\sim}{s}_2$. The author has found, following a different approach, that other spurious solutions can also appear, these solutions being inadmissible in the sense of yielding, for instance, point Q of the second solution identical to point G of the first one, thus making impossible the construction of the linkage.

The solution to the present problem is now obtained following a different procedure as that proposed by Roth and Tsai. In fact, although the algorithm proposed by these authors can be implemented even with a desk calculator (not necessarily a programmable one), the author could not see how to extend it to overdetermined problems, i.e. problems that involve the guidance of a rigid body through more than three different configurations, thus not allowing for an exact solution, in which case the best approximation (See Section 6.3) should be sought. The procedure followed here is based on the use of SUBROUTINE NRDAMP. First the synthesis equations are simplified considering that vector $\underset{\sim}{g}-\underset{\sim}{q}$ is perpendicular to both $\underset{\sim}{f}$ and $\underset{\sim}{m}$. Thus, the synthesis equations reduce to

For i=1,2:

$$F_i=\tan\frac{\theta_i}{2}\underset{\sim}{f}^T(\underset{\sim}{I}-\underset{\sim}{s}_i\underset{\sim}{s}_i^T)\underset{\sim}{m}+\det(\underset{\sim}{f},\underset{\sim}{m},\underset{\sim}{s}_i)=0 \tag{5.3.53a}$$

$$F_{i+2}=\underset{\sim}{f}^T(\underset{\sim}{I}-\underset{\sim}{s}_i\underset{\sim}{s}_i^T)(\underset{\sim}{a}_i-\underset{\sim}{g})=0 \tag{5.3.53b}$$

$$F_{i+4}=\underset{\sim}{m}^T(\underset{\sim}{I}-\underset{\sim}{s}_i\underset{\sim}{s}_i^T)(\underset{\sim}{a}_i-\underset{\sim}{q})=0 \tag{5.3.53c}$$

$$F_{i+6}=\underset{\sim}{s}_i^T(\underset{\sim}{g}-\underset{\sim}{q})-\frac{t_i}{2}=0 \tag{5.3.53d}$$

$$F_9=\underset{\sim}{f}^T\underset{\sim}{f}-1=0 \tag{5.3.53e}$$

$$F_{10}=\underset{\sim}{m}^T\underset{\sim}{m}-1=0 \tag{5.3.53f}$$

$$F_{11}=\underset{\sim}{f}^T(\underset{\sim}{g}-\underset{\sim}{q})=0 \tag{5.3.53g}$$

$$F_{12}=\underset{\sim}{m}^T(\underset{\sim}{g}-\underset{\sim}{q})=0 \tag{5.3.53h}$$

The Jacobian matrix of the system, which is needed by NRDAMP, is now formed by computing the partial derivatives of the above functions with respect to the unknowns $\underset{\sim}{f},\underset{\sim}{m},\underset{\sim}{g}$ and $\underset{\sim}{q}$. Thus,

For i=1,2:

$$\frac{\partial F_i}{\partial \underset{\sim}{f}} = \tan\frac{\theta_i}{2}(\underset{\sim}{I}-\underset{\sim}{s}_i\underset{\sim}{s}_i^T)\underset{\sim}{m}+\underset{\sim}{m}\times\underset{\sim}{s}_i \tag{5.3.54a}$$

$$\frac{\partial F_i}{\partial \underset{\sim}{m}} = \tan\frac{\theta_i}{2}(\underset{\sim}{I}-\underset{\sim}{s}_i\underset{\sim}{s}_i^T)\underset{\sim}{f}-\underset{\sim}{f}\times\underset{\sim}{s}_i \tag{5.3.54b}$$

$$\frac{\partial F_i}{\partial \underset{\sim}{g}} = \underset{\sim}{0},\ \frac{\partial F_i}{\partial \underset{\sim}{q}} = \underset{\sim}{0} \tag{5.3.54c}$$

$$\frac{\partial F_{i+2}}{\partial \underset{\sim}{f}} = (\underset{\sim}{I}-\underset{\sim}{s}_i\underset{\sim}{s}_i^T)(\underset{\sim}{a}_i-\underset{\sim}{g}),\ \frac{\partial F_{i+2}}{\partial \underset{\sim}{m}} = \underset{\sim}{0} \tag{5.3.54d}$$

$$\frac{\partial F_{i+2}}{\partial g} = -(I-s_i s_i^T)f, \quad \frac{\partial F_{i+2}}{\partial q} = 0 \tag{5.3.54e}$$

$$\frac{\partial F_{i+4}}{\partial f} = 0, \quad \frac{\partial F_{i+4}}{\partial m} = (I-s_i s_i^T)(a_i-q) \tag{5.3.54f}$$

$$\frac{\partial F_{i+4}}{\partial g} = 0, \quad \frac{\partial F_{i+4}}{\partial q} = -(I-s_i s_i^T)m \tag{5.3.54g}$$

$$\frac{\partial F_{i+6}}{\partial f} = 0, \quad \frac{\partial F_{i+6}}{\partial m} = 0, \quad \frac{\partial F_{i+6}}{\partial g} = s_i, \quad \frac{\partial F_{i+6}}{\partial q} = -s_i \tag{5.3.54h}$$

$$\frac{\partial F_9}{\partial f} = 2f, \quad \frac{\partial F_9}{\partial m} = 0, \quad \frac{\partial F_9}{\partial g} = 0, \quad \frac{\partial F_9}{\partial q} = 0 \tag{5.3.54i}$$

$$\frac{\partial F_{10}}{\partial f} = 0, \quad \frac{\partial F_{10}}{\partial m} = 2m, \quad \frac{\partial F_{10}}{\partial g} = 0, \quad \frac{\partial F_{10}}{\partial q} = 0 \tag{5.3.54j}$$

$$\frac{\partial F_{11}}{\partial f} = g-q, \quad \frac{\partial F_{11}}{\partial m} = 0, \quad \frac{\partial F_{11}}{\partial g} = f, \quad \frac{\partial F_{11}}{\partial q} = -f \tag{5.3.54k}$$

$$\frac{\partial F_{12}}{\partial f} = 0, \quad \frac{\partial F_{12}}{\partial m} = g-q, \quad \frac{\partial F_{12}}{\partial g} = m, \quad \frac{\partial F_{12}}{\partial q} = -m \tag{5.3.54l}$$

Subroutines FUN and DFDX, required by NRDAMP, compute functions F_i (i=1,...12) and the Jacobian matrix. For this purpose, a vector x of twelve components is defined in the following manner:

$x_i=f_i$, i=1,2,3

$x_{i+3}=m_i$, i=1,2,3

$x_{i+6}=g_i$, i=1,2,3

$x_{i+9}=q_1$, i=1,2,3

Analogously, a vector p of given parameters is defined as

For i=1,2,3,

$$p_i = s_{1i}$$

$$p_{i+3} = s_{2i}$$

$$p_{i+6} = a_{1i}$$

$$p_{i+9} = a_{2i}$$

for $i=1,2,$

$$p_{i+12} = \theta_i$$

$$p_{i+14} = t_i$$

In order to automate the calculations, the following arrays are defined:

VE(I,J), for I=1,2,...,7 and J=1,2,3, is the J<u>th</u> component of the I<u>th</u> vector of the array, where

$\mathbf{f}$= 1<u>st</u> vector of the array

$\mathbf{m}$= 2<u>nd</u> vector of the array

$\mathbf{a}_1-\mathbf{g}$= 3<u>rd</u> vector of the array

$\mathbf{a}_2-\mathbf{g}$= 4<u>th</u> vector of the array

$\mathbf{a}_1-\mathbf{q}$= 5<u>th</u> vector of the array

$\mathbf{a}_2-\mathbf{q}$= 5<u>th</u> vector of the array

$\mathbf{g}-\mathbf{q}$ = 7<u>th</u> vector of the array

MA(I,J,K), for I=1,2, and J,K=1,2,3, is the (J,K)<u>th</u> entry of the I<u>th</u> matrix of the array, where

$\mathbf{I}-\mathbf{s}_i\mathbf{s}_i^T$= the i<u>th</u> matrix of the array.

Other subsidiary subroutines are the following:

SUBROUTINE MAVE (I,MA,J,VE,PROV) computes the product of matrix I of MA times vector J of VE and stores this product in array PROV.

SUBROUTINE VEVE(I,VE,V,PROS) computes the inner product of vector I of VE times V and stores this product in PROS.

```
      SUBROUTINE FUN(X,F,P,N)
      REAL PROS,X(13),P(16),F(12),PROV(3),CROS(3),V(3)
      REAL MA(2,3,3),VE(7,3)
      COMMON/ONE/MA
      COMMON/TWO/VE
      CALL VECTOR(P,X,VE)
      DO 1 I=1,2
           CALL MAVE(I,MA,2,VE,PROV)
           CALL VEVE(1,VE,PROV,PROS)
           F(I)=TAN(P(12+I)/2.)*PROS
           CALL CROSS(2,X,I,P,CROS)
           CALL VEVE(1,VE,CROS,PROS)
1     F(I)=F(I)+PROS
      L=2
      DO 2 I=1,2
           DO 2 J=1,2
                L = L+1
                CALL MAVE(J,MA,L,VE,PROV)
                CALL VEVE(I,VE,PROV,PROS)
2     F(L)=PROS
      DO 4 I=1,2
           DO 3 J=1,3
3          V(J)=P(3*(I-1)+J)
           CALL VEVE(7,VE,V,PROS)
4     F(I+6)=PROS-P(14+I)/2.
      DO 6 I=1,2
           DO 5 J=1,3
5          V(J)=VE(I,J)
           CALL VEVE(I,VE,V,PROS)
6     F(8+I)=PROS-1.
      DO 8 I=1,2
           DO 7 J=1,3
7          V(J)=VE(I,J)
           CALL VEVE(7,VE,V,PROS)
8     F(10+I)=PROS
      RETURN
      END
#
```

Fig. 5.3.4 Listing of SUBROUTINE FUN

```
      SUBROUTINE DFDX(X,DF,P,N)
      REAL X(13),P(16),PROV(3),V(3),CROS(3),
   &  DF(12,12),MA(2,3,3),VE(7,3)
      COMMON/ONE/MA
      COMMON/TWO/VE
      DO 1 I =1,2
           SG=-1.
           DO 1 J=1,2
                SG=-SG
                JJ=3-J
                CALL MAVE(I,MA,JJ,VE,PROV)
                CALL CROSS(JJ,X,I,P,CROS)
                DO 1 K=1,3
 1    DF(I,3*(J-1)+K)=TAN(P(I+12)/2.)*PROV(K)+SG*CROS(K)
      DO 2 I=1,2
           DO 2 J=7,12
 2    DF(I,J)=0.
      DO 3 I=1,2
           DO 3 J=1,2
                JJ=3*(2-J)
                DO 3 K=1,3
                     DF(I+J+J,JJ+K)=0.
 3    DF(I+J+J,JJ+K+6)=0.
      L=2
      DO 6 I=1,2
           DO 6 J=1,2
                L=L+1
                CALL MAVE(J,MA,L,VE,PROV)
                DO 4 K=1,3
 4              DF(L,K+3*(I-1))=PROV(K)
                CALL MAVE(J,MA,I,VE,PROV)
                DO 5 K=1,3
 5              DF(L,K+3*(I+1))=-PROV(K)
 6    CONTINUE
      DO 7 I=7,8
           DO 7 J=1,6
 7    DF(I,J)=0.
      SG=-1.
      DO 8 I=1,2
           DO 8 J=1,2
                SG=-SG
                DO 8 K=1,3
 8    DF(I+6,6+K+3*(J-1))=SG*P(3*(I-1)+K)
      DO 9 I=9,12
           DO 9 J=1,12
 9    DF(1,J)=0.
      DO 10 I=1,2
            DO 10 J=1,3
10    DF(I+8,3*(I-1)+J)=VE(I,J)+VE(I,J)
      DO 11 I=1,2
           DO 11 J=1,3
11    DF(I+10,J+3*(I-1))=VE(7,J)
```

Fig. 5.3.5 Listing of SUBROUTINE DFDX

```
      SG=-1.
      DO 12 I=1,2
            DO 12 J=1,2
                  SG=-SG
                  DO 12 K=1,3
   12 DF(I+10,3*(J+1)+K)=VE(I,K)*SG
      RETURN
      END
#
```

Fig. 5.3.5 Listing of SUBROUTINE DFDX (Continued)

```
      SUBROUTINE MAVE(I,MA,J,VE,PROV)
      REAL MA(2,3,3),VE(7,3),PROV(3)
      DO 1 K=1,3
           PROV(K)=0.
           DO 1 L=1,3
    1 PROV(K)=PROV(K)+MA(I,K,L)*VE(J,L)
      RETURN
      END
#
```

Fig. 5.3.6 Listing of SUBROUTINE MAVE

```
      SUBROUTINE VEVE(I,VE,V,PROS)
      REAL PROS,VE(7,3),V(3)
      PROS=0.
      DO 1 J=1,3
    1 PROS=PROS+VE(I,J)*V(J)
      RETURN
      END
#
```

Fig. 5.3.7 Listing of SUBROUTINE VEVE

```
      SUBROUTINE CROSS(I,X,J,P,CROS)
      REAL X(13),P(16),CROS(3)
      DO 1 K=1,3
           CROS(K)=0.
           DO 1 L=1,3
                DO 1 M=1,3
                     N=(L-K)*(M-L)*(K-M)
    1 CROS(K)=CROS(K)-N*X(3*(I-1)+L)*P(3*(J-1)+M)/2.
      RETURN
      END
#
```

Fig. 5.3.8 Listing of SUBROUTINE CROSS

SUBROUTINE CROSS (I,X,J,P,CROS) computes the cross product of either $\underset{\sim}{f}$(I=1) or $\underset{\sim}{m}$(I=2), stored in X, times either $\underset{\sim}{s}_1$(J=1) or $\underset{\sim}{s}_2$(J=2), both stored in P. It stores the product in array CROS. The listings of the aforementioned subroutines appear in Figs 5.3.4, 5.3.5, 5.3.6, 5.3.7 and 5.3.8.

The two meaningful solutions that were obtained, using the Newton-Raphson method, are

$$\underset{\sim}{f}_1 = \begin{pmatrix} 0 \\ 0 \\ -1 \end{pmatrix} \quad \underset{\sim}{m}_1 = \begin{pmatrix} +0.707 \\ 0.0 \\ -0.707 \end{pmatrix} \quad \underset{\sim}{q}_1 = \begin{pmatrix} -0.5 \\ 1.5 \\ 0.5 \end{pmatrix} \quad \underset{\sim}{q}_1 = \begin{pmatrix} -0.5 \\ 1.0 \\ 0.5 \end{pmatrix}$$

$$\underset{\sim}{f}_2 = \begin{pmatrix} -0.7071 \\ 0 \\ 0.7071 \end{pmatrix} \quad \underset{\sim}{m}_2 = \begin{pmatrix} -0.666 \\ -0.666 \\ 0.333 \end{pmatrix} \quad \underset{\sim}{q}_2 = \begin{pmatrix} -0.5 \\ 1.0 \\ 0.5 \end{pmatrix} \quad \underset{\sim}{q}_2 = \begin{pmatrix} -0.1666 \\ 0.8333 \\ 0.8333 \end{pmatrix}$$

The conditions for the above linkage to constitute a Bennett mechanism (5.20) are, according to Fig 5.3.9,

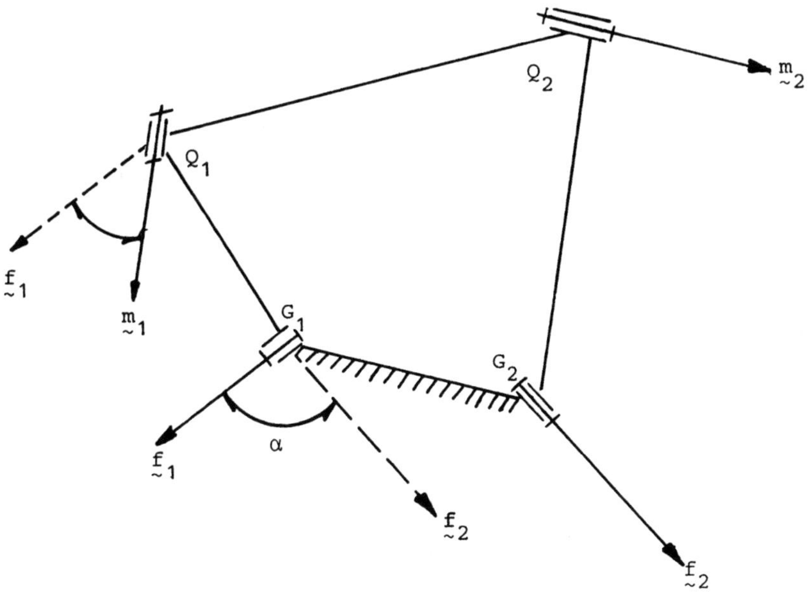

Fig 5.3.9 A Bennett Mechanism

$$\overline{G_1Q_1}=\overline{G_2Q_2}=a$$

$$\overline{G_1G_2}=\overline{Q_1Q_2}=b$$

$$|\underset{\sim}{f}_1^T\underset{\sim}{f}_2|=|\underset{\sim}{m}_1^T\underset{\sim}{m}_2|$$

$$|\underset{\sim}{f}_1^T\underset{\sim}{m}_1|=|\underset{\sim}{f}_2^T\underset{\sim}{m}_2|$$

$$\frac{\sin\alpha}{a}=\frac{\sin\beta}{b}$$

It can be readily proved that the linkage just designed satisfies the above conditions and thus constitutes a Bennett mechanism.

The algorithm of Roth and Tsai requires referring the problem to a particular set of coordinate axes, in such a way that, letting $\underset{\sim}{i},\underset{\sim}{j},\underset{\sim}{k}$ denote unit vectors along new X,Y and Z axes, respectively, these vectors are defined as

$$\underset{\sim}{k}=\underset{\sim}{s}_1$$

$$\underset{\sim}{j}=\frac{\underset{\sim}{s}_1\times\underset{\sim}{s}_2}{||\underset{\sim}{s}_1\times\underset{\sim}{s}_2||}$$

$$\underset{\sim}{i}=\underset{\sim}{j}\times\underset{\sim}{k}$$

Furthermore, the origin of coordinates is to be placed at the intersection of the common perpendicular to the axes of screws $\underset{\sim}{S}_1$ and $\underset{\sim}{S}_2$ with the axis of $\underset{\sim}{S}_1$. This way, in the new coordinates,

$$\underset{\sim}{s}_1=\begin{pmatrix}0\\0\\1\end{pmatrix},\underset{\sim}{a}_1=\begin{pmatrix}0\\0\\0\end{pmatrix},\underset{\sim}{s}_2=\begin{pmatrix}u\\0\\v\end{pmatrix},\underset{\sim}{a}_2=\begin{pmatrix}0\\h\\0\end{pmatrix}$$

where h is the distance between the axes of $\underset{\sim}{S}_1$ and $\underset{\sim}{S}_2$ times the sign of $\underset{\sim}{s}_1\times\underset{\sim}{s}_2\cdot(\underset{\sim}{a}_2-\underset{\sim}{a}_1)$. The solution obtained using the data in the new coordinate axes must be then transformed into the original coordinates. This is done via the following affine transformation:

$$(\underset{\sim}{f})_o=(\underset{\sim}{R})_o(\underset{\sim}{f})_n$$

$$(\mathbf{m})_o = (\mathbf{R})_o (\mathbf{m})_n$$

$$(\mathbf{g})_o = (\mathbf{a}_1)_o + (\mathbf{R})_o (\mathbf{g})_n$$

$$(\mathbf{g})_o = (\mathbf{a}_1)_o + (\mathbf{R})_o (\mathbf{g})_n$$

where subscripts o and n refer to original and new coordinates, respectively. Matrix $(\mathbf{R})_o$ can be obtained via Definition 1.2.1, as

$$(\mathbf{R})_o = \left(\mathbf{j}\times\mathbf{k} \,\middle|\, \frac{\mathbf{s}_1\times\mathbf{s}_2}{\|\mathbf{s}_1\times\mathbf{s}_2\|} \,\middle|\, \mathbf{s}_1\right)^T$$

where the transpose of the given matrix should be taken because the matrix taken as such (without transposing) represents the rotation of the original axes the new ones.

<u>Exercise 5.3.1</u> Check the solution given above to Example 5.3.1, applying Tsai and Roth's algorithm (5.12).

5.4 A DIFFERENT APPROACH TO THE SYNTHESIS PROBLEM FOR RIGID-BODY GUIDANCE.

For dyads containing any combination of revolute and spherical pairs, this problem can be formulated in an alternative way, which formally resembles the plane linkage synthesis problem (5.17) discussed in Appendix 4. Let $X_0Y_0Z_0$ be coordinate axes attached to the rigid body intended to guide, in its reference configuration, and $X_jY_jZ_j$ those attached to it in its jth configuration. Furthermore, let S_{12} and S_{23} be spherical pairs connecting the dyad to synthesize to the frame of the mechanism (linkage 1) and to the rigid body (linkage 3). Then, from Fig 5.4.1, the next relationship follows

$$\tilde{r}_j+\tilde{a}_j+\tilde{b}_j=\tilde{a}_0+\tilde{b}_o \quad (5.4.1)$$

$$\tilde{a}_j=\tilde{Q}_j\tilde{a}_0 \quad (5.4.2)$$

$\tilde{Q}_j$ being the rotation matrix of the screw carrying the rigid body from its reference to its jth configuration.

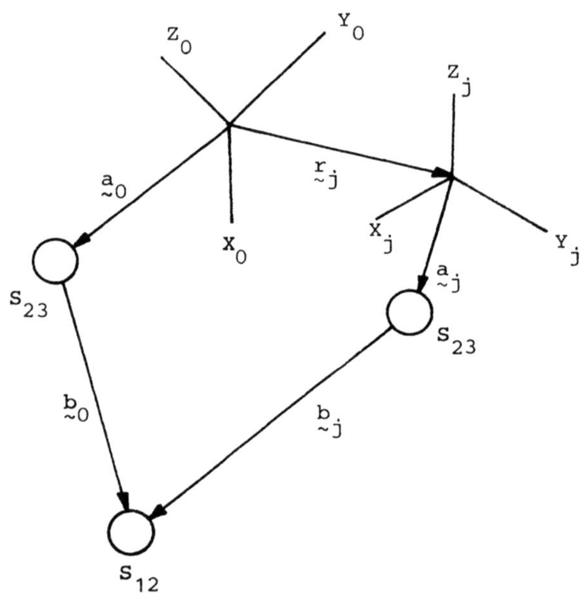

Fig 5.4.1 S-S dyad conducting a rigid body

Introducing eq. (5.4.2) into eq. (5.4.1) and solving for $\underset{\sim}{b}_j$ one obtains

$$\underset{\sim}{b}_j=(\underset{\sim}{I}-\underset{\approx}{Q}_j)\underset{\sim}{a}_0+\underset{\sim}{b}_0-\underset{\sim}{r}_j \qquad (5.4.3)$$

Since the dyad is rigid,

$$||\underset{\sim}{b}_j||=||\underset{\sim}{b}_0|| \qquad (5.4.4)$$

Substituting eq. (5.4.3) into eq. (5.4.4) and squaring both sides of the resulting equation yields

$$||(\underset{\sim}{I}-\underset{\approx}{Q}_j)\underset{\sim}{a}_0+\underset{\sim}{b}_0-\underset{\sim}{r}_j||^2=||\underset{\sim}{b}_0||^2, j=1,2,\ldots,n \qquad (5.4.5)$$

which constitute, in general, a nonlinear system of n synthesis equations to solve for the unknowns $\underset{\sim}{a}_0$ and $\underset{\sim}{b}_0$. Since these unknowns contain 6 scalar components, for exact synthesis the largest value that n can attain is 6, and hence, with an S-S dyad, a rigid body can be conducted through 7 configurations, counting the reference configuration as the seventh one.

Exercise 5.4.1 If it is intended to conduct the rigid body through 4 configurations only, then 3 unknowns must be specified to render the problem determined. These specified unknowns can be any combination of the six involved coordinates. Show that, if the three components of $\underset{\sim}{a}_0$ are specified, the system of equations becomes linear.

Regarding the four-configuration synthesis problem discussed in Exercise 5.4.1, if $\underset{\sim}{a}_0$ happens to be specified in such a way that for the kth configurations the next holds.

$$\underset{\sim}{r}_k=(\underset{\sim}{I}-\underset{\approx}{Q}_k)\underset{\sim}{a}_0 \qquad (5.4.6)$$

then the kth synthesis equation becomes an identity, thus leading to an underdetermined system of equations, which can be rendered determined by either specifying one fifth configuration or by specifying one of the components of $\underset{\sim}{b}_0$. An alternative way of handling this situation, amply discussed in Ch. 6, is to leave the system of equations undertermined but imposing some optimality condition like, for instance, to minimize $||\underset{\sim}{b}_0||$, thus enabling the designer to obtain the minimum-length dyad.

From the above discussion it is apparent that it is highly desirable to be able to specify $\underset{\sim}{a}_0$ so as to satisfy eq. (5.4.6). However, it is not possible to solve for $\underset{\sim}{a}_0$ from this equation, for the matrix involved is singular, as discussed in Section 2.6. In that section it was also mentioned that in a system of equations like the one "defining" $\underset{\sim}{a}_0$ in eq. (5.4.6), two of the three equations are linearly dependent. Hence, this system is underdetermined and, properly speaking, <u>does not</u> actually <u>define</u> $\underset{\sim}{a}_0$. This vector can, however, be defined by imposing the condition that its magnitude be a minimum, which can be done via the pseudoinverse matrix of the system of 2 (linearly independent) equations in three unknowns obtained from eq. (5.4.6).

<u>Exercise 5.4.2</u> Show that if $\underset{\sim}{r}_k$ is parallel to the axis of the screw carrying the rigid body from its reference configuration to its $k^{\underline{th}}$ configuration, then it is not possible to find a vector a_0 satisfying eq. (5.4.6).

If vector $\underset{\sim}{r}_k$ is not on the null space of $\underset{\sim}{I}-\underset{\sim}{Q}_k$ (Sect. 1.3), i.e., if it is not parallel to the axis of rotation of $\underset{\sim}{Q}_k$, then a minimum-norm vector $\underset{\sim}{a}_0$ can be obtained from the underdetermined system of the two linearly independent equations contained in eq. (5.4.6). This is a very useful result, because, as the length of vector $\underset{\sim}{a}_0$ is made the shortest possible, the weight and the dimensions of the linkage are minimized, thus allowing to construct a linkage with the smallest amount of material, driving it with the least possible amount of power and placing it in the narrowest space. More on linkage optimization will be discussed in Ch. 6

<u>Synthesis of R-S dyads.</u> In this case a configuration analogous to the one shown in Fig 5.4.1 is obtained, except that pair S_{12} is replaced by pair R_{12}. Similarly, for n+1 rigid-body configurations, the same synthesis equations (5.4.5) follow. Since this dyad can only rotate abount the axis

of R_{12}, suitable constraints must be introduced in order to insure the normality between vectors $\underset{\sim}{b}_j$ ($j=0,1,\ldots,n$) and the unit vector $\underset{\sim}{c}$, directed along the axis of R_{12}. This normality condition is equivalent to the coplanarity condition among all vectors $\underset{\sim}{b}_j$. Algebraically, the latter condition can be stated as

$$\det(\underset{\sim}{b}_0,\underset{\sim}{b}_1,\underset{\sim}{b}_s)=0,\ s=2,\ldots,n \tag{5.4.7}$$

thus obtaining n-1 additional synthesis equations. The number of synthesis parameters (unknowns) in this case is again six, i.e. the three scalar components of vectors $\underset{\sim}{a}_0$ and $\underset{\sim}{b}_0$. Hence, in order to match the number of equations to that of unknowns, one must have

$$2n-1=6$$

i.e.

$$n=\frac{7}{2}$$

which, unfortunately is not an integer.

From the foregoing discussion it follows that some parameters should be specified beforehand. If m parameters are specified, then n must have the value

$$n=\frac{7-m}{2}$$

Thus, letting $N=n+1$ be the number of prescribed rigid body configurations, the following table is obtained

m	N
1	4
3	3
5	2

It is pointed out, again, that if three parameters are specified, these can be chosen to be the three components of $\underset{\sim}{a}_0$ satisfying the two linearly independent equations of (5.4.6), for one k, such that the magnitude of

$\mathbf{a}_0$ is a minimum. If no constraint equations (5.4.7) were to be present in this case, this would allow one to prescribe one fourth configuration of the rigid body under guidance. The presence of the said equations, however, does not prevent the introduction of such an extra configuration, for, from equations (5.4.3) and (5.4.6), $\mathbf{b}_k=\mathbf{b}_0$, for that k for which (5.4.6) holds, the kth equation of (5.4.7) thereby being identically satisfied. In conclusion, the R-S dyad can be synthesised for a maximum of 4 prescribed configurations of a rigid body in such a way that $||\mathbf{a}_0||$ is minimised, therefore allowing for a minimum-weight synthesis. The latter result could also be concluded from the discussions of the same problem appearing in (5.11) and (5.21), where the authors show that the said rigid-body guidance problem leads to an underdetermined system of equations. The synthesis of S-R dyads is essentially the same as that of R-S dyads

Synthesis of R-R dyads. Again the picture shown in Fig 5.4.1 relating the reference-and the jth configurations for an S-S dyad, still holds for R-R dyads, except that pairs S_{12} and S_{23} are replaced by R_{12} and R_{23}, respectively, as shown in Fig 5.4.2. In addition to the synthesis equations (5.4.5), constraint equations for these pairs should be introduced. In fact, if both revolute pairs are connected by means of a link of axis parallel to the common perpendicular to the axis of both revolute pairs, this perpendicularity should be observed throughout the different specified configurations.

The perpendicularity of the axis of the link with the axis of R_{12} can be stated as

$$\mathbf{u}^T\mathbf{b}_j=0, \ j=0,1,\ldots,n \qquad (5.4.8)$$

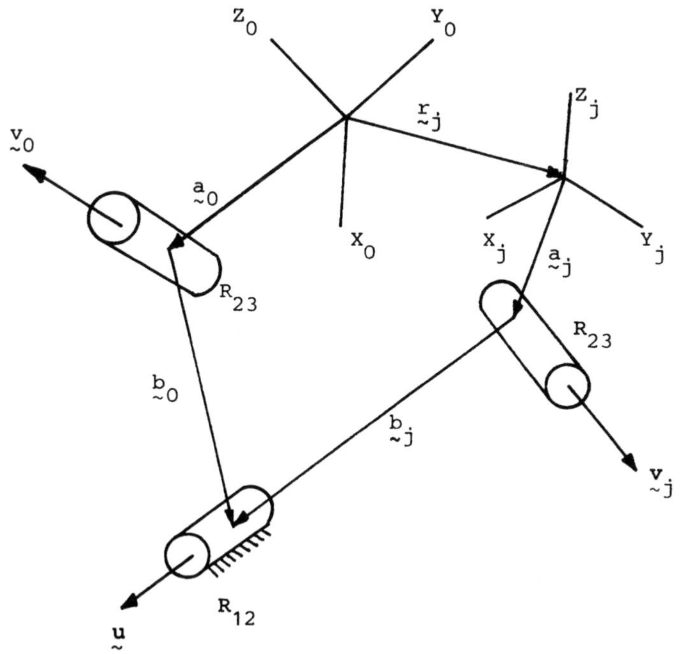

Fig 5.4.2 Rigid-body guidance by means of an R-R dyad

Additionally, the perpendicularity of the said link axis with the axis of R_{23} can be stated as

$$\underset{\sim}{v}^T \underset{\sim}{b}_j = 0, j=0,1,\ldots,n$$

or, introducing eq. (5.4.3), and the fact that

$$\underset{\sim}{v}_j = \underset{\sim}{Q}_j \underset{\sim}{v}_0, j=0,1,\ldots,n$$

the latter perpendicularity condition appears as

$$\underset{\sim}{v}_0^T \underset{\approx}{Q}^T \left((\underset{\sim}{I} - \underset{\approx}{Q}\ ) \underset{\sim}{a}_0 + \underset{\sim}{b}_0 - \underset{\sim}{r}_j \right) = 0, j=0,1,\ldots,n \tag{5.4.9}$$

where

$$\underset{\sim}{Q}_0 \equiv \underset{\sim}{I}, \underset{\sim}{r}_0 \equiv 0$$

From Fig 5.4.2, it is clear that vectors $\underset{\sim}{u}$ and $\underset{\sim}{v}$ belong to the same rigid body, namely the R-R link to be synthesised.

Equations (5.4.8) and (5.4.9), however, do not guarantee that these vectors undergo a rigid body motion from the reference to the $j^{\underline{th}}$ configuration, In fact, a 180-degree rotation of vector $\underset{\sim}{v}_j$ about the link axis (vector $\underset{\sim}{b}_j$) -thus producing a reflexion of the link- could not be detected from eq. (5.4.9), for this would still hold after such a reflection. Hence, an additional constraint should be imposed, namely the constancy of the angle between vectors $\underset{\sim}{u}$ and $\underset{\sim}{v}$, i.e.

$$u^T v_j = u^T v_0, j=1,2,\ldots,n \tag{5.4.10}$$

Finally, the normality of vectors $\underset{\sim}{u}$ and $\underset{\sim}{v}_0$ leads to

$$||\underset{\sim}{u}||=1, ||\underset{\sim}{v}_0||=1 \tag{5.4.11}$$

Summarizing, eqs. (5.4.5), (5.4.8-11) constitute a system of 4n+4 equations to compute the 12 unknown components of vectors $\underset{\sim}{a}_0, \underset{\sim}{b}_0, \underset{\sim}{u}$ and $\underset{\sim}{v}_0$. To match the number of equations to that of unknowns, one must have

$$4n+4=12, \text{i.e. } n=2$$

Hence, by means of an R-R dyad, a rigid body can be guided through 3 configurations (counting the reference as the third configuration), a result which was mentioned in Section 5.3. As in the two foregoing cases, it is possible to realize an optimal synthesis of this linkage. In fact, instead of obtaining the unknown vectors from the 12 derived equations, find $\underset{\sim}{a}_0$ from eq. (5.4.6) such that its magnitude be a minimum. Hence, the number of unknowns reduces to nine, but the number of equations to satisfy is nine, too, thus making it possible to obtain one or several exact solutions to the problem. In fact, find $\underset{\sim}{a}_0$ from eq. (5.4.6) for $\underset{\sim}{r}_j=\underset{\sim}{r}_2$; then the system of equations is: eq. (5.4.5) for j=1; additional equations are eq. (5.4.9a)- next derived-for j=0,1, eq. (5.4.10) for j=1,2 and the two eqs. (5.4.11). Eqs. (5.4.5) (5.4.8) and (5.4.9a) are identically satisfied for j=2 if $\mathbf{a}_0$ is obtained from eq. (5.4.6). Eq. (5.4.9a) is an alternate form of the perpendicularity condition between $\mathbf{b}_j$ and $\mathbf{v}_j$ -eq. (5.4.9)-, which can be stated as the perpendicularity between $\mathbf{b}_j$ and $\mathbf{v}_0$, i.e.

$$\mathbf{v}_j^T\left((\mathbf{I}-\mathbf{Q}_j)\mathbf{a}_0+\mathbf{b}_0-\mathbf{r}_j\right)=0, j=0,1,\ldots,n \qquad (5.4.9a)$$

Exercise 5.4.3 Show that, if $\mathbf{a}_0$ is chosen so as to satisfy eq. (5.4.6), with $\mathbf{r}_j=\mathbf{r}_2$, then eqs. (5.4.5), (5.4.8) and (5.4.9a) are identically satisfied for j=2.

Using the latter approach to the synthesis problem under study would yield two of the specified configurations of the body corresponding to two conjugate configurations of the linkage, as can be readily proved. Regarding this approach, a word of caution is in order, however: As is proposed to show in Exercise 5.4.2, it is possible that situations arise where it is not possible to find $\mathbf{a}_0$ from eq. (5.4.6)

Example 5.4.1 Synthesis an R-R dyad to conduct a rigid body -to which axes $X_iY_iZ_i$ (i=0,1,2) are attached- throught the three positions shown in Fig 5.4.3

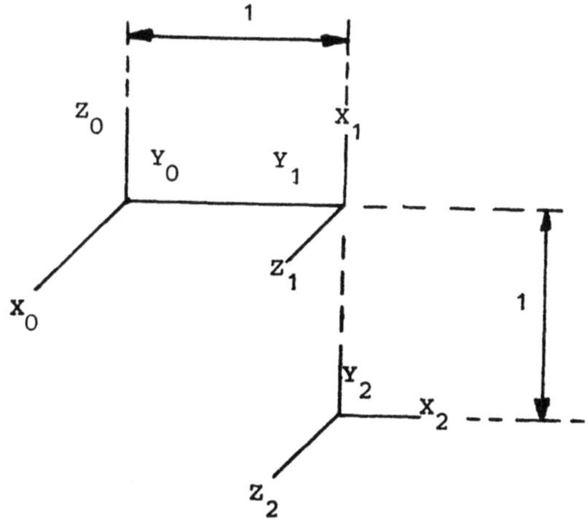

Fig 5.4.3 Rigid-body guidance throught three configurations.

Solution:

In what follows all vectors and matrices are referred to $X_iY_iZ_i$ axes. Matrices $\underset{\sim}{Q}_1$ and $\underset{\sim}{Q}_2$ rotating axes from the reference configuration 0 to configurations 1 and 2 are obtained using Definition 1.2.1, as is shown in Section 2.3. Hence,

$$\underset{\sim}{Q}_1=\begin{pmatrix} 0 & 0 & 1 \\ 0 & -1 & 0 \\ 1 & 0 & 0 \end{pmatrix}, \quad \underset{\sim}{Q}_2=\begin{pmatrix} 0 & 0 & 1 \\ 1 & 0 & 0 \\ 0 & 1 & 0 \end{pmatrix} \tag{5.4.12}$$

Then $\underset{\sim}{a}_0$ is determined from the following equation, so that its lenght is a minimum

$$(\underset{\sim}{I}-\underset{\sim}{Q}_2)\underset{\sim}{a}_0=\underset{\sim}{r}_2 \tag{5.4.13}$$

As shown in Section 2.6, $\underset{\sim}{I}-\underset{\sim}{Q}_2$ is singular of rank 2; hence eq. (5.4.13) contains exactly 2 linearly independent equations, these being

$$\begin{pmatrix} 1 & 0 & -1 \\ -1 & 1 & 0 \end{pmatrix}\begin{pmatrix} a_{01} \\ a_{02} \\ a_{03} \end{pmatrix}=\begin{pmatrix} 0 \\ 1 \end{pmatrix}$$

whose minimum norm solution is readily obtained via the Moore-Penrose pseudo-inverse (See Section 1.12). The computed solution is

$$\underset{\sim}{a}_0=\left(-\frac{1}{3},\frac{2}{3},-\frac{1}{3}\right)^T$$

Let the unknown vectors $\underset{\sim}{b}_0$, $\underset{\sim}{u}$ and $\underset{\sim}{v}_0$ have the following components:

$$\underset{\sim}{b}_0=\begin{pmatrix} b_{01} \\ b_{02} \\ b_{03} \end{pmatrix}, \quad \underset{\sim}{u}=\begin{pmatrix} u_1 \\ u_2 \\ u_3 \end{pmatrix}, \quad \underset{\sim}{v}_0=\begin{pmatrix} v_{01} \\ v_{02} \\ v_{03} \end{pmatrix} \tag{5.4.16}$$

Eq. (5.4.5) for j=1 takes on the form

$$\frac{2}{3}\, b_{02} - \frac{4}{3}\, b_{03} = -\frac{5}{9} \qquad (5.4.17a)$$

Vectors $\underset{\sim}{b}_1$ and $\underset{\sim}{b}_2$ are computed from eq. (5.4.3), from which it can readily be checked that $\underset{\sim}{b}_2=\underset{\sim}{b}_0$, thereby satisfying eq. (5.4.5) for j=2 identically. Eq. (5.4.8) for j=0,1 then, takes on the forms

$$u_1 b_{01}+u_2 b_{02}+u_3 b_{03}=0 \qquad (5.4.17b)$$

$$u_1 b_{01}+u_2\left(\frac{1}{3}+b_{02}\right)+u_3\left(-\frac{2}{3} + b_{03}\right)=0 \qquad (5.4.17c)$$

Eq. (5.4.9a), for j=0,1 leads to

$$u_{01} b_{01}+v_{02} b_{02}+v_{03} b_{03}=0 \qquad (5.4.17d)$$

$$v_{01} b_{01}+v_{02}\left(\frac{1}{3} +b_{02}\right)+v_{03}\left(-\frac{2}{3} +b_{03}\right)= 0 \qquad (5.4.17e)$$

Eq. (5.4.10), for j=1,2 yields

$$u_1 v_{03}-u_2 v_{02}+u_3 v_{01}=u_1 v_{01}+u_2 v_{02}+u_3 v_{03} \qquad (5.4.17f)$$

$$u_1 v_{03}+u_2 v_{01}+u_3 v_{02}=u_1 v_{01}+u_2 v_{02}+u_3 v_{03} \qquad (5.4.17g)$$

where $\underset{\sim}{v}_j$ was computed performing the product $\underset{\sim}{Q}_j \underset{\sim}{v}_0$. Finally, eqs. (5.4.11) take on the form

$$u_1^2+u_2^2+u_3^2=1 \qquad (5.4.17h)$$

$$v_{01}^2+v_{02}^2+v_{03}^2=1 \qquad (5.4.17i)$$

The solutions to eqs. (5.4.17) provide the different possible dyads that guide the rigid body through the different configurations shown in Fig 5.4.2. This system of nine equations in nine unknowns, though nonlinear, can be solved without the aid of a digital computer, as is shown next. To refer to the foregoing equations, only the letter attached to each is mentioned, for shortness.

Subtracting (b) from (c),

$$u_2-2u_3=0 \qquad \text{(a')}$$

Subtracting (f) from (g),

$$u_2(v_{01}+v_{02})+v_3(v_{02}-v_{01})=0 \qquad \text{(b')}$$

Substituting (a') into (b'),

$$u_3(v_{01}+3v_{02})=0 \qquad \text{(c')}$$

First solution:

To satisfy eq. (c') there are two possibilities. One is

$$u_3=0 \qquad \text{(d')}$$

(d') in (a') leads to

$$u_2=0 \qquad \text{(e')}$$

(d') and (c') in h yields

$$u_1=\pm 1 \qquad \text{(f')}$$

In what follows, this solution is developed taking only the plus sign of eq. (f'). Substitution of (f') into (f) yields

$$v_{03}=v_{01} \qquad \text{(g')}$$

Subtracting (d) from (e),

$$v_{02}-2v_{03}=0 \qquad \text{(h')}$$

Solving for v_{01} and v_{02} in (g') and (h'), in terms of v_{03} and substituting the resulting expressions into (i), one obtains

$$6v_{03}^2=1$$

from which

$$v_{03}=\frac{\sqrt{6}}{6} \qquad (i')$$

and

$$v_{01}=\frac{\sqrt{6}}{6} \qquad (j')$$

$$v_{02}=\frac{2\sqrt{6}}{6} \qquad (k')$$

Substituting (d'),(e') and (f') into eq. (b) leads to

$$b_{01}=0 \qquad (l')$$

Substituting (i'),(j') and (k') into eq. (d) leads to

$$2b_{02}+b_{03}=0 \qquad (m')$$

Solving for b_{02} and b_{03} from eqs. (a) and (m'),

$$b_{02}=-\frac{1}{6} \qquad (n')$$

$$v_{03}=\frac{1}{3} \qquad (p')$$

Hence, the first solution to this problem is

$$\underset{\sim}{b}_0=\begin{pmatrix}0\\-1/6\\1/3\end{pmatrix},\ \underset{\sim}{u}=\begin{pmatrix}1\\0\\0\end{pmatrix},\underset{\sim}{v}_0=\begin{pmatrix}\sqrt{6}/6\\\sqrt{6}/3\\\sqrt{6}/6\end{pmatrix} \qquad (5.4.18)$$

<u>Second solution:</u>

If now u_3 is assumed to have a nonzero value, it is possible to divide eq. (c') by it and obtain

$$v_{01}+3v_{02} = 0 \qquad (a'')$$

From (h') and (a") it is possible to solve for v_{01} and v_{03} in terms of v_{02},

$$v_{01}=-3v_{02} \tag{b"}$$

$$v_{03}=\frac{1}{2}v_{02} \tag{c"}$$

Substitution of (b") and (c") into (i) yields

$$v_{02}=\frac{2\sqrt{41}}{41} \tag{d"}$$

Hence

$$v_{01}=-\frac{6\sqrt{41}}{41} \tag{e"}$$

$$v_{03}=\frac{\sqrt{41}}{41} \tag{f"}$$

Substituting (d"),(e") and (f") into eqs. (f) and (g) leads to

$$u_1-2u_2-6u_3=-6u_1+2u_2+u_3 \tag{g"}$$

$$u_1=6u_2+2u_3=-6u_1+2u_2+u_3 \tag{h"}$$

from which one can solve for u_1 and u_2 in terms of u_3, thus obtaining

$$u_1=\frac{15}{7}u_3 \tag{i"}$$

$$u_2=2u_3 \tag{j"}$$

Substitution of (i") and (j") into eq. (h) leads to

$$u_3=\frac{7\sqrt{470}}{470} \tag{k"}$$

Hence,

$$u_1=\frac{15\sqrt{470}}{470} \tag{l"}$$

$$u_2=\frac{14\sqrt{470}}{470} \tag{m"}$$

Substituting (k"), (l") and (m") into eq. (b) one obtains

$$15b_{01}+14b_{02}+7b_{03}=0 \tag{n"}$$

Performing a similar substitution into eq. (d), one obtains

$$-6b_{01}+2b_{02}+b_{03}=0 \qquad (p'')$$

Solving eqs. (n") and (p") for b_{01} and b_{02} in terms of b_{03},

$$b_{01}=0 \qquad (q'')$$

$$b_{02}=\frac{1}{2}\, b_{03} \qquad (r'')$$

Substitution of (r") into eq. (a) leads finally to

$$b_{02}=\frac{5}{18} \qquad (s'')$$

$$b_{03}=\frac{5}{9} \qquad (t'')$$

Thus, the second solution is

$$\underset{\sim}{b}_0=\begin{pmatrix} 0 \\ 5/18 \\ 5/9 \end{pmatrix}, \quad \underset{\sim}{u}=\begin{pmatrix} 7 \\ 15 \\ 14 \end{pmatrix}\alpha, \underset{\sim}{v}_0=\begin{pmatrix} -6 \\ 2 \\ 1 \end{pmatrix}\beta \qquad (5.4.19)$$

where

$$\alpha=\frac{\sqrt{470}}{470}, \quad \beta=\frac{\sqrt{41}}{41}$$

Both dyads defined by $\underset{\sim}{a}_0$ as given by (5.4.13) and both expressions for $\underset{\sim}{b}_0$, (5.4.16) and (5.4.17), are optimal in the sense that $\underset{\sim}{a}_0$ is of minimum length. However, with respect to $\underset{\sim}{b}_0$, the first solution is the best.

In conclusion, regarding R-R dyad synthesis, it has been shown how to obtain an optimal dyad for a problem usually known to have only two real solutions, thus not allowing for any optimisation.

5.5 LINKAGE SYNTHESIS FOR PATH GENERATION. This problem can be stated as: "Synthesise a linkage of a given topology such that a point R of one of its links attains successively the specified positions R_0, $R_1 \ldots, R_n$". This problem can be regarded as one of rigid-body guidance with incompletely specified displacements [5.22], for the motion of one single point of a rigid body does not define the body motion (its orientation thus remains undefined). To formulate the synthesis equations, use is made of eq. (5.3.12), written in the form

$$r_j = r_0 + w_j f + u_j Q_{2j} m + (Q_{3j} - I)(r_0 - q) + (Q_{2j} - I)(q - g), j=1,2,\ldots,n \qquad (5.5.1)$$

where r_0, r_j (j=1,...,n) are the position vectors of points R_0, R_j (j=1,...,n), respectively. Other variables indexed with j are parameters of the involved screws -defined in Section 5.3- relating the jth configuration of the rigid body under study to its reference configuration, indexed with 0. Notice that vectors f, m, q and g are not indexed, this being due to the fact that they remain the same throughout the rigid-body motion, except when dealing with spherical pairs, as is shown later.

Differently from what occurs with the rigid-body guidance problem, in this case the dyads, constituting the linkage intended to design, are not synthesised separately; the linkage is synthesised as a whole, instead. In fact, since in this case the rigid-body rotations involved are not specified, the composed-screw-axis directions, s_j, as well as the corresponding rotations, θ_j, are unknown. These are not independent, once the directions of the screw-axes f and m are specified, for then θ_j can be obtained from eq. (5.3.9), in terms of s_j, f and m. Hence, only the n s_j vectors remain as independent unknowns. These vectors, however, remain the same, no matter what dyad is being synthesised, i.e. although in general the variables

w_j, f, u_j, m, q, and g change from one dyad to the other, the s_j vectors are common. The synthesis procedure is illustrated with one example.

Example 5.5.1 Synthesis of an RRSS path generator linkage.

With no loss of generality, it can be assumed that $r_0 = 0$ in eqs. (5.5.1). Then, the R-R dyad synthesis is first formulated, although, as mentioned previously, this synthesis cannot be performed independently from the one of the S-S dyad. This is due to the fact that the synthesis equations for both dyads are coupled via the s_j vectors. In performing the R-R dyad synthesis, the different variables have the following meanings (See Fig 5.3.1 and its description): u_j and w_j are the displacements associated with the fixed and the moving R pairs of the dyad, respectively. Hence, they vanish.

f and m are unit vectors pointing in the direction of the axes of the fixed and the moving R pairs of the dyad. s_j is the unit vector pointing in the direction of the axis of the screw carrying the body from its reference- to its jth configuration.

g and q are position vectors of points located along the axes of the screws F_j and M_j, respectively. As in Section 5.3, those points will be located at the intersections of the said axes with their common normal.

Under these cirucumstances, eqs. (5.5.1) take on the form

$$r_j = -(Q_{3j} - I)q + (Q_{2j} - I)(q - g) = r_j(f, m, s_j, q, g) \qquad (5.5.2)$$

where, clearly

$$Q_{2j} = Q_{2j}(f, \gamma_j), \gamma_j = \gamma_j(f, m, s_j)$$

$$Q_{3j} = Q_{3j}(s_j, \theta_j), \theta_j = \theta_j(f, m, s_j)$$

The normality of vectors f, m and $s_j (j=1,2,\ldots,n)$ is expressed as

$$||f|| = 1, \; ||m|| = 1, \; ||s_j|| = 1, \; j = 1, \ldots, n \qquad (5.5.3)$$

whereas the specific locations of points G and Q along the screw axes are expressed as

$$(\underset{\sim}{q}-\underset{\sim}{g})^{T}\underset{\sim}{f}=0,\quad (\underset{\sim}{q}-\underset{\sim}{g})^{T}\underset{\sim}{m}=0 \tag{5.5.4}$$

Thus, the synthesis of this dyad leads to a system of 3n+12 unknowns: $\underset{\sim}{f}$, $\underset{\sim}{m}$, $\underset{\sim}{s}_j$ (3n unknown scalars), $\underset{\sim}{q}$ and $\underset{\sim}{g}$, in the 4n+4 equations: 3n equations (5.5.2) (one for each direction of the physical space)+ n+2 equations (5.5.3) + 2 equations (5.5.4).

The synthesis of the S-S dyad is formulated in the following way: Referring to Fig 5.5.1, let S_{12} and S_{23} be the spherical pairs of the dyad under consideration. Let B_0 and B_j be the reference- and the $j^{\underline{th}}$ configurations of body B, and B_j' an intermediate configuration. Now, the guidance of B from B_0 through B_j is performed in two stages: first, B_0 is rotated about point Q through a rotation of axis $\underset{\sim}{m}_j$ and angle α_j (it will be seen next that neither $\underset{\sim}{m}_j$ nor α_j appear explicitly in the resulting synthesis equations for this dyad). Let B_j' be the configuration of B after this rotation. Next, fixing B_j to the dyad, rotate the whole assembly as one single rigid body about point G through an axis $\underset{\sim}{f}_j$ and an angle γ_j, B_j being the final configuration of the rigid body.

Since in this case also $\underset{\sim}{u}_j=w_j=0$, and $\underset{\sim}{r}_0$ can be equated to 0, eq. (5.5.1) takes on the form

$$\underset{\sim}{r}_j=-(\underset{\approx}{Q}_{3j}-\underset{\sim}{I})\underset{\sim}{g}+(\underset{\approx}{Q}_{2j}-\underset{\sim}{I})(\underset{\sim}{q}-\underset{\sim}{g})=\underset{\sim}{r}_j(\underset{\sim}{f}_j,\underset{\sim}{m}_j,\underset{\sim}{s}_j,\underset{\sim}{q},\underset{\sim}{g}) \tag{5.5.5}$$

where

$$\underset{\sim}{Q}_{2j}=\underset{\sim}{Q}_{2j}(\underset{\sim}{f}_j,\gamma_j),\gamma_j=\gamma_j(\underset{\sim}{f}_j,\underset{\sim}{m}_j,\underset{\sim}{s}_j)$$

$$\underset{\sim}{Q}_{3j}=\underset{\sim}{Q}_{3j}(\underset{\sim}{s}_j,\theta_j),\theta_j=\theta_j(\underset{\sim}{f}_j,\underset{\sim}{m}_j,\underset{\sim}{s}_j)$$

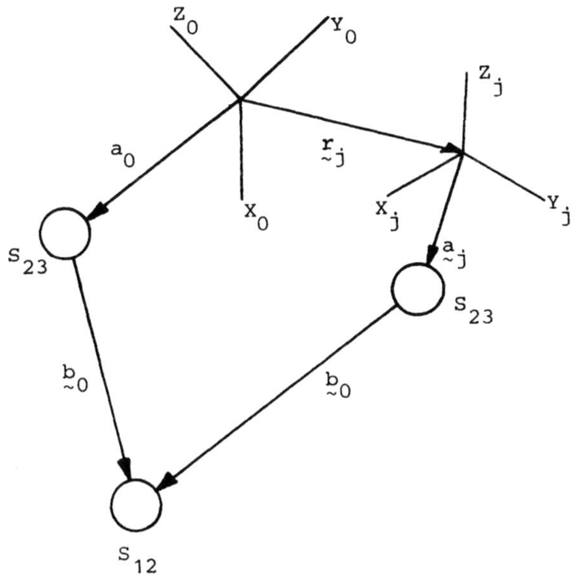

Fig 5.5.1 Displacement of a rigid by means of an S-S dyad

Clearly, vectors $\mathbf{f}_j$ and $\mathbf{m}_j$ are different from $\mathbf{f}$ and $\mathbf{m}$ appearing in eqs. (5.5.2)-(5.5.4), which is made clear because the latter vectors are not indexed. Vectors $\mathbf{s}_j$, appearing in eqs. (5.5.2)-(5.5.3), however, are identical to those appearing in eqs. (5.5.5); hence they do not add extra unknowns to the synthesis problem of the overall linkage.

Other synthesis equations are the following:

The normality of vectors $\mathbf{f}_j$ and $\mathbf{m}_j$ is expressed as

$$||\mathbf{f}_j||=1, \quad ||\mathbf{m}_j||=1, \quad j=1,2,\ldots,n \tag{5.5.6}$$

Since the S-S dyad can rotate about axis GQ without changing either configuration B_j' or B_j, it can be assumed that the axes of rotation, given by unit vectors $\mathbf{f}_j$ and $\mathbf{m}_j$, are both perpendicular to axis GQ, i.e.

$$(\underset{\sim}{q}-\underset{\sim}{g})^T \underset{\sim}{f}_j = 0, \quad j=1,2,\ldots,n \qquad (5.5.7)$$

$$(\underset{\sim}{q}-\underset{\sim}{g})^T \underset{\sim}{m}_j = 0, \quad j=1,2,\ldots,n \qquad (5.5.8)$$

Thus, the synthesis of the S-S dyad adds the following unknowns to the overall linkage synthesis problem: $\underset{\sim}{f}_j$, $\underset{\sim}{m}_j$ $(j=1,\ldots,n)$, $\underset{\sim}{q}$ and $\underset{\sim}{g}$, i.e. 6n+6 additional unknowns (Vectors $\underset{\sim}{s}_j$ are not counted as new unknowns, since these were already taken into account when discussing the synthesis of the R-R dyad). The additional equations are: 3n equations (5.5.5) + 2n equations (5.5.6) +n equations (5.5.7)+n equations (5.5.8). Summarizing, then, the RRSS-linkage-synthesis problem for path generation gives rise to system of 11n+4 equations in 9n+18 unknowns. Matching the number of equations to that of unknowns, it is soon realized that n should have the value 7, i.e. it is possible to generate a path passing through 8 specified points (counting the reference- as the $8^{\underline{th}}$ position), which coincides with the result obtained by Suh [5.23].

Remarks on the formulation of the RRSS linkage synthesis problem for path generation.

i) In the previous linkage-synthesis problem angles γ and θ appear in eq. (5.5.1) and angles γ_j and θ_j do in eq. (5.5.2). These angles, however, are not counted as additional unknowns, since they can be computed, via eqs. (5.5.4) and (5.5.9), with the values of $\underset{\sim}{f}$, $\underset{\sim}{m}$ (or $\underset{\sim}{f}_j$ and $\underset{\sim}{m}_j$) and $\underset{\sim}{s}_j$.

ii) The problem, as just stated for 8-point synthesis, leads to a system of 81 equations in 81 unknowns which, because of being nonlinear, is diffecult to handle numerically. Suh's formulation [5.23] leads to a system of 39 equations in 39 unknowns, in this instance. Roth and Tsai

(5.11) recommend that, for dyads containing revolute, prismatic, helical or cylindrical pairs, one point on the pair axis (G or Q of the previous formulation) be taken as the intersection of the axis with, say, the X-Y plane, thus diminishing the number of unknowns (one component of the $\underset{\sim}{f}$ and of the $\underset{\sim}{m}$ vectors is thus zero). This way, the number of equations and of unknowns is reduced to 65 which, anyway, is about twice that of Suh's. Thus, it might seem more advantageous to use the latter approach.

Since Suh's formulation is amply discused in (5.23), it is not presented here. An alternative approach, based on Section 5.4, is presented instead.

<u>An alternative approach to the problem formulation for the RRSS linkage synthesis problem.</u>

Referring to Figs 5.4.1 and 5.4.2, the synthesis equations are next written. To distinguish the position vectors of points A and B, in their reference configurations, belonging to the R-R dyad, from those belonging to the S-S dyad, the latter are starred. The synthesis equations for the R-R dyad are those derived in Section (5.4), i.e.

$$||(\underset{\sim}{I}-\underset{\sim}{Q}_j)\underset{\sim}{a}_0+\underset{\sim}{b}_0-\underset{\sim}{r}_j||^2=||\underset{\sim}{b}_0||^2, j=1,\ldots,n \tag{5.5.9}$$

where it is convenient to express $\underset{\sim}{Q}_j$ as appearing in eq. (2.5.4), i.e.

$$\underset{\sim}{Q}_j=-\underset{\sim}{P}_j^2\cos\theta_j+\underset{\sim}{P}_j\sin\theta_j+\underset{\sim}{R}_j \tag{5.5.10}$$

Calling $\underset{\sim}{e}_j$ the unit vector parallel to the axis of $\underset{\sim}{Q}_j$ and $(\alpha_j,\beta_j,\gamma_j)^T$ its components (in the $X_0Y_0Z_0$ axes), the matrices appearing at the right-hand side of eq. (5.5.10) are, from eq. (2.5.4a),

$$\underset{\sim}{P}_j=\begin{pmatrix} 0 & -\gamma_j & \beta_j \\ \gamma_j & 0 & -\alpha_j \\ -\beta_j & \alpha_j & 0 \end{pmatrix}, \quad \underset{\sim}{R}_j=\begin{pmatrix} \alpha_j^2 & \alpha_j\beta_j & \alpha_j\gamma_j \\ \alpha_j\beta_j & \beta_j^2 & \beta_j\gamma_j \\ \alpha_j\gamma_j & \beta_j\gamma_j & \gamma_j^2 \end{pmatrix} \tag{5.5.11}$$

Additional equations for the R-R dyad are:

$$\underset{\sim}{u}^T\underset{\sim}{b}_j=0,\ j=1,\ldots,n \tag{5.5.12}$$

$$\underset{\sim}{v}_0^T\underset{\sim}{b}_j=0,\ j=1,\ldots,n \tag{5.5.13}$$

$$\underset{\sim}{u}^T\underset{\sim}{v}_j=\underset{\sim}{u}^T\underset{\sim}{v}_0,\ j=1,2,\ldots,n \tag{5.5.14}$$

$$||\underset{\sim}{u}||=1,\ ||\underset{\sim}{v}_0||=1 \tag{5.5.15}$$

$$||\underset{\sim}{e}_j||=1,\ j=1,2,\ldots,n \tag{5.5.16}$$

The synthesis equations for the S-S dyad are

$$||(\underset{\sim}{I}-\underset{\sim}{Q}\ )\underset{\sim}{a}_0^* + \underset{\sim}{b}_0^* - \underset{\sim}{r}_j||^2= ||\underset{\sim}{b}_0^*||^2,\ j=1,\ldots,n \tag{5.5.17}$$

Summarizing, the problem of an RRSS linkage synthesis for path generation leads to a system of 6n + 4 equations in 4n + 18 unknowns. The equations are: n equations (5.5.9)+(n+1) equations (5.5.12)+(n+1) equations (5.5.13)+n equations (5.5.14)+2 equations (5.5.15)+n equations (5.5.516)+n equations (5.5.17). The unknowns are:

$$\underset{\sim}{a}_0,\underset{\sim}{b}_0,\underset{\sim}{e}_j\,(j=1,\ldots,n),\theta_j,\underset{\sim}{u},\underset{\sim}{v}_0,\underset{\sim}{a}_0^* \text{ and } \underset{\sim}{b}_0^*.$$

If the number of equations is to be matched with that of unknowns, it is readily concluded that the linkage thus synthesised can trace a spatial path passing through 8 prescribed points, which result coincides with that of Suh's Via the latter approach, however, the number of equations and unknowns (for 8-point synthesis) is 46, a number still greater than that of Suh's, yet smaller than that obtained using Roth and Tsai's equations.

5.6. EPILOGUE. The subject of linkage synthesis is far more extense than has been presented here. Topics that were not covered are, amongst others, linkage synthesis for rigid-body guidance and for path generation with infinitesimally-separated positions, i.e. to meet prescribed conditions on velocities and higher derivatives. The subject is treated in (5.9), (5.10) and (5.11), but was not included here due to space limitations.

REFERENCES

5.1 Denavit J. and R.S. Hartenberg, Kinematic Synthesis of Linkages, McGraw-Hill Book Co., N. York, 1964. p. 346

5.2 Mohan Rao A.V., G.N. Sandor, D. Kohli and A.H. Soni, "Closed form synthesis of spatial function generating mechanisms for the maximum number of precision points", Journal of Engineering for Industry, Trans. ASME, Series B, vol. 95, Aug. 1973, pp. 725-736.

5.3 Luck K., "Computersynthese des viergliedrigen räumlichen Koppelgetriebes vom Typ RSSR", Mechanism and Machine Theory, vol. 11, 1976, pp. 213-225

5.4 Björck Å. and G. Dahlquist, Numerical Methods, Prentice-Hall, Inc., Englewood Cliffs, 1974, pp. 131-136.

5.5 Suh C.H. and C.W. Radcliffe, "Synthesis of spherical linkages with use of the displacement matrix", Journal of Engineering for Industry, Trans. ASME, Series B, vol. 89, 1967, pp. 215-222

5.6 Suh C.H., "Design of space mechanisms for function generation", Journal of Engineering for Industry, Trans. ASME, Series B, vol. 90, 1968, pp. 507-512

5.7 Roth B., "The kinematics of motion through finitely separated positions", Journal of Applied Mechanics, Trans, ASME, Series E, vol. 89, 1967, pp. 591-598

5.8 Roth B., "Finite position theory applied to mechanism synthesis", Journal of Applied Mechanics, Trans. ASME, Series E, vol. 89, 1967, pp. 599-605

5.9 Chen P. and B. Roth, "A unified theory for the finitely and infinitesimally separated position problems of kinematic synthesis", Journal of Engineering for Industry, Trans. ASME, Series B, vol. 91, 1969, pp. 203-208

5.10 Chen P. and B. Roth, "Design equations for the finitely and infinitesimally separated position synthesis of binary links and combined chains", Journal of Enginnering for Industry, Trans. ASME, Series B, vol. 91, 1969, pp. 209-219

5.11 Tsai L.W. and B. Roth, "Design of dyads with helical, cylindrical, spherical and prismatic joints", Mechanism and Machine Theory, vol. 7, 1972, pp. 85-102

5.12 Tsai L.W. and B. Roth, "A note on the design of revolute-revolute cranks", Mechanism and Machine Theory, vol. 8, 1973, pp. 23-31

5.13 Fox E.A., Mechanics, Harper & Row, N. York, 1967, p. 8

5.14 Bisshopp K.E., "Rodrigues' Formula and the screw matrix", Journal of Engineering for Industry, Trans. ASME, Series B., vol. 91, 1969, pp. 179-185.

5.15 Ericksen J.L."Tensor Fields" in Flügge S., ed. Handbuch der Physik, vol. III/1, Springer-Verlag, Berlin, 1960, pp. 794-796.

5.16 Suh C.H. and C.W. Radcliffe "Synthesis of plane linkages with use of the displacement matrix", Journal of Engineering for Industry, Trans. ASME, Series B, vol. 89, 1967, pp. 206-214

5.17 Angeles J., Análisis y Síntesis Cinemáticos de Sistemas Mecánicos, (in Spanish) Editorial Limusa, Mexico City, 1978.

5.19 Roth B., "Finite position theory applied to mechanism synthesis", Journal of Applied Mechanics, Trans. ASME, Series E, vol. 34, 1967, pp. 599-605

5.19 Suh C.H., "On the duality in the existence of R-R links for three positions", Journal of Engineering for Industry, Trans. ASME, Series B, vol. 91, 1969, 129-134.

5.20 Bennett G.T., "A new mechanism", Engineering, vol. 76, 1903, pp. 777-778

5.21 Suh C.H., "Design of space mechanisms for rigid-body guidance", Journal of Engineering for Industry, Trans. ASME, Series B, vol. 90, No. 3, 1968, pp. 499-506.

5.22 Tsai L.W. and B. Roth, "Incompletely specified displacements: Geometry and spatial linkage synthesis", ASME paper No. 72-Mech-13.

5.23 Suh C.H. and C.W. Radcliffe, Kinematics and Mechanisms Design, John Wiley & Sons, Inc., N. York, 1978, pp. 201-202.

6. An Introduction to the Optimal Synthesis of Linkages

6.1 INTRODUCTION. The problems of linkage synthesis outlined in Ch. 5 lead in general to nonlinear systems of algebraic equations. As discussed in Section 1.13, these systems can admit multiple solutions or none. The said solutions satisfy all the equations exactly, if roundoff errors are disregarded, which is done in this case for the numerical accuracy of the solution will be taken for granted in the current discussion. The resulting solutions, however, were not evaluated as to the "quality" of the linkage they produced, once this quality is defined. The quality under consideration could be the overall size of the linkage (this size could be in turn defined as the sum of the lengths of all the links, for example), its minimum mechanical advantage or the maximum angular acceleration attained by one of its links.

The several possibilities that could arise, different to those studied in Ch. 5 are:

i) the number of prescribed conditions exceeds the number of parameters available, thus producing an overdetermined system of equations which in general cannot be solved exactly.

ii) the number of prescribed conditions is smaller than that of parameters available, thus making it possible for the designer to select the best linkage (in some sense) from a variety of alternatives. In this case, two possibilities can arise, namely, either the designer has absolute freedom (unconstrained problem) to choose the best alternative, or else his alternatives are to lie within a certain allowable set (constrained problems). Such restrictions arise whenever questions of cost, availability of space, weight, etc. are considered.

In what follows, more attention will be paid to the question of evaluating the quality of a linkage by properly defining an objective function to be optimised. The foregoing problems will be dealt with in this chapter.

6.2 THE OPTIMISATION PROBLEM. The most general optimisation problem can be stated as follows:

"Given a vector space V and a subset Ω of it, find the vector $\underset{\sim}{x}^{*}\epsilon\Omega$ such that a (also given) functional f defined over V, attains its optimal value at $\underset{\sim}{x}^{*}$"

In the above statement, a funtional is nothing but a mapping of V into the field over which it is defined, i.e. a scalar.

The problem thus defined has not yet been solved in all its generality, but solutions have been found to some special cases of it, many of which are applicable to linkage synthesis problems.

To begin with, the kind of problems that usually arise in the realm of linkage optimisation are defined over a vector space V of finite dimension, i.e. the problems that will be handled in this context contain a finite number of independent variables. Problems containing an infinite (even more, a non-denumerable) set of unknowns arise in areas such as optimal control systems (6.1) where the vector space V is a Banach space (6.2), i.e. a vector space whose elements are continuous functions of one real variable (time) with some special properties that will not be listed here, for they will not be needed.

Furthermore, the subset Ω of the finite-dimensional vector spaces V that will be dealt with in this chapter, is defined by a set of either equality or inequality algebraic relations (in the case of Banach spaces, the said subset is defined by differential equations). In addition, the objective function to optimise is a functional over V, i.e. a scalar algebraic function (as opposed to integral functional in the case of Banach spaces) of the n independent variables corresponding to the dimension of V. Hence, the optimisation problem arising in linkage synthesis can be stated in the general form:

$$\begin{aligned} &\text{optimise} && f=f(\tilde{x}) \\ &\text{subject to} && \tilde{g}(\tilde{x})=\tilde{0} \\ &\text{and} && \tilde{h}(\tilde{x})\geq\tilde{0} \end{aligned} \qquad (6.2.1)$$

where $\tilde{x}$ is an n-dimensional vector of space V, f is a vector-valued scalar function, $\tilde{g}$ is an m-dimensional (m<n) vector-valued vector function and $\tilde{h}$, a p-dimensional vector-valued vector function.

Remark: Vector spaces are non-ordered sets [6.3], hence no inequality relation can be associated to them. Therefore, the inequality appearing in (6.2.1) should be understood as a compact symbol to mean

$$h_i(\tilde{x})\geq 0 \qquad i=1,\ldots,p \qquad (6.2.2)$$

h_i being the $i^{\underline{th}}$ component of vector $\tilde{h}$. A similar symbology is used to express positive definite and positive semidefinite square matrices. Thus, $\tilde{A}>\tilde{0}$ and $\tilde{A}\geq\tilde{0}$ designate positive definite and positive semidefinite matrices, respectively.

The modern approach to the linkage optimisation problem places it within the frame of mathematical programming, as discussed in [6.4]. In the abstract of this paper the author states: "...the synthesis of mechanisms thus ceases to be a narrow and isolated field but becomes part of a broad sphere of science encompassing areas obviously very distant such as economy, military science, automation, cybernetics and others".

The foregoing approach will be followed in this chapter. In the first place the overdetermined linkage synthesis problems will be presented and solved.

6.3 <u>OVERDETERMINED PROBLEMS OF LINKAGE SYNTHESIS</u>. It was shown in Section 5.2 that the synthesis of the RSSR linkage for function generation leads to a system of up to eight linear equations in eight unknowns, thus making it possible to satisfy up to eight conditions of the type $\phi_i = \phi(\psi_i)$ between the input and the output angles, ψ_i and ϕ_i, respectively*. In Section 5.3 it was shown that a rigid body can be guided through up to three successive configurations by means of an R-R dyad and reference was made to results appearing in the literature [5.11,12,17-19] showing that the said three-position-rigid-body guidance problem, when solvable, has two real solutions, which constitute a Bennett (RRRR-single-degree-of-freedom-spatial) linkage. The resulting synthesis equations were shown to be twelve nonlinear equations in twelve unknowns. Further on, in Section 5.5 it was shown that, by means of an RRSS linkage, a spatial path can be traced that passes through up to eight points in space, the resulting equations being 39 (according to Suh [5.22]) nonlinear equations in 39 unknowns.

There can arise technical problems, however, that require to synthesise either; i) an RSSR linkage that satisfies a discrete input-output function over more than eight points, or ii) an RR dyad that carries a rigid body through more than three configurations, or iii) an RRSS linkage that traces a spatial path passing through more than eight points. In all these cases the number of resulting synthesis equations will exceed that of available linkage parameters (unknowns), thus leading to an overdetermined problem. This problem could involve a system of linear equations, as in case i) above, or a system of nonlinear ones, as in the remaining two previously discussed cases.

---

* It was also mentioned in that Section that, as Mohan Rao et al. [5.4] claim, it is possible to extend this synthesis problem to 10 precision points if scale factors between the synthesised function and input and output angles are not specified; but this possibility will not be discussed here because the synthesised function is assumed to be defined only at a discrete set of points.

The overdetermined arising problem does not have a solution in the "usual" sense, i.e. in the sense of satisfying the equations exactly. As was discussed in Section 1.11 for linear systems and in Section 1.13 for nonlinear ones, a solution vector $\underset{\sim}{x}_0$ is sought that yields the minimum error (in the approximation to the said equations), in this instance. The most common way of measuring such error is through the Euclidean norm of the vector whose components are the involved equations, i.e. if this error is denoted by e and the system of equations is $\underset{\sim}{f}(\underset{\sim}{x})=0$ ($\underset{\sim}{A}\underset{\sim}{x}-\underset{\sim}{b}=\underset{\sim}{0}$ in the linear case), then

$$e^2=||\underset{\sim}{f}(\underset{\sim}{x})||^2=\underset{\sim}{f}^T\underset{\sim}{f} \tag{6.3.1a}$$

or

$$e^2=||\underset{\sim}{A}\underset{\sim}{x}-\underset{\sim}{b}||^2=(\underset{\sim}{A}\underset{\sim}{x}-\underset{\sim}{b})^T(\underset{\sim}{A}\underset{\sim}{x}-\underset{\sim}{b}) \tag{6.3.1b}$$

depending upon whether the system of equations is nonlinear or linear. The arising optimisation problem can be stated as

$$\text{"Minimize } \underset{\sim}{f}^T\underset{\sim}{f} \text{ over } \underset{\sim}{x}\text{"} \tag{6.3.2}$$

If the system of equations is linear, $\underset{\sim}{f}$ should be replaced by $\underset{\sim}{A}\underset{\sim}{x}-\underset{\sim}{b}$ in (6.3.2), of course.

Notice that problem (6.3.2) is a particular case of problem (6.2.1), in which no constraints are present, i.e. this is a so-called "unconstrained optimisation problem"; furthermore, due to its particular quadratic nature, it is referred to specifically as a least-square problem. The optimum $\underset{\sim}{x}_0$ is found, as outlined in section 1.12 for the linear case, by triangularisation of matrix $\underset{\sim}{A}$ via Householder reflections and "back substitution", both of which are implemented in subroutines HECOMP and HOLVE appearing in Figs. 1.12.4 and 1.12.5. The nonlinear case, as discussed in Section 1.13, is solved iteratively by the Newton-Raphson method, which requires the computation of the optimal correction at each stage, via the solution of a linear least-square problem, which is done, as already mentioned, by means of subroutines HECOMP and HOLVE. SUBROUTINE NRDAMC, whose listing appears in Fig. 1.13.2, combines HECOMP and HOLVE to solve the nonlinear least-square problem (6.3.2).

The least-square problem arising in linkage syntheses has been already dealt with in the litterature. Suh and Mecklenburg (6.5) solved such problem by application of Powell's method. The said method is one of the so-called "direct search methods" and specifically it performs uni-directional searches, i.e. it optimises the objective function once at each iteration, keeping fixed all the variables but one, at each time. More on direct search methods will be discussed in Section 6.5.

The linear least-square solution via Householder reflections is next illustrated with an example taken from (6.5) for comparisson purposes.

Example 6.3.1. Synthesise an RSSR linkage so that its input ψ and its output ϕ be related according to Table 6.3.1. Referring to the nomenclature of Fig 5.2.1, assume

$$a_4=1,\ \alpha_4=90°$$

Solution:

Use is made of eq. (5.2.18) and definitions (5.2.19), thus obtaining a system of 19 linear equations in the six unknowns $k_1, k_2, \ldots, k_6$, of the form

$$\underset{\sim}{A}\underset{\sim}{x}=\underset{\sim}{b} \qquad (6.3.3)$$

where $\underset{\sim}{A}$ is a 19x6 matrix, $\underset{\sim}{x}$ is a 6-dimensional vector whose i^{th} component is k_i and $\underset{\sim}{b}$ is a 19-dimensional vector whose i^{th} component is $c\phi_i c\psi_i + c\alpha_4 s\psi_i s\phi_i$. Both matrix $\underset{\sim}{A}$ and vector $\underset{\sim}{b}$ appear in Table 6.3.2.

Subroutines HECOMP and HOLVE were used to obtain the least-square solution $\underset{\sim}{x}_0$, i.e. the six values $k_1, \ldots, k_6$ that approximate the 19 equations (6.3.3) with the minimum error, in the sense of the Euclidean norm. The optimal resulting values and the corresponding linkage parameters appear in Table 6.3.3

The error obtained with Powell's method, reported in (6.5) , as well as that obtained with Householder reflections, are shown in Table 6.3.4. The overall error, for comparisson purposes, is taken as

$$e = \sqrt{\frac{1}{n}\Sigma(\phi_i^{*}-\phi_i^{**})^2}$$

where ϕ_i^* and ϕ_i^{**} are the generated and the specified values of the output variable, respectively, corresponding to the value ψ_i of the input variable. The nonlinear case solution is illustrated with the following example.

Example 6.3.2 Synthesis of a plane RRRR linkage for rigid-body guidance through 17 specified configurations.

The specified configurations are shown in Table 6.3.5. The synthesis equations are those derived in Appendix 4, namely, eqs. (A.4.6) which are next rewritten as

$$f_j=||(1-e^{i\theta_j})a_0+b_0-r_j||^2-||b_0||^2=0, j=1,\ldots,16 \qquad (6.3.4)$$

with $i\equiv\sqrt{-1}$

The objetive function is, then

$$\phi(a_0,b_0)=\underset{\sim}{f}^T\underset{\sim}{f}=\sum_1^{16} f_j^2$$

Minimizing ϕ, as given by eq. (6.3.5) is, then, a nonlinear least-square problem, already discussed in case ii) of section 1.13. Thus, it can be solved using SUBROUTINE NRDAMC. The 17 specified configurations are shown in Table 6.3.5.

A program was written to obtain the least-square solution to the problem, which turned out to have two different solutions, one for each side of the linkage, thus enabling the designer to construct the whole linkage. The two solutions are shown in Table 6.3.6

TABLE 6.3.1

THESE ARE THE SPECIFIED POINTS

	PSI(DEGREES)	PHI(DEGREES)		PSI(DEGREES)	PHI(DEGREES)
1	0.	0.	11	-.50000000E+02	.37200000E+02
2	-.50000000E+01	.24000000E+01	12	-.55000000E+02	.42300000E+02
3	-.10000000E+02	.51000000E+01	13	-.60000000E+02	.47500000E+02
4	-.15000000E+02	.82000000E+01	14	-.65000000E+02	.53000000E+02
5	-.20000000E+02	.11500000E+02	15	-.70000000E+02	.58700000E+02
6	-.25000000E+02	.15200000E+02	16	-.75000000E+02	.64600000E+02
7	-.30000000E+02	.19100000E+02	17	-.80000000E+02	.70900000E+02
8	-.35000000E+02	.23300000E+02	18	-.85000000E+02	.78000000E+02
9	-.40000000E+02	.27700000E+02	19	-.90000000E+02	.90000000E+02
10	-.45000000E+02	.32300000E+02			

TABLE 6.3.2

MATRIX A						VECTOR B
1.000000	0.000000	-1.000000	0.000000	0.000000	1.000000	1.000000
0.999123	0.041876	-0.996195	-0.087156	0.041716	1.000000	0.995321
0.996041	0.088894	-0.984808	-0.173648	0.087544	1.000000	0.980909
0.989776	0.142629	-0.965926	-0.258819	0.137769	1.000000	0.956050
0.979925	0.199368	-0.939693	-0.342020	0.187345	1.000000	0.920828
0.965016	0.262189	-0.906308	-0.422618	0.237624	1.000000	0.874602
0.944949	0.327218	-0.866025	-0.500000	0.283379	1.000000	0.818350
0.918446	0.395546	-0.819152	-0.573576	0.324012	1.000000	0.752347
0.885394	0.464842	-0.766044	-0.642788	0.356090	1.000000	0.678251
0.845262	0.534352	-0.707107	-0.707107	0.377844	1.000000	0.597690
0.796530	0.604599	-0.642788	-0.766044	0.388629	1.000000	0.512000
0.739631	0.673013	-0.573576	-0.819152	0.386024	1.000000	0.424235
0.675590	0.737277	-0.500000	-0.866025	0.368639	1.000000	0.337795
0.601815	0.798636	-0.422618	-0.906308	0.337518	1.000000	0.254338
0.519519	0.854459	-0.342020	-0.939693	0.292242	1.000000	0.177686
0.428935	0.903335	-0.258819	-0.965926	0.233800	1.000000	0.111017
0.327218	0.944949	-0.173648	-0.984808	0.164089	1.000000	0.056821
0.207912	0.978148	-0.087156	-0.996195	0.085251	1.000000	0.018121
-0.000000	1.000000	0.000000	-1.000000	-0.000000	1.000000	0.000000

TABLE 6.3.3.

LINKAGE PARAMETERS		THESE ARE PARAMETERS K	
a_1	= 0.91126871	k_1	= 0.081873
a_2	= 2.62059792	k_2	= -3.489375
a_3	= 0.80357731	k_3	= 1.176454
a_4	= 1.00000000	k_4	= -1.395556
s_1	= -1.18624029	k_5	= 0.386427
s_4	= -2.41755601	k_6	= 2.093812
ϕ_0	= 21.12786811°		

TABLE 6.3.4.

POINT NO.	APPROXIMATION ERROR USING POWELL'S METHOD (DEGREES)	APPROXIMATION ERROR USING HOUSEHOLDER'S METHOD (DEGREES)
1	0.0000	0.0142
2	-0.0012	0.0010
3	-0.0306	-0.0351
4	0.0233	0.0159
5	-0.0268	-0.0345
6	0.0326	0.0260
7	0.0160	0.0108
8	0.0383	0.0344
9	0.0152	0.0120
10	-0.0379	-0.0411
11	-0.0064	-0.0096
12	0.0227	0.0193
13	-0.0422	-0.0450
14	-0.0002	-0.0015
15	0.0355	0.0343
16	0.0106	0.0132
17	-0.0002	0.0014
18	-0.0125	-0.0197
19	0.0298	0.0041

ROOT MEAN SQUARE ERROR

USING POWELL'S METHOD (DEGREES)	USING HOUSEHOLDER'S METHOD (DEGREES)
0.0018	0.0018

TABLE 6.3.5 Specified configurations for rigid-body guidance

THESE ARE THE SPECIFIED CONFIGURATIONS

	X (CM)	Y (CM)	THETA (DEGREES)
0	7.880	-0.260	313.720
1	8.490	-7.290	332.330
2	7.680	2.820	349.930
3	6.300	4.210	353.180
4	4.580	4.950	359.870
5	2.740	5.010	355.840
6	1.010	4.410	356.300
7	0.259	3.880	3.900
8	-0.400	-3.090	3.670
9	0.250	-3.760	3.690
10	1.000	-4.290	4.150
11	2.730	-4.890	5.120
12	4.560	-4.830	6.810
13	6.280	-4.090	10.000
14	7.660	-2.700	13.000
15	8.440	-0.610	18.000
16	7.790	-2.690	46.270

TABLE 6.3.6 Solutions to overdetermined rigid-body guidance problem of linkage synthesis

First solution	Second solution
A_0(5.124,2.255)	A_0(1.444,-6.705)
B_0(0.549,-0.703)	B_0(6.371,-9.315)

Thus, the linkage is composed of one fixed link with input and output links hinged to it at each of both points B shown in Table 6.3.6. The coupler link is hinged to the input and output links at each of both points A_0 also shown in that table.

Since the problem is overdetermined, the solutions obtained do not zero each of the 16 functions f_j shown in (6.3.4). They minimize the quadratic norm (6.3.5), instead. Since each f_j is in turn quadratic in a_0 and b_0, ϕ is quartic in these complex numbers. Thus, ϕ has units of lenght raised to the fourth power. In order to obtain a dimensionless measure of the error in the approximation, this is computed as

$$e = \frac{1}{16|b_0|} \sum_{1}^{16} \sqrt{|f_j|} \tag{6.3.6}$$

This way, the error of the first solution resulted to be 38.74%, whereas that of the second one, 60.77%. More examples considering different numbers of specified configurations appear in [6.6].

An overdetermined problem of the RR spatial dyad for rigid-body guidance is presented in the next example.

Example 6.3.3. Synthesis of the RR spatial dyad for rigid-body guidance.

In this case, one fourth position of points A, B and C is added to those three already specified in Example 5.3.1. All four positions are the following:

$A_0(0, 0, 1)$ $B_0(0, 1, 0)$ $C_0(1, 0, 0)$

$A_1(0, 1, 1)$ $B_1(-1, 1, 0)$ $C_1(0, 0, 0)$

$A_2(-1, 1, 1)$ $B_2(-1, 2, 0)$ $C_2(-2, 1, 0)$

$A_3(0, 1, 0)$ $B_3(1, 0, 0)$ $C_3(0, 9, 1)$

The synthesis equations are those of Example 5.3.1, eqs. (5.3.53a) to (5.3.53h), except that eqs. (5.3.53a) to (5.3.53d) are now written for i=1,2,3, thus obtaining a system of 16 equations in 12 unknowns. The first two screws were already obtained in that example. The third one, corresponding to the fourth configuration, was also obtained via SUBROUTINE SCREW. All three screws are then

$$\underset{\sim}{s}_1 = \begin{pmatrix} 0.707 \\ -0.707 \\ 0.000 \end{pmatrix}, \quad \underset{\sim}{a}_1 = \begin{pmatrix} 0.250 \\ 0.250 \\ 0.500 \end{pmatrix}, \quad t_1 = -0.707, \quad \theta_1 = -70.529°$$

$$\underset{\sim}{s}_2 = \begin{pmatrix} 0.000 \\ -0.707 \\ 0.707 \end{pmatrix}, \quad \underset{\sim}{a}_2 = \begin{pmatrix} -0.250 \\ 1.000 \\ 1.000 \end{pmatrix}, \quad t_2 = -0.707, \quad \theta_2 = -109.471°$$

$$\underset{\sim}{s}_3 = \begin{pmatrix} 0.577 \\ 0.577 \\ 0.577 \end{pmatrix}, \quad \underset{\sim}{a}_3 = \begin{pmatrix} 0.000 \\ 0.000 \\ 0.000 \end{pmatrix}, \quad t_3 = 0.000, \quad \theta_3 = -120.000°$$

The nonlinear least-square problem arising from this dyad-synthesis problem was solved using the method of Newton-Raphson, as outlined in Section 1.13. Damping, however, proved to lead to divergence in this case, for which reason an undamped version of SUBROUTINE NRDAMC was used. This is SUBROUTINE NLLS (Nonlinear Least-Squares), whose listing appears in Fig. 6.3.1. The dimensionless error of the approximation was computed as

$$e = \sqrt{\frac{1}{n}\sum_1^n h_i^2}$$

with h_i defined as a normalised F_i , which is in turn defined in eqs. (5.3.53), i.e.

For i=1,2,3,

```
      SUBROUTINE NLLS(X,FUN,DFDX,P,TOL,N,M,ITER,MAX)
      REAL X(N),F(M),DF(M,N),P(1),U(M),DELTA(M)
C
C   THIS PROGRAM OBTAINS THE LEAST SQUARE SOLUTION TO THE NONLINEAR
C   SYSTEM F(X)= O, WHERE F AND X ARE M-AND N-DIMENSIONAL VECTORS, M
C   BEING GREATER THAN N. THE PROCEDURE IS ITERATIVE, AND AT EACH
C   ITERATION FINDS THE LEAST SQUARE SOLUTION TO THE LINEAR SYSTEM
C   DF*DELTA= -F, WHERE DF IS THE JACOBIAN M X N MATRIX OF THE ORIGINAL
C   SYSTEM, COMPUTED AT THE CURRENT VALUE OF X. THE LINEAR LEAST SQUARE
C   SOLUTION AT EACH ITERATION IS FOUND VIA HOUSEHOLDER REFLECTIONS (
C   BJOERCK A. AND G. DAHLQUIST, NUMERICAL METHODS, PRENTICE HALL,
C   ENGLEWOOD CLIFFS, N.J., 1974, PP. 201-206, 443-444).
C   PARAMETERS :
C       X     ..= N-DIMENSIONAL VECTOR OF UNKNOWNS.
C       F     ..= M-DIMENSIONAL VECTOR OF FUNCTIONS WHOSE EUCLIDEAN NORM
C                 IS TO BE MINIMIZED.
C       DF    ..= M X N JACOBIAN MATRIX OF F WITH RESPECT TO X.
C       P     ..= VECTOR OF PARAMETERS APPEARING IN FUN AND/OR DFDX. ITS
C                 DIMENSION VARIES FROM PROBLEM TO PROBLEM.
C       TOL   ..= A REAL POSITIVE "SMALL" VARIABLE DENOTING THE IMPOSED
C                 TOLERANCE. IT IS SUPPLIED BY THE USER.
C       ITER  ..= AN INTEGER VARIABLE DENOTING THE NUMBER OF THE CURRENT
C                 ITERATION.
C       MAX   ..= AN INTEGER VARIABLE DENOTING THE MAXIMUM NUMBER OF
C                 ALLOWED ITERATIONS.
C   SUBSIDIARY SUBROUTINES :
C
C   HECOMP  = TRIANGULARIZES A RECTANGULAR MATRIX BY HOUSEHOLDER
C             REFLECTIONS (MOLER C. B., MATRIX EIGENVALUE AND LEAST-
C             SQUARE COMPUTATIONS, COMPUTER SCIENCE DEPARTAMENT,
C             STANFORD UNIVERSITY. MARCH, 1973.)
C   HOLVE   = SOLVES TRIANGULARIZED SYSTEM BY BACK-SUBSTITUTION (MOLER
C             C. B., OP. CIT.)
C   FUN     = COMPUTES F.
C   DFDX    = COMPUTES DF.
C   FNORM   = COMPUTES THE MAXIMUM NORM OF A VECTOR.
C
C
      ITER=0
    1 ITER=ITER+1
      CALL FUN(X,F,P,M,N)
      FNOR=FNORM(F,M)
      IF(ITER.GT.MAX) GO TO 5
C
C   FORMS LINEAR LEAST SQUARE PROBLEM
      CALL DFDX(X,DF,P,M,N)
      CALL HECOMP(M,M,N,DF,U)
      CALL HOLVE(M,M,N,DF,U,F)
C
#
```

Fig 6.3.1 Listing of SUBROUTINE NLLS

```
C     COMPUTES CORRECTION BETWEEN TWO SUCCESSIVE ITERATIONS
          DO 2 I=1,M
               DELTA(I)=F(I)
   2      CONTINUE
          DELNOR=FNORM(DELTA,N)
          IF(DELNOR.LT.TOL) GO TO 6
C
C     IF DELNOR IS STILL LARGE, PERFORMS CORRECTION TO VECTOR X
   3      DO 4 I=1,N
               X(I)=X(I)-DELTA(I)
   4      CONTINUE
          GO TO 1
   5      WRITE(6,101) ITER,FNOR
          GO TO 7
   6      WRITE(6,102) ITER,FNOR
   7      DO 8 I=1,N
   8      WRITE(6,103) I,X(I)
          RETURN
 101      FORMAT(5X,"PROCEDURE DIVERGES AT ITERATION NUMBER",I3/5X,
     -            "THE NORM OF THE FUNCTION IS : ",E20.7/5X,
     -            "THE CURRENT VALUE OF X IS :"/)
 102      FORMAT(5X,"AT ITERATION NUMBER ",I3/5X,"THE NORM OF THE ",
     -           "FUNCTION IS ",E20.7/5X,"PROCEDURE CONVERGED, THE ",
     -           "SOLUTION BEING :"/)
 103      FORMAT(5X,2HX(I2,3H)= F15.5)
          END
#
```

Fig 6.3.1 Listing of SUBROUTINE NLLS (continued)

$$h_i = \tan\frac{\theta i}{2}\,\mathbf{f}^T(\mathbf{I}-\mathbf{s}_i\mathbf{s}_i^T)\mathbf{m}+\det(\mathbf{f},\mathbf{m}\ \mathbf{s}_i) \qquad (6.3.7a)$$

$$h_{i+3} = \frac{\mathbf{f}^T(\mathbf{I}-\mathbf{s}_i\mathbf{s}_i^T)(\mathbf{a}_i-\mathbf{g})}{||\mathbf{a}_i-\mathbf{g}||} \qquad (6.3.7b)$$

$$h_{i+6} = \frac{\mathbf{m}^T(\mathbf{I}-\mathbf{s}_i\mathbf{s}_i^T)(\mathbf{a}_i-\mathbf{q})}{||\mathbf{a}_i-\mathbf{q}||} \qquad (6.3.7c)$$

$$h_{i+q} = \begin{cases} \dfrac{2\mathbf{s}_i^T(\mathbf{g}_i-\mathbf{q}_i)-t_i}{t_i}, & \text{if } t_i \neq 0 \\[2ex] \dfrac{2\mathbf{s}_i(\mathbf{g}_i-\mathbf{q}_i)}{||g_i-q_i||}, & \text{if } t_i = 0 \end{cases} \qquad (6.3.7d)$$

$$h_{13} = \mathbf{f}^T\mathbf{f} - 1 \qquad (6.3.7e)$$

$$h_{14} = \mathbf{m}^T\mathbf{m} - 1 \qquad (6.3.7f)$$

$$h_{15} = \frac{\mathbf{f}^T(\mathbf{g}-\mathbf{q})}{||\mathbf{g}-\mathbf{q}||} \qquad (6.3.7g)$$

$$h_{16} = \frac{\mathbf{m}^T(\mathbf{g}-\mathbf{q})}{||\mathbf{g}-\mathbf{q}||} \qquad (6.3.7h)$$

The computer program written to solve the foregoing problem led to three different solutions, which appear in Table 6.3.7. It is pointed out that, since the given equations are not zeroed, $\mathbf{f}$ and $\mathbf{m}$ fail to be unit vectors, which is at any rate irrelevant, because they are meant to indicate only the directions of the axes of the involved revolute pairs. Moreover, the three different RRRR spatial mechanisms arising from the three different combinations that can be formed out of the three different solutions obtained, do not result to be Bennett mechanisms, as is the case in dealing with exact synthesis, already discussed in Section 5.3. Thus, to complete a single-loop-single-degree-of-fredom linkage to guide a rigid body for over 3 configurations, a second dyad of a different type (RS, SR, CC, for instance) should be resorted to.

Table 6.3.7 Least-square solutions to the overdetermined problem of the synthesis of an R-R dyad for rigid-body guidance through 4 specified configuration.

	First solution	Second solution	Third solution
$\underset{\sim}{f}$	0.565 0.474 0.599	-0.497 -0.690 -0.457	-0.902 0.341 -0.149
$\underset{\sim}{m}$	-0.158 0.310 0.898	-0.382 0.858 0.229	-0.649 -0.520 -0.503
$\underset{\sim}{g}$	-0.367 1.829 0.106	-0.425 -0.027 1.936	-0.225 -0.620 2.446
$\underset{\sim}{q}$	-0.155 1.534 0.282	-0.355 -0.273 2.189	-0.200 -0.929 2.651
e	0.483	0.235	0.208

6.4 UNDERDETERMINED PROBLEMS OF LINKAGE SYNTHESIS SUBJECT TO NO INEQUALITY CONSTRAINTS. The problems that will be discussed in the present Section are the opposite of those presented in Section 6.3, i.e. problems that will be treated here contain an excess of the number of unknowns over that of equations; hence, they admit infinitely many solutions. If the system is linear, then a unique solution exists whose norm is a global minimum and can be found via the Moore-Penrose (See Section 1.11) pseudo-inverse matrix of the system. If the said system is nonlinear, then all local minima can be found, as discussed in Section 1.13 via the introduction of Lagrange multipliers. As Fox and Gupta (6.7) point out, linear optimisation problems arise rarely in mechanism synthesis, that is to say, although (as was shown in Section 5.2) linkage synthesis problems for function generation can be formulated as linear problems, in this case the linkage parameters do not appear directly in the synthesis equations linearly. Hence, only the non-linear case will be discussed in this Section and the solution procedure will be illustrated with two examples.

Example 6.4.1. Synthesise an R-R dyad that guides a rigid body through the first two configurations shown in Fig 5.4.3. Since the general number of equations for this problem is 4n+4 (See Section 5.4), and for two configurations n=1, the resulting number of equations is 8, but the number of unknowns is 12, thus obtaining an underdetermined system of equations.

Solution:

The data are

$$\underset{\sim}{Q}_1 = \begin{pmatrix} 0 & 0 & 1 \\ 0 & -1 & 0 \\ 1 & 0 & 0 \end{pmatrix}, \quad \underset{\sim}{r}_1 = \begin{pmatrix} 0 \\ 1 \\ 0 \end{pmatrix} \tag{6.4.1}$$

where components are referred to the $X_0Y_0Z_0$ coordinate frame.

The unknowns are:

$$\underset{\sim}{a}_0 = \begin{pmatrix} a_{01} \\ a_{02} \\ a_{03} \end{pmatrix}, \quad \underset{\sim}{b}_0 = \begin{pmatrix} b_{01} \\ b_{02} \\ b_{03} \end{pmatrix}, \quad \underset{\sim}{u} = \begin{pmatrix} u_1 \\ u_2 \\ u_3 \end{pmatrix}, \quad \underset{\sim}{v}_0 = \begin{pmatrix} v_{01} \\ v_{02} \\ v_{03} \end{pmatrix} \tag{6.4.2}$$

Equation (5.4.5), written for j=1, is

$$(a_{01}-a_{03})^2+4a_{02}^2+(-a_{02}+a_{03})^2+2b_{01}(a_{01}-a_{03})+$$

$$+4(b_{02}-1)a_{02}+2b_{03}(-a_{01}+a_{03})-2b_{02}+1=0 \tag{6.4.3a}$$

Equation (5.4.8), written for j=0,1 leads to

$$u_1b_{01}+u_2b_{02}+u_3b_{03}=0 \tag{6.4.3b}$$

$$u_1(a_{01}-a_{03})+2u_2a_{02}+u_3(-a_{01}+a_{03})-u_2=0 \quad (6.4.3c)$$

Equation (5.4.9), written for j=0, 1 leads to

$$v_{01}b_{01}+v_{02}b_{02}+v_{03}b_{03}=0 \quad (6.4.3d)$$

$$v_{01}(a_{01}-a_{03})+2v_{02}a_{02}+v_{03}(-a_{01}+a_{03})-v_{02}=0 \quad (6.4.3e)$$

Equation (5.4.10), written for j=1 is

$$u_1v_{03}-u_2v_{02}+u_3v_{01}=u_1v_{01}+u_2v_{02}+u_3v_{03} \quad (6.4.3f)$$

Finally, eqs. (5.4.11) lead to

$$u_1^2+u_2^2+u_3^2=1 \quad (6.4.3g)$$

$$v_{01}^2+v_{02}^2+v_{03}^2=1 \quad (6.4.3h)$$

Equations (6.4.3a-h) constitute a system of 8 equations in the 12 unknowns: a_{0i}, b_{0i}, u_i, v_{0i} (i=1, 2, 3). The objective function that is now introduced, meant to be minimized, is a "length" of the dyad defined by the sum of the magnitudes of vectors $\underset{\sim}{a}_0$ and $\underset{\sim}{b}_0$. The objetive function is, then

$$\phi(a_0, b_0) = a_{01}^2 + a_{02}^2 + a_{03}^2 + b_{01}^2 + b_{02}^2 + b_{03}^2 \quad (6.4.4)$$

Thus, the optimisation problem can be stated as:

"Minimize $\phi(\underset{\sim}{a}_0,\underset{\sim}{b}_0)$ as given by eq. (6.4.4), subject to eqs. (6.4.3a-h)"

The solution to the foregoing problem proceeds as follows: Let $\underset{\sim}{f}$ be an 8-dimensional vector whose components are the left-hand sides of eqs. (6.4.3) when all the right-hand sides are set equal to zero, the non-zero terms having been transfered to the left hand side, of course. Let $\underset{\sim}{x}$ be a 12-dimensional vector whose components are $(a_{01},a_{02},a_{03},b_{01},b_{02},b_{03},u_1,u_2,u_3, v_{01},v_{02},v_{03})^T$. Define a new objetive function ψ as

$$\psi=\phi+\underset{\sim}{\lambda}^T\underset{\sim}{f} \quad (6.4.5)$$

where $\underset{\sim}{\lambda}=(\lambda_1,\lambda_2,\lambda_3,\lambda_4,\lambda_5,\lambda_6,\lambda_7,\lambda_8)^T$ is a vector containing the Lagrange multipliers of the system. The condition for function ψ to have a stationary value (See Section 1.10) is

$$\frac{\partial \psi}{\partial \underset{\sim}{x}} \equiv \frac{\partial \phi}{\partial \underset{\sim}{x}} + \left(\frac{\partial \underset{\sim}{f}}{\partial \underset{\sim}{x}}\right)^T \underset{\sim}{\lambda} = \underset{\sim}{0} \qquad (6.4.6)$$

which yields a system of 12 additional equations. System (6.4.6), together with system (6.4.3), constitutes a system of 20 equations in 20 unknowns, which are the 12 components of vector $\underset{\sim}{x}$ + the 8 components of vector $\underset{\sim}{\lambda}$. This is a determined system whose roots can be found via the method of Newton-Raphson.

Exercise 6.4.1 Solve system (6.4.3a-h), (6.4.6)

Another example, of a slightly different nature, is next presented

Example 6.4.2 Synthesise a plane RRRR linkage that generates the input-output values appearing in Table 6.4.1 and possessing the best transmision over the interval $\psi_1 \leq \psi \leq \psi_3$

Table 6.4.1 Prescribed input-output pairs

	input	output
i	ψ_i (degrees)	ϕ_i (degrees)
1	30	45
2	150	60
3	270	90

Solution:

The first step towards the solution is to define an objetive function whose optimisation yields the desired operation. In this case this operation is specified as two conditions. These are; i) a very precisely defined one specifying that the linkage generate the three pairs of values appearing in Table 6.4.1, and ii) a less precisely defined condition stating only that the linkage have the best possible transmission throughout the interval $\psi_1 \leq \psi \leq \psi_3$. The first condition can be given a mathematical representation very easily, namely, introducing Freudenstein's equation (6.8), with reference to Fig 6.4.1. This equation is:

$$k_1 - k_2\cos\phi + k_3\cos\psi + \cos(\phi-\psi) = 0 \quad (6.4.7)$$

where

$$k_1 = \frac{a_3^2 - a_2^2 - a_4^2}{2a_2a_4}, \quad k_2 = \frac{a_1}{a_2}, \quad k_3 = \frac{a_1}{a_4} \quad (6.4.8)$$

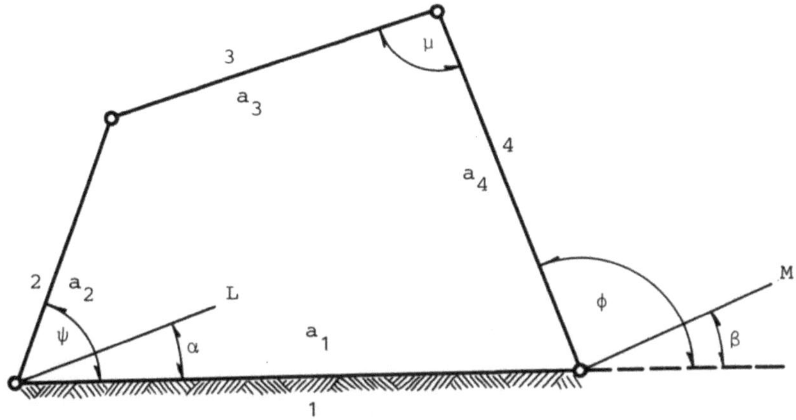

Fig 6.4.1 Plane RRRR function-generating linkage

Freudenstein's equation states the relationship that the linkage dimensions should observe in order to meet a set of input-output prescribed pairs, as given in Table 6.4.1. Notice, however, that Freudenstein's equation, as given in eqs. (6.4.7) and (6.4.8), contains only three undetermined parameters. Hence, in order to determine them, three input-output relationships should be prescribed, thus obtaining a system of three linear equations in the three unkowns k_1, k_2 and k_3. The problem at hand, this way, will have exactly one solution, if the involved system of equations is not singular. Thus, whether its transmission is good or bad, nothing else can be done about it. Nevertheless, in no part of the problem has it been required that the input and the output angles be measured from the line defined by the axes of the pairs R_{12} and R_{41}. One can, therefore, measure those angles from lines $R_{12}L$ and $R_{41}M$, respectively, making angles of α and β with line $R_{12}R_{41}$. This way, two additional undetermined parameters have been introduced, which allow, by a proper selection, to obtain the linkage that not only generates the pairs of values of Table 6.4.1, but also does it with the best transmission. What is still left to be defined is how to measure this transmission and so decide whether a given mechanism has a "good" or a "bad" one. It is widely known that the transmission of a linkage, i.e. its ratio of output torque to input torque, is measured by its transmission angle μ. The said transmission attains its best value for $\mu = 90°$ or $270°$, whereas its worst, for $\mu = 0°$ or $180°$. Hence a proper way to measure the "goodness" of the transmission of a linkage is measuring how much μ deviates either from $90°$ or from $270°$. Of course, the smaller this deviation is, the better the transmission of the linkage will be. Hence, a reasonable way to measure this deviation, z, is through the root-mean square (r.m.s.) value of $\cos\mu$, over the interval of interest, i.e.

$$z = \frac{1}{\Delta\psi}\int_{\psi_1}^{\psi_3}\cos^2\mu d\psi \tag{6.4.9}$$

where

$$\Delta\psi \equiv \psi_3 - \psi_1 \tag{6.4.10}$$

and

$$\cos\mu = \frac{2a_1a_2\cos(\psi+\alpha) - a_1^2 - a_2^2 + a_3^2 + a_4^2}{2a_3a_4} \tag{6.4.11}$$

Inverting eqs. (6.4.8), with $a_1 \equiv 1$,

$$a_2 = \frac{1}{k_2}, \quad a_3 = \frac{\sqrt{r}}{k_2k_3}, \quad a_4 = \frac{1}{k_3} \tag{6.4.12a}$$

where

$$r \equiv k_2^2 + k_3^2 + k_2^2\ k_3^2 + 2k_1k_2k_3 \tag{6.4.12b}$$

Substituting eqs. (6.4.12) into eq. (6.4.11),

$$\cos\mu = A\cos(\psi+\alpha) - B \tag{6.4.13}$$

$$A \equiv \frac{k_3^2}{\sqrt{r}} \quad , \quad B = \frac{k_3^2-r^2-k_2^2+k_2^2k_3^2}{2k_2\sqrt{r}} = -\frac{k_2+k_1k_3}{\sqrt{r}} \tag{6.4.14}$$

Thus,

$$\cos^2\mu = \frac{k_3^4}{r}\cos^2(\psi+\alpha) + \frac{2(k_2+k_1k_3)k_3^2}{r}\cos(\psi+\alpha) + \frac{(k_2+k_1k_3)^2}{r} \tag{6.4.15}$$

Substituting eq. (6.4.15) into (6.4.9),

$$z = \frac{k_3^4}{r\Delta\psi}\{\frac{1}{2}\Delta\Psi + \frac{1}{4}\left(\sin2(\psi_3+\alpha)-\sin2(\psi_1+\alpha)\right)\}$$
$$+ \frac{2k_3(k_2+k_1k_3)}{r\Delta\psi}\left(\sin(\psi_3+\alpha)-\sin(\psi_1+\alpha)\right) + \frac{(k_2+k_1k_3)^2}{r} = \frac{N}{r} \tag{6.4.16}$$

where

$$N \equiv k_3^4\{\frac{1}{2} + \frac{1}{4\Delta\psi}\left(\sin2(\psi_3+\alpha)-\sin2(\psi_1+\alpha)\right)\} +$$
$$+ \frac{2k_3^2(k_2+k_1k_3)}{\Delta\psi}\left(\sin(\psi_3+\alpha)-\sin(\psi_1+\alpha)\right) + (k_2+k_1K_3)^2 \tag{6.4.17}$$

To determine minima of z its stationary values should first be found. These are found by computing the zeros of its gradient

$$\underset{\sim}{g} = \frac{\partial z}{\partial \underset{\sim}{x}} \tag{6.4.18}$$

where $\underset{\sim}{x} \equiv (\alpha,\beta)^T$ is the decision vector. Now, since z does not depend directly upon α and β , but through vector $\underset{\sim}{k} \equiv (k_1,k_2,k_3)^T$, $\underset{\sim}{g}$ is computed applying the "chain rule",

$$\frac{\partial z}{\partial \underset{\sim}{x}} = \left(\frac{\partial \underset{\sim}{k}}{\partial x}\right)^T \frac{\partial z}{\partial \underset{\sim}{k}} \tag{6.4.19}$$

Clearly,

$$\frac{\partial z}{\partial \underset{\sim}{k}} = \frac{1}{r}\left\{\frac{\partial N}{\partial \underset{\sim}{k}} - \frac{\partial r}{\partial \underset{\sim}{k}} z\right\} \tag{6.4.20}$$

and*

---

* c ≡cos , s ≡ sin

$$\frac{\partial N}{\partial \underset{\sim}{k}} = \begin{bmatrix} \frac{2k_3^3}{\Delta\psi}\left(s(\psi_3+\alpha) - s(\psi_1+\alpha)\right) + 2(k_2+k_1k_3)k_3 \\ \frac{2k_3^2}{\Delta\psi}\left(s(\psi_3+\alpha) - s(\psi_1+\alpha)\right) + 2(k_2+k_1k_3) \\ a \end{bmatrix} \tag{6.4.21a}$$

$$a \equiv 4k_3^3\left\{\frac{1}{2} + \frac{1}{4\Delta\psi}\left(\sin 2(\psi_3+\alpha) - \sin 2(\psi_1+\alpha)\right)\right\} + \\ + \frac{1}{\Delta\psi}(4k_2k_3+6k_1k_3^3)\left(\sin(\psi_3+\alpha) - \sin(\psi_1+\alpha)\right) + \\ + 2(k_2+k_1k_3)k_1 \tag{6.4.21b}$$

$$\frac{\partial r}{\partial \underset{\sim}{k}} = \begin{bmatrix} 2k_2k_3 \\ 2k_1k_3+2k_2+2k_2k_3^2 \\ 2k_1k_2+2k_3+2k_2^2k_3 \end{bmatrix} \tag{6.4.22}$$

Vector $\underset{\sim}{k}$ is computed by solving Freudenstein's equation (6.4.7) written 3 times, once for each pair of input-output values. This leads to the linear system of equations

$$\underset{\sim}{A}\,\underset{\sim}{k} = \underset{\sim}{b} \tag{6.4.23}$$

where

$$\underset{\sim}{A} = \begin{bmatrix} 1 & -c(\phi_1+\beta) & c(\psi_1+\alpha) \\ 1 & -c(\phi_2+\beta) & c(\psi_2+\alpha) \\ 1 & -c(\phi_3+\beta) & c(\psi_3+\alpha) \end{bmatrix} \equiv \begin{bmatrix} 1 & -c_1 & \tilde{c}_1 \\ 1 & -c_2 & \tilde{c}_2 \\ 1 & -c_3 & \tilde{c}_3 \end{bmatrix} \tag{6.4.24a}$$

and

$$\underset{\sim}{b} = \begin{bmatrix} -c(\phi_1-\psi_1-\alpha+\beta) \\ -c(\phi_2-\psi_2-\alpha+\beta) \\ -c(\phi_3-\psi_3-\alpha+\beta) \end{bmatrix} \equiv \begin{bmatrix} -c_1^* \\ -c_2^* \\ -c_3^* \end{bmatrix} \tag{6.4.24b}$$

Solving eq. (6.4.23) explicitly by application of Gauss's elimination scheme or, equivalently, of the LU de composition (See Section 1.12), one obtains

$$k_i = m_{ij}b_j \text{ or } \underset{\sim}{k} = \underset{\sim}{M}\underset{\sim}{b} \tag{6.4.25}$$

where the usual index convention holds, i.e. a sum over repeated indices follows, with

$$q \equiv \tilde{c}_1(c_2-c_3)+\tilde{c}_2(c_3-c_1)+\tilde{c}_3(c_1-c_2) \tag{6.4.26a}$$

$$m_{1i} = \frac{c_{i-1}\tilde{c}_{i+1}-c_{i+1}\tilde{c}_{i-1}}{q}, \quad m_{2i} = \frac{\tilde{c}_{i+1}-\tilde{c}_{i-1}}{q}, \quad m_{3i} = \frac{c_{i+1}-c_{i-1}}{q} \tag{6.4.26b}$$

Thus, from eq. (6.4.25), one obtains

$$\frac{\partial \underset{\sim}{k}}{\partial \underset{\sim}{x}} = \underset{\sim}{M}\frac{\partial \underset{\sim}{b}}{\partial \underset{\sim}{x}} + \frac{\partial \underset{\sim}{M}}{\partial \underset{\sim}{x}}\underset{\sim}{b} \tag{6.4.27}$$

where, from eq. (6.4.24b),

$$\frac{\partial \underset{\sim}{b}}{\partial \underset{\sim}{x}} = \begin{pmatrix} -s(\phi_1-\psi_1-\alpha+\beta) & s(\phi_1-\psi_1-\alpha+\beta) \\ -s(\phi_2-\psi_2-\alpha+\beta) & s(\phi_2-\psi_2-\alpha+\beta) \\ -s(\phi_3-\psi_3-\alpha+\beta) & s(\phi_3-\psi_3-\alpha+\beta) \end{pmatrix} = \begin{pmatrix} -s_1^* & s_1^* \\ -s_2^* & s_2^* \\ -s_3^* & s_3^* \end{pmatrix} \tag{6.4.28}$$

and

$$\frac{\partial m_{1i}}{\partial x_j} = \frac{1}{q}\frac{\partial}{\partial x_j}(c_{i-1}\tilde{c}_{i+1}-c_{i+1}\tilde{c}_{i-1}) - \frac{m_{1i}}{q}\frac{\partial q}{\partial x_j} \tag{6.4.29a}$$

$$\frac{\partial m_{2i}}{\partial x_j} = \frac{1}{q}\frac{\partial}{\partial x_j}(\tilde{c}_{i+1}-\tilde{c}_{i-1}) - \frac{m_{2i}}{q}\frac{\partial q}{\partial x_j} \tag{6.4.29b}$$

$$\frac{\partial m_{3i}}{\partial x_j} = \frac{1}{q}\frac{\partial}{\partial x_j}(c_{i+1}-c_{i-1}) - \frac{m_{3i}}{q}\frac{\partial q}{\partial x_j} \tag{6.4.29c}$$

$$\frac{\partial}{\partial x_1}(c_{i-1}\tilde{c}_{i+1}-c_{i+1}\tilde{c}_{i-1}) = -c_{i-1}\tilde{s}_{i+1}+c_{i+1}\tilde{s}_{i-1} \tag{6.4.29d}$$

$$\frac{\partial}{\partial x_2}(c_{i-1}\tilde{c}_{i+1}-c_{i+1}\tilde{c}_{i-1}) = -s_{i-1}\tilde{c}_{i+1}+s_{i+1}\tilde{c}_{i-1} \tag{6.4.29e}$$

$$\frac{\partial q}{\partial x_1} = -\tilde{s}_1(c_2-c_3)-\tilde{s}_2(c_3-c_1)-\tilde{s}_3(c_1-c_2) \tag{6.4.29f}$$

$$\frac{\partial q}{\partial x_2} = -\tilde{c}_1(s_2-s_3)-\tilde{c}_2(s_3-s_1)-\tilde{c}_3(s_1-s_2) \tag{6.4.29g}$$

$$\frac{\partial}{\partial x_1}(\tilde{c}_{i+1}-\tilde{c}_{i-1}) = -(\tilde{s}_{i+1}-\tilde{s}_{i-1}) \tag{6.4.29h}$$

$$\frac{\partial}{\partial x_2}(\tilde{c}_{i+1}-\tilde{c}_{i-1}) = 0 \tag{6.4.29i}$$

$$\frac{\partial}{\partial x_1}(c_{i+1}-c_{i-1}) = 0 \tag{6.4.29j}$$

$$\frac{\partial}{\partial x_2}(c_{i+1}-c_{i-1}) = -(s_{i+1}-s_{i-1}) \tag{6.4.29k}$$

Substituting expressions (6.4.29d-k) into (6.4.29a-c), one obtains

$$\frac{\partial m_{1i}}{\partial x_1} = \frac{1}{q}(-c_{i-1}\tilde{s}_{i+1}+c_{i+1}\tilde{s}_{i-1}) + \frac{m_{1i}}{q}\left(\tilde{s}_1(c_2-c_3)+\tilde{s}_2(c_3-c_1)+\tilde{s}_3(c_1-c_2)\right) \tag{6.4.30a}$$

$$\frac{\partial m_{1i}}{\partial x_2} = \frac{1}{q}(-s_{i-1}\tilde{c}_{i+1}+s_{i+1}\tilde{c}_{i-1}) + \frac{m_{1i}}{q}\left(\tilde{c}_1(s_2-s_3)+\tilde{c}_2(s_3-s_1)+\tilde{c}_3(s_1-s_2)\right) \tag{6.4.30b}$$

$$\frac{\partial m_{2i}}{\partial x_1} = -\frac{1}{q}(\tilde{s}_{i+1}-\tilde{s}_{i-1}) + \frac{m_{2i}}{q}\left(\tilde{s}_1(c_2-c_3)+\tilde{s}_2(c_3-c_1)+\tilde{s}_3(c_1-c_2)\right) \tag{6.4.30c}$$

$$\frac{\partial m_{2i}}{\partial x_2} = \frac{m_{2i}}{q}\left(\tilde{c}_1(s_2-s_3)+\tilde{c}_2(s_3-s_1)+\tilde{c}_3(s_1-s_2)\right) \tag{6.4.30d}$$

$$\frac{\partial m_{3i}}{\partial x_1} = \frac{m_{3i}}{q}\left(\tilde{s}_1(c_2-c_3)+\tilde{s}_2(c_3-c_1)+\tilde{s}_3(c_1-c_2)\right) \tag{6.4.30e}$$

$$\frac{\partial m_{3i}}{\partial x_2} = -\frac{1}{q}(s_{i+1}-s_{i-1}) + \frac{m_{3i}}{q}\left(\tilde{c}_1(s_2-s_3)+\tilde{c}_2(s_3-s_1)+\tilde{c}_3(s_1-s_2)\right) \tag{6.4.30f}$$

The (i,j) element of matrix $\frac{\partial \underset{\sim}{M}}{\partial \underset{\sim}{x}} \underset{\sim}{b}$ is computed as

$$\left(\frac{\partial \underset{\sim}{M}}{\partial \underset{\sim}{x}} \underset{\sim}{b}\right)_{ij} = \frac{\partial m_{i\ell}}{\partial x_j} b_\ell \tag{6.4.31}$$

In order to compute the zeros of $\underset{\sim}{g}$, the Newton-Raphson method is applied. This requires the computation of the Hessian matrix, $\underset{\sim}{J}$, of z, i.e.

$$\underset{\sim}{J} \equiv \frac{\partial \underset{\sim}{g}}{\partial \underset{\sim}{x}} = \frac{\partial^2 z}{\partial \underset{\sim}{x}^2} \tag{6.4.32}$$

Again, $\underset{\sim}{J}$ is computed applying the "chain" rule, as

$$\frac{\partial \underset{\sim}{g}}{\partial \underset{\sim}{x}} = \frac{\partial \underset{\sim}{g}}{\partial \underset{\sim}{k}} \frac{\partial \underset{\sim}{k}}{\partial \underset{\sim}{x}} \tag{6.4.33}$$

where

$$\frac{\partial \underset{\sim}{g}}{\partial \underset{\sim}{x}} = \left(\frac{\partial^2 \underset{\sim}{k}}{\partial \underset{\sim}{k} \partial \underset{\sim}{x}}\right)^T \frac{\partial z}{\partial \underset{\sim}{k}} + \left(\frac{\partial \underset{\sim}{k}}{\partial \underset{\sim}{x}}\right)^T \frac{\partial^2 z}{\partial \underset{\sim}{k}^2} \qquad (6.4.34)$$

But

$$\frac{\partial^2 \underset{\sim}{k}}{\partial \underset{\sim}{k} \partial \underset{\sim}{x}} = \frac{\partial}{\partial \underset{\sim}{x}} \quad \frac{\partial \underset{\sim}{k}}{\partial \underset{\sim}{k}} = \frac{\partial \underset{\sim}{I}}{\partial \underset{\sim}{x}} = \underset{\sim}{O} \qquad (6.4.35)$$

$\underset{\sim}{I}$ being the 3x3 identity matrix. Thus,

$$\frac{\partial \underset{\sim}{g}}{\partial \underset{\sim}{k}} = \left(\frac{\partial \underset{\sim}{k}}{\partial \underset{\sim}{x}}\right)^T \frac{\partial^2 z}{\partial \underset{\sim}{k}^2} \qquad (6.4.36)$$

and $\underset{\sim}{J}$ is finally obtained as

$$\underset{\sim}{J} = \frac{\partial \underset{\sim}{g}}{\partial \underset{\sim}{x}} = \left(\frac{\partial \underset{\sim}{k}}{\partial \underset{\sim}{x}}\right)^T \frac{\partial^2 z}{\partial \underset{\sim}{k}^2} \frac{\partial \underset{\sim}{k}}{\partial \underset{\sim}{x}} \qquad (6.4.37)$$

where

$$\frac{\partial^2 z}{\partial \underset{\sim}{k}^2} = \frac{1}{r} \left\{ \frac{\partial^2 N}{\partial \underset{\sim}{k}^2} - z \frac{\partial^2 r}{\partial \underset{\sim}{k}^2} - \left[\frac{\partial z}{\partial \underset{\sim}{k}} \left(\frac{\partial r}{\partial \underset{\sim}{k}} \right)^T + \frac{\partial r}{\partial \underset{\sim}{k}} \left(\frac{\partial z}{\partial \underset{\sim}{k}} \right)^T \right] \right\} \qquad (6.4.38)$$

From eqs. (6.4.21) and (6.4.22),

$$\frac{\partial 2_N}{\partial \underset{\sim}{k}^2} = \begin{bmatrix} 2k_3^2 & 2k_3 & \frac{6k_3^2}{\Delta\psi}\left(s(\psi_3+\alpha)-s(\psi_1+\alpha)\right) + b \\ 2k_3 & 2 & \frac{4k_3}{\Delta\psi}\left(s(\psi_3+\alpha)-s(\psi_1+\alpha)\right) + 2k_1 \\ c & d & e \end{bmatrix} \qquad (6.4.39)$$

$b \equiv 2k_2+4k_1k_3$

$$c \equiv \frac{6k_3^2}{\Delta\psi}\left(\sin(\psi_3+\alpha)-\sin(\psi_1+\alpha)\right)+2k_2+4k_1k_3$$

$$d \equiv \frac{4k_3}{\Delta\psi}\left(\sin(\psi_3+\alpha)-\sin(\psi_1+\alpha)\right)+2k_1$$

$$e \equiv 12k_3^2\left\{ \frac{1}{2} + \frac{1}{4\Delta\psi}\left(\sin 2(\psi_3+\alpha)-\sin 2(\psi_1+\alpha)\right)\right\} +$$

$$+ \frac{1}{\Delta\psi}(4k_2+12k_1k_3)\left(\sin(\psi_3+\alpha)-\sin(\psi_1+\alpha)\right) +2k_1^2$$

$$\frac{\partial^2 r}{\partial \underset{\sim}{k}^2} = \begin{bmatrix} 0 & 2k_3 & 2k_2 \\ 2k_3 & 2(1+k_3^2) & 2k_1+4k_2k_3 \\ 2k_2 & 2k_1+4k_2k_3 & 2+2k_2^2 \end{bmatrix} \qquad (6.4.40)$$

With expressions (6.4.18)-(6.4.22) to compute the gradient $\underset{\sim}{g}$ of the objective function and (6.4.32)-(6.4.40) to compute its Hessian $\underset{\sim}{J}$, the Newton Raphson algorithm, discussed in Section 1.13, can now be applied, as:

Given a random starting value $\underset{\sim}{x}_0 = (\alpha_0, \beta_0)^T$, generate the sequence $\underset{\sim}{x}_1, \underset{\sim}{x}_2, \ldots, \underset{\sim}{x}_k, \underset{\sim}{x}_{k+1}, \ldots$, that, if convergence is reached, will approach the solution $\underset{\sim}{x}^*$ that cancels $\underset{\sim}{g}$. The said sequence is generated as:

$$\underset{\sim}{x}_{k+1} = \underset{\sim}{x}_k - \underset{\sim}{J}^{-1}(\underset{\sim}{x}_k)\underset{\sim}{g}(\underset{\sim}{x}_k)$$

The computer subprogram NRDAMP, introduced in section 1.13, implements the foregoing iterative procedure with damping, as discussed in that Section. This subprogram was applied to solve the problem at hand, the results obtained being:

$$\underset{\sim}{x}^* \equiv \begin{pmatrix} \alpha^* \\ \beta^* \end{pmatrix} = \begin{pmatrix} 1.7367 \\ 3.4714 \end{pmatrix} \text{ rad}$$

which was reached after 19 iterations and 31 evaluations of the gradient $\underset{\sim}{g}$. The resulting value of $\underset{\sim}{g}$ was

$$\underset{\sim}{g}(x^*) = (0.000004,\ 0.000001)^T \text{ (dimensionless)}$$

which corresponds to a value of z equal to 0.1082 (dimensionless). Thus the r.m.s. value of $\cos\mu$ is 0.3289. The linkage thus obtained was analysed with the aid of a program written ad hoc and the mean value of μ was obtained, within the interval $\psi_1 \le \psi \le \psi_3$, as 89.25°, very close to 90°, as expected.

6.5 LINKAGE OPTIMISATION SUBJECT TO INEQUALITY CONSTRAINTS. PENALTY FUNCTION METHODS.

The class of optimisation problems discussed in Section 6.3 does not include any constraints defined by inequality relations, thus lending themselves to relatively easy handling. One problem involving equality constraints was introduced in Section 6.4. This problem was formulated by application of Langrange multipliers, which led to the solution of a nonlinear system of algebraic equations containing more uknowns than decision variables. The number of unknowns increases because of the introduction of the unknown Langrage multipliers. If the arising nonlinear system of equations can be easily differentiated, then its solution can be attempted by the method of Newton-Raphson, as was already discussed. If, however, that system either cannot be differentiated, or the computation of its derivatives requires prohibitively large execution times, then other techniques should be resorted to. The penalty function methods introduced herewith will be discussed only in connection with inequality constraints. Should the problem at hand contain also equality constraints, these can be easily handled, as shown in (6.9). Optimisation problems subject to inequality constraints arise very often in linkage synthesis. For example, if the RRRR plane linkage shown in Fig 6.4.1 is to be driven by a uniform -velocity electrical motor whose torque is applied at link 2, then this link should be a crank, i.e. it should be able to rotate through a 360° angle. The conditions under which link 2 is a crank can be readily written from Grashoff's conditions. These conditions are derived by writting the conditions of existence of the triangle arising for the positions corresponding to $\psi = 0, \pi$ of the linkage shown in Fig 6.4.1. These are

$$|a_1 - a_2| > a_3 + a_4 \qquad (6.5.1a)$$

$$|a_1-a_2| > |a_4-a_3| \quad (6.5.1b)$$

$$a_1+a_2 < a_3+a_4 \quad (6.5.1c)$$

$$a_1+a_2 > |a_4-a_3| \quad (6.5.1d)$$

As another example, assume the transmision angle of the same linkage, μ, defined in Section 6.4, is constrained to lie between μ_a and μ_b, then

$$\mu_a \leq \mu \leq \mu_b \quad (6.5.2)$$

would be the said inequality constraint.

As a "rule of thumb", the transmission angle is usually constrained to lie between 40° and 140°, in order to ensure a "good" force transmission. This class of problems then, is of the general form (6.2.1), where the inequalities define a region of the decision-variable space, Ω , called "the allowable region" or, equivalently, "the feasible region".

An extensive account of different methods to solve this problem appears in [6.7]. Fox and Gupta in this paper classify the different methods of solution to this problem as; i) penaly function methods and ii) direct methods. Penalty function methods are dealt with in this section, whereas direct methods are discussed in section 6.6.

The idea behind penalty function methods is to approach the optimal solution asymptotically by extrapolation of a sequence of optimal solutions to unconstrained problems. In this respect there are two possibilities: either the optimum is approached within the feasible region or from outside of it. The penalty function is accordingly referred to as either interior or exterior. Given an objective function $\phi(\underset{\sim}{x})$ subject to equality constraints that will not be taken into account here (These can be thought of as having already been used to solve for the corresponding set of variables, thus dealing only with independent decision variables), and to inequality constraints $\underset{\sim}{h}(\underset{\sim}{x}) \geq 0$, a sequence $\{\psi_k\}$ of <u>interior penalty functions</u> is next constructed. Let

$$\psi_k(\underset{\sim}{x};r_k) = \phi(x) + r_k \sum_{1}^{p} \frac{1}{h_i(\underset{\sim}{x})}, \quad k=1,2,\ldots \tag{6.5.3}$$

where all r_k's are positive and observe a decreasing order, i.e.

$$r_1 > r_2 > r_3 > \ldots \tag{6.5.4}$$

and

$$h_i(\underset{\sim}{x}) \geq 0, \quad i = 1, \ldots, p \tag{6.5.5}$$

This way, when trying to minimize $\phi(\underset{\sim}{x})$, the penalty term in ψ_k is positive and increases repidly as vector $\underset{\sim}{x}$ approaches one of the constraints. The solution to each unconstrained problem, $\underset{\sim}{x}^*(r_k)$, is forced then to lie within the feasible region if $h_i(\underset{\sim}{x})$ is redefined as zero when the given constraints (6.5.5) are violated. The solution to the original constrained problem, $\underset{\sim}{x}^*$, is thus found as

$$\underset{\sim}{x}^* = \lim_{r_k \to 0} \underset{\sim}{x}(r_k)$$

By allowing

$$r_{k+1} = r_k/t_k \qquad 10 \leq t_k \leq 50 \tag{6.5.6}$$

a good extrapolation value $\underset{\sim}{x}^*$ can be computed by fitting the $x_j^*(r_k)$ values for each component of the vector $\underset{\sim}{x}$ to a continuous function of r, as suggested by Fiacco and McCormick {6.10}. If i values $x_j(r_k)$, $j = 1, \ldots, n$; $k=1, \ldots, i$ are available, the suggested function takes on the form

$$x_j(r) = A_j + \sum_{k=1}^{i-1} B_{jk} r^{k/2}, \quad j=1, \ldots, n \tag{6.5.7}$$

This way, then

$$x_j^* = A_j, \quad j = 1, \ldots, n \tag{6.5.8}$$

The question of how to assign r_1 is dealt with by specifying that the penalty term be a given fraction, f, of the objetive function when initialising the iteration process, i.e. r_1 is solved for from the following equation

$$r_1 \sum_1^p \frac{1}{h_i(\underset{\sim}{x}_0)} = f\ \phi(\underset{\sim}{x}_0) \qquad (6.5.9)$$

$\underset{\sim}{x}_0$ being the initial guess, lying of course within the feasible region. The value of f is recommended to be taken between 0.01 and 0.5

Exterior penalty functions have the form

$$\psi(x;r_k) = \phi(x) + r_k \sum_1^p \left(\min\{0,h_i(x)\}\right)^2, \quad k=1,2,\ldots \qquad (6.5.10)$$

where all r_k's are again positive and observe now an increasing order , i.e.

$$r_1 < r_2 < r_3 < \ldots$$

This way, whenever vector $\underset{\sim}{x}$ leaves the feasible region, the penalty function increases, thus forcing the solution to approach this region. Notice that, within the feasible region, the penalty term vanishes.

Both interior and exterior penalty functions are shown in Fig 6.5.1, for one-dimensional problems, with a feasible region defined by $a \leq x \leq b$.

The question that arises naturally is how to solve the corresponding unconstrained problem, for each value of r. In this respect, there are three possibilities:

i) The objetive function and the penalty term have first and second derivatives readily available.

ii) Only first derivatives are available.

iii) No derivatives are available

The solution to each of these problems is next outlined.

Case i) An obvious way to proceed is to compute the roots of the gradient of the penalty function via the Newton-Raphson method.

These roots are found from the system

$$\psi'(\underset{\sim}{x}) = \underset{\sim}{0}$$

which is nonlinear. In section 1.13 the numerical solution to this problem was extensively discussed. According to that section, the Newton-Raphson

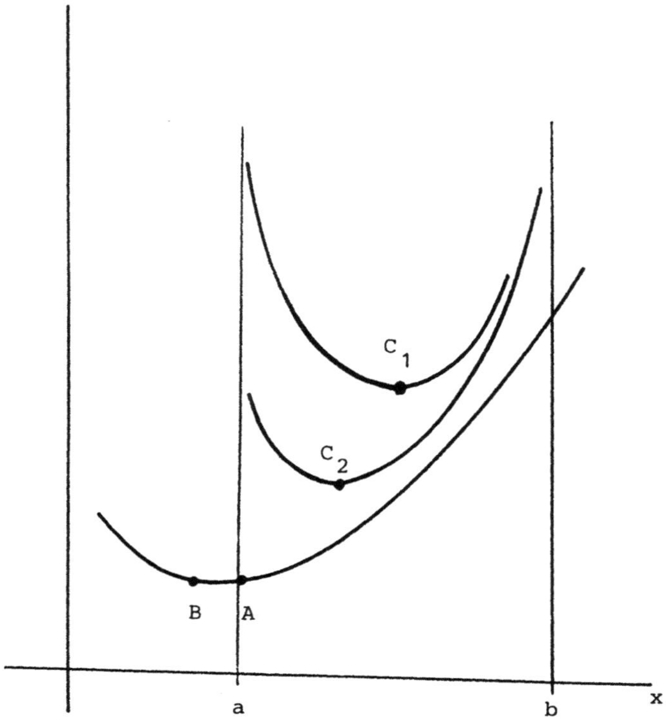

a) Interior penalty function

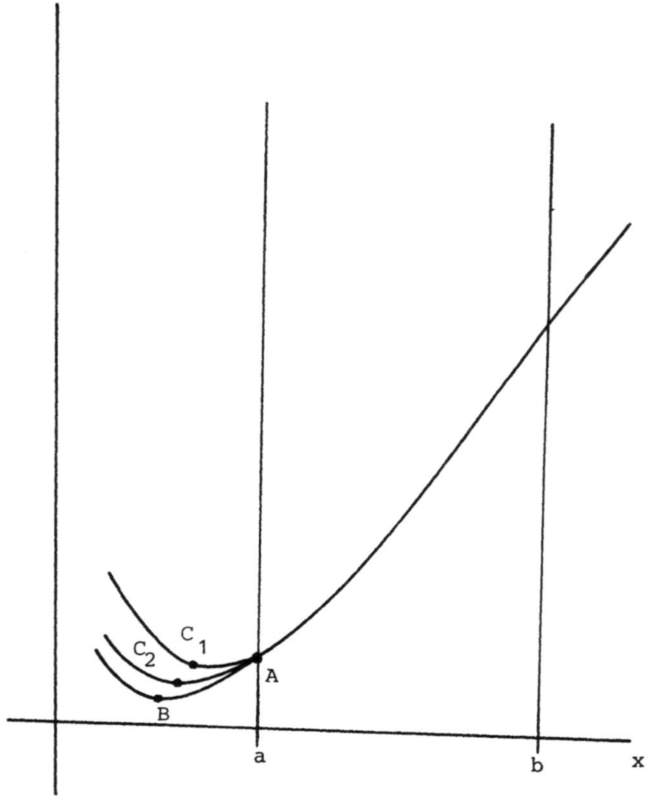

b) Exterior penalty function

A: Constrained minimum
B: Unconstrained minimum
C_i: Successive minima of unconstrained problems

Fig 6.5.1. Penalty functions

method requires the computation of the Jacobian matrix of the system, i.e. the Hessian matrix $\psi''(\underset{\sim}{x})$ which, as assumed, is readily available. This procedure can then be carried out with the aid of subroutines NRDAMP, DECOMP and SOLVE, whose listings appear in Section 1.

Case ii) Since no second derivatives, or at least not all of them, are available, the Newton-Raphson method cannot be applied in this instance. One can resort, however, to gradient methods, i.e. methods that require only first derivatives, like the one of Fletcher and Reeves (6.11). This procedure is next outlined: Let $f(\underset{\sim}{x})$ be a scalar function which is intended to minimize, and $\underset{\sim}{g}(\underset{\sim}{x})$ be its gradient. Then

a) Assign a starting value $\underset{\sim}{x}_0$ arbitrarily.

b) Define: $\underset{\sim}{g}_0(\underset{\sim}{x}) = \underset{\sim}{g}(\underset{\sim}{x}_0)$, $\underset{\sim}{p}_0 = -\underset{\sim}{g}_0$

c) Compute the minimum $\underset{\sim}{x}_{i+1}$ of $f(\underset{\sim}{x})$ on the line through $\underset{\sim}{x}_i$ in the direction of $\underset{\sim}{p}_i$. Notice that this minimization need only be performed in one dimension, which anyway could be an extremely time-consuming task in itself, if not impossible to execute. In order to make their method efficient, the authors estimate this minimum by interpolating the function through three points.

d) Let $\underset{\sim}{g}_{i+1} = \underset{\sim}{g}(\underset{\sim}{x}_{i+1})$

e) Define: $\beta_i = ||\underset{\sim}{g}_{i+1}||^2 / ||\underset{\sim}{g}_i||^2$

f) Let $\underset{\sim}{p}_{i+1} = -\underset{\sim}{g}_{i+1} + \beta_i \underset{\sim}{p}_i$

g) After every n+1 iterations, verify if $|f(\underset{\sim}{x}_{k+n+1}) - f(\underset{\sim}{x}_k)| \leq \varepsilon$, ε being a small prescribed quantity. If the inequality holds, then one minimum has been found; if it does not, then repeat the procedure starting at step c.

This procedure guarantees the computation of the minimum of a quadratic function in at most n iterations. For non-quadratic functions the process should be stopped whenever the prescribed convergence criterion has been met. Step c, which is crucial in the foregoing algorithm, is performed in three stages: first, the order of magnitude of the distance between the maximum x_{i+1} and the previous one, $\underset{\sim}{x}_i$, is estimated; next, bounds for this distance are established and, finally, the location of $\underset{\sim}{x}_{i+1}$ is estimated through interpolation. Reference (6.11) includes an ALGOL procedure implementing the foregoing algorithm.

Case iii) In this case the most suitable method to solve the problem is that of Powell (6.12) . The procedure is based upon searches along a set of linearly independent directions $\underset{\sim}{\xi}_1, \underset{\sim}{\xi}_2, \ldots, \underset{\sim}{\xi}_n$, as outlined in the procedure appearing in the cited reference, which is next reproduced:

a) For r = 1, 2, ..., n, calculate λ_r so that $f(\underset{\sim}{p}_{r-1} + \lambda_r \underset{\sim}{\xi}_r)$ is a minimum, where $f(\underset{\sim}{x})$ is the objetive function. Moreover, define

$$\underset{\sim}{p}_r = \underset{\sim}{p}_{r-1} + \lambda_r \underset{\sim}{\xi}_r$$

b) Find the integer m, $1 \leq m \leq n$, so that $\{f(\underset{\sim}{p}_{m-1}) - f(\underset{\sim}{p}_m)\}$ is a maximum and define $\Delta = f(\underset{\sim}{p}_{m-1}) - f(\underset{\sim}{p}_m)$

c) Calculate $f_3 = f(2\underset{\sim}{p}_n - \underset{\sim}{p}_0)$ and define $f_1 = f(\underset{\sim}{p}_0)$ and $f_2 = f(\underset{\sim}{p}_n)$

d) If either $f_3 \geq f_1$ and or $(f_1 - 2f_2 + f_3)(f_1 - f_2)^2 \geq \frac{1}{2}\Delta(f_1 - f_3)^2$, use the old directions $\underset{\sim}{\xi}_1, \underset{\sim}{\xi}_2, \ldots, \underset{\sim}{\xi}_n$ for the next iteration and use $\underset{\sim}{p}_n$ for the next $\underset{\sim}{p}_0$; otherwise,

e) Defining $\underset{\sim}{\xi} = \underset{\sim}{p}_n - \underset{\sim}{p}_0$, calculate λ so that $f(\underset{\sim}{p}_n + \lambda\underset{\sim}{\xi})$ is a minimum, use $\underset{\sim}{\xi}_1, \ldots \underset{\sim}{\xi}_{m-1}, \underset{\sim}{\xi}_{m+1}, \ldots, \underset{\sim}{\xi}_n, \underset{\sim}{\xi}$ as the directions and $\underset{\sim}{p}_n + \lambda\underset{\sim}{\xi}$ as the starting point for the next iteration.

For quadratic functions

$$f(\underset{\sim}{x}) = \underset{\sim}{x}^T \underset{\sim}{A}\underset{\sim}{x} + \underset{\sim}{b}^T \underset{\sim}{x} + c$$

the search directions are conjugate, i.e., for $i \neq j$,

$$\underset{\sim}{\xi}_i^T A \underset{\sim}{\xi}_j = 0$$

Theoretically Powell's algorithm finds the minimum of such a quadratic function in n iterations. In practice, however, two conjugate directions may be quasi-dependent, the procedure thus failing to converge. This situation is taken into account in the foregoing algorithm at step d. The said algorithm, then, may require more than n iterations to find the minimum of a quadratic function.

The method used to minimize a function of one variable without calculating derivatives is based upon an interpolation formula to fit three points on a quadratic function.

The application of the penalty function method is next illustrated with two simple examples.

Example 6.5.1. One-dimensional optimisation subject to inequality constraints.

Let $\phi = 1 + x^2$ be the function to be minimized subject to the constraints

$1 \leq x \leq 2$

Solution: The minimum of the function in the absence of constraints is 1, which is attained at $x = 0$. Due to the constraints, the function attains its minimum at $x = 1$, the corresponding minimum being equal to 2. Although this problem thus can be solved by inspection, it is presented here to illustrate the method of the penalty function.

An interior penalty function is constructed as

$$\psi(x;\, r_k) = 1 + x^2 + r_k\left(\frac{1}{x-1} - \frac{1}{x-2}\right),\ k = 1, 2, \ldots$$

which is then minimized for

$$r_1 = 1.0,\ r_2 = 0.1,\ r_3 = 0.01$$

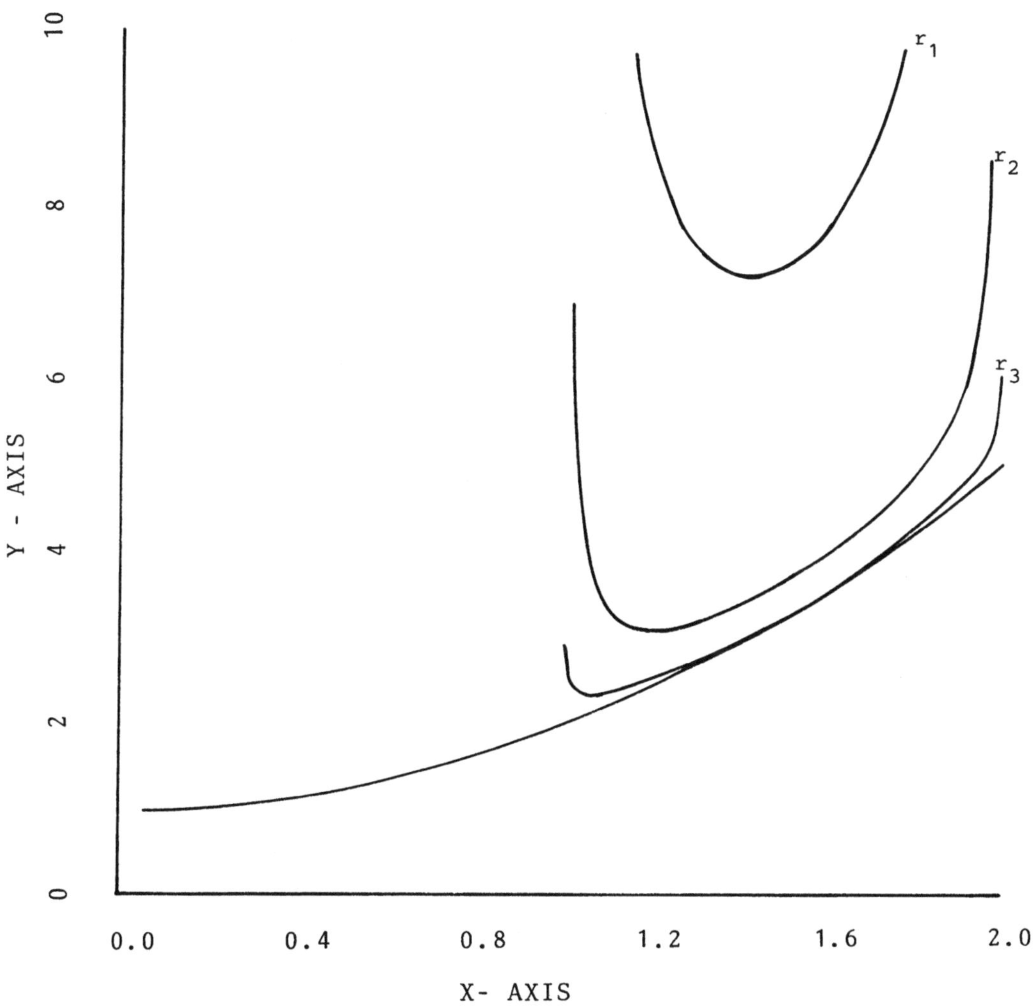

Fig 6.5.2 One -dimensional minimization using an interior penalty function.

This is done by computing the roots of the derivative of ψ with respect to x , which leads to the solution of the following nonlinear equation

$$\psi'(x) = 2x - r_k \left(\frac{1}{(x-1)^2} - \frac{1}{(x-2)^2}\right) = 0$$

Proceeding to solve the foregoing equation via the method of Newton-Raphson, the following minimizing values were found:

$x^*(r_1) = 1.416$, $x^*(r_2) = 1.198$, $x^*(r_3) = 1.068$

which clearly show the trend towards the minimizing value $x^* = 1.0$ of the original constrained problem. The above values are next fit to a parabola as

$$x = A + B_1 r^{1/2} + B_2 r$$

thus obtaining

$$A = 0.998, \; B_1 = 0.733, \; B_2 = -0.315$$

i.e.

$$x^* = \lim_{r \to 0} x^*(r) = A = 0.998$$

The objetive function, as well as the corresponding penalty functions are shown in Fig 6.5.2.

Example 6.5.2. Two-dimensional minimization subject to inequality constraints.

Let

$$\phi = x_1^2 + 2x_2^2$$

be the function to be minimized, subject to the following constraints:

$$x_1 \geq 0, \; x_2 \geq 0, \; x_1 + x_2 \geq 1$$

The following penalty function is then constructed:

$$\psi(x;\, r_k) = x_1^2 + 2x_2^2 + r_k\left(\frac{1}{x_1} + \frac{1}{x_2} + \frac{1}{x_1+x_2-1}\right), \; k = 1,2, \ldots$$

for

$$r_1 = 0.1, \; r_2 = 0.01 \text{ and } r_3 = 0.001$$

The gradient of ψ has the components

$$\frac{\partial\psi}{\partial x_1} = 2x_1 + r_k \left(- \frac{1}{x_1^2} - \frac{1}{(x_1+x_2-1)^2}\right)$$

$$\frac{\partial\psi}{\partial x_2} = 4x_2 + r_k \left(- \frac{1}{x_2^2} - \frac{1}{(x_1+x_2-1)^2}\right)$$

which vanishes at an extremum. The roots of the foregoing nonlinear system of algebraic equations were then computed using cubroutines NRDAMP, DECOMP and SOLVE, which yielded the following values:

$$x_1(r_1) = 0.7941, \; x_2(r_1) = 0.4704$$

$$x_1(r_2) = 0.7140, \; x_2(r_2) = 0.3703$$

$$x_1(r_3) = 0.6836, \; x_2(r_3) = 0.3434$$

which were reached in at most 6 iterations of the Newton-Raphson algorithm, thus showing very quick convergence.

The obtained optimising values were then fit to a parabola for each component, as follows

$$x_i(r) = A_i + B_i r^{1/2} + C_i, \; i = 1,2$$

After performing the corresponding substitutions in the given representation of $\underset{\sim}{x}^*$, a linear system of equations was obtained, whose roots were found to be

$$A_1 = 0.6687, \; B_1 = 0.4790, \; C_1 = - 0.2605$$

$$A_2 = 0.3317, \; B_2 = 0.3612, \; C_2 = 0.2443$$

Thus,

$$x_1 = A_1 = 0.6687$$

$$x_2 = A_2 = 0.3317$$

the corresponding exact optimising values being

$$x_1 = \frac{2}{3}, \; x_2 = \frac{1}{3}$$

as can be verified from the condition that, at the minimum, the gradient of the ellipses $x_1^2 + 2x_2^2$ = constant, must be normal to the line $x_1 + x_2 = 1$, defining the third inequality constraint. An algorithm called "Sequential Unconstrained Minimization Technique" (SUMT), implementing the interior penalty function method, due to Fiacco and McCormick, has been published and extensively tested. The corresponding program appears in (6.13).
Amongst the methods mentioned to solve the unconstrained minimization problem arising in connection with penalty function methods for constrained minimization, they only provide local minima. An algorithm called the "tunneling algorithm", developed by Velasco-Levy and Montalvo at the University of Mexico (UNAM) (6.14) finds the global minimum of an unconstrained minimization problem.
The interested reader can find examples of linkage optimisation problems by application of penalty functions either in (6.9) or in (6.15).

6.6 LINKAGE OPTIMISATION SUBJECT TO INEQUALITY CONSTRAINTS. DIRECT METHODS.

Direct methods are those that handle inequalities as given, i.e. without resorting to auxiliary unconstrained problems, as is the case when using penalty function methods. There are several methods of this kind, but the ones that will be discussed here are:

i) the method of the feasible directions,
ii) the gradient projection method, and
iii) the complex method

The method of the feasible directions is applicable to linear constraints. In the presence of nonlinear constraints, a previous linearisation should be performed. It requires first derivatives of both the objetive function and of the inequality constraints. The gradient projection method, on the other hand, considers nonlinear constraints as such, but also requires first derivatives of the objetive function and of the constraints. The complex method, however, does not require ther linearity nor differentiability at all and hence, results attractive when dealing with nonlinear problems whose derivatives are not available.

In what follows, let $\phi(\underset{\sim}{x})$ denote the objetive function to be minimized (for maximizing problems, simply change the sign of the objetive function), $\underset{\sim}{x}$, the decision vector and $\underset{\sim}{h}(\underset{\sim}{x})\geq\underset{\sim}{0}$, the inequality constraints. Equality constraints will be disregarded for the same reasons as in Section 6.5

Each of the aforementioned methods is next outlined.

The method of the feasible directions, developed by Zoutendijk [6.16], assumes linear inequality constraints. It requires, first, a search direction $\underset{\sim}{e}$ and, second, a search distance, s, so that when both are determined, given a current value $\underset{\sim}{x}_k$ of the vector argument, the next value, $\underset{\sim}{x}_{k+1}$, is given by

$$\underset{\sim}{x}_{k+1}=\underset{\sim}{x}_k+s_k\underset{\sim}{e}_k \tag{6.6.1}$$

where

$$||\underset{\sim}{e}_k||=1.$$

To find the direction $\underset{\sim}{e}_k$, assume that the inequality constraints are linear, i.e. of the form

$$\underset{\sim}{A}\underset{\sim}{x}+\underset{\sim}{b}\geq\underset{\sim}{0} \tag{6.6.2}$$

$\underset{\sim}{A}$ being an mxn matrix (m<n). There are two possibilities, namely, either $\underset{\sim}{x}_k$ lies within the feasible region or it lies on its boundary. If it lies within this region, then $\underset{\sim}{e}_k$ is just taken as the gradient of the objective function, i.e.

$$\underset{\sim}{e}_k = \pm \frac{\nabla\phi}{||\nabla\phi||} \tag{6.6.3}$$

where the + sign is taken if ϕ is to be maximized; otherwise, the - sign should be chosen. If, however, $\underset{\sim}{x}_k$ lies on the boundary of the feasible region, then some, say m' , of the m inequalities (6.6.2) become equations. If the new value $\underset{\sim}{x}_{k+1}$ is not going to violate the constraints (6.6.1), then the search direction and distance should satisfy the inequalities

$$\underset{\sim}{A}\underset{\sim}{x}_k+s_k\underset{\sim}{A}\underset{\sim}{e}_k+\underset{\sim}{b}\geq\underset{\sim}{0} \tag{6.6.4}$$

which constitute a set of m inequalities. However, m' of the components of vector $\underset{\sim}{A}\underset{\sim}{x}_k+\underset{\sim}{b}$ vanish, as assumed before. Hence, taking s_k positive, if eq. (6.6.4) is to be satisfied for all its components, then the search direction $\underset{\sim}{e}_k$ should satisfy the following inequality

$$\underset{\sim}{A}\underset{\sim}{e}_k>\underset{\sim}{0} \tag{6.6.5}$$

Since vector $\underset{\sim}{e}_k$ must satisfy a linear inequality, it is most convenient to use a linear norm for it, instead of the Euclidean one, i.e. let

$$||\underset{\sim}{e}_k||_1=\max_i\{(e_{ik})\}=1 \tag{6.6.6}$$

where e_{ik} is the $i^{\underline{th}}$ component of vector $\underset{\sim}{e}_k$. In order to turn the finding of vector $\underset{\sim}{e}_k$ into a standard linear programming problem, which requires the sought vector to meet a nonnegativity condition, the following change of variable is introduced

$$\underset{\sim}{t}_k=\underset{\sim}{e}_k+\underset{\sim}{f} \tag{6.6.7}$$

where $\underset{\sim}{f}$ is an m-dimensional vector whose components are all unity.

Thus, $\underset{\sim}{e}_k$ is found from the following linear program:

$$\text{Max } (\nabla\phi_k)^T(\underset{\sim}{t}_k-\underset{\sim}{f})$$

subject to

$$-\underset{\sim}{t}_k+2\underset{\sim}{f}\geq\underset{\sim}{0}$$

and

$$\underset{\sim}{t}_k\geq 0$$

where the 2n involved inequality constraints arise from eq. (6.6.6). The foregoing procedure to determine $\underset{\sim}{e}_k$ is due to Glass and Cooper (6.17).
If the inequality constraints are nonlinear, then a local linearisation should be performed to find matrix $\underset{\sim}{A}$ of (6.6.2) as

$$\underset{\sim}{A} = \frac{\partial \underset{\sim}{h}}{\partial \underset{\sim}{x}} \bigg|_{\underset{\sim}{x}=\underset{\sim}{x}_k}$$

which yields good results if the feasible region is convex.
Once the search direction is found, the search distance s_k is determined performing a unidirectional optimisation in the direction of $\underset{\sim}{e}_k$. There can arise two possibilities: the first one is that the optimising s_k carries $\underset{\sim}{x}$ into a point within the feasible region, in which case no difficulty is present; the other possibility appears when $\underset{\sim}{x}_{k+1}$ happens to fall outside of the feasible region. In this case, inequality (6.6.4) does not hold any longer for all its components. Assuming that m' of the m inequalities (6.6.4) are violated for the previously found optimising s_k, then there exist s_{ik} values smaller than s_k (i=1,...,m') for which the corresponding relations (6.6.4) become equations, i.e. for which

$$a_{ij}x_{jk}+s_{ik}a_{ij}e_{jk}+b_i=0, \quad i=1,2,\ldots,m' \qquad (6.6.8)$$

where a renumbering of the said relations could have been performed.
Solving for s_{ik} from (6.6.8),

$$s_{ik} =- \frac{a_{ij}x_{jk}+b_i}{a_{ij}e_{jk}} \qquad (6.6.9)$$

Hence, the search distance is then taken as

$$s_k=\min_i\{s_{ik}\}_1^{m'}$$

thus completing one iteration of the whole procedure.

<u>The gradient projection method</u>, developed by Rosen (6.18, 19), performs the search (for the optimum) initially along the boundaries of the feasible region. The procedure is, as outlined by Beveridge and Schechter (6.20), the following:

1. Given a point $\underset{\sim}{x}_p$ where r of the inequality constraints become equations, evaluate $\nabla\phi$ at $\underset{\sim}{x}_p$. The said set of equations is then written as

$$g_i = h_i(x_p) = 0, \quad i = 1, \ldots, r \qquad (6.6.10)$$

2. Calculate the projection of the gradient onto the plane tangent to the surface (6.6.10). This projection defines a direction $\underset{\sim}{e}$ in space, lying in the said tangent plane. Vector $\underset{\sim}{e}$ is determined as follows:

   Let $\frac{d\phi}{ds}$ be the directional derivative of ϕ along direction $\underset{\sim}{e}$, thus

$$\frac{d\phi}{ds} = (\nabla\phi)^T \underset{\sim}{e} \qquad (6.6.11)$$

   Since $\underset{\sim}{e}$ is contained in a plane tangent to (6.6.10),

$$\frac{d\underset{\sim}{g}}{ds} (\nabla \underset{\sim}{g})^T \underset{\sim}{e}=0 \qquad (6.6.12)$$

   vector $\underset{\sim}{e}$ being defined as of unit magnitude, i.e.

$$\underset{\sim}{e}^T \underset{\sim}{e}=1 \qquad (6.6.13)$$

   Define the objective function

$$\psi= \frac{d\phi}{ds} + \lambda_o \;(1-\underset{\sim}{e}^T\underset{\sim}{e})+ \underset{\sim}{\lambda}^T(\nabla \underset{\sim}{g})^T \underset{\sim}{e} \qquad (6.6.14)$$

   which accounts for constraints (6.6.11) and (6.6.12). Substitution of (6.6.10) into (6.6.13) together with the condition for ψ to be stationary yield

$$\frac{d\psi}{d\underset{\sim}{e}}=\nabla\phi+\nabla \underset{\sim}{g}\underset{\sim}{\lambda}-2\lambda_o\underset{\sim}{e}=\underset{\sim}{0} \qquad (6.6.15)$$

   from which

$$\underset{\sim}{e} = \frac{\nabla \underset{\sim}{g}\underset{\sim}{\lambda}+\nabla\phi}{2\lambda_o} \qquad (6.6.16)$$

which, when substituted into (6.6.11), yields an equation for $\underset{\sim}{\lambda}$, namely

$$(\nabla \underset{\sim}{g})^T \nabla \underset{\sim}{g} \underset{\sim}{\lambda} = -(\nabla \underset{\sim}{g})^T \nabla \phi \tag{6.6.17}$$

λ_o being computed from

$$\lambda_o = \frac{1}{2}\sqrt{(\nabla \phi)^T \nabla \underset{\sim}{g} \underset{\sim}{\lambda} + ||\nabla \phi||^2} \tag{6.6.18}$$

which follows from the substitution of eq. (6.6.15) into eq. (6.6.12). Hence, $\underset{\sim}{e}$ is given as

$$\underset{\sim}{e} = \frac{\nabla \underset{\sim}{g} \underset{\sim}{\lambda} + \nabla \phi}{\sqrt{(\nabla \phi)^T \nabla g \lambda + ||\nabla \phi||^2}} \tag{6.6.19}$$

3. At this stage two possibilities can arise: either $\underset{\sim}{e}$ vanishes or it does not. If it does not vanish, initiate a search in this direction until a constraint boundary is found. Let $\underset{\sim}{x}'_p$ be the intersection of the line going from $\underset{\sim}{x}_p$ in the direction of $\underset{\sim}{e}$, with the boundary. Then, if $\phi(\underset{\sim}{x}'_p)$ is better than $\phi(\underset{\sim}{x}_p)$, one iteration is completed and the procedure is restarted. If, on the contrary, $\phi(\underset{\sim}{x}'_p)$ is not better than $\phi(\underset{\sim}{x}_p)$, perform a unidirectional search along the line connecting $\underset{\sim}{x}_p$ and $\underset{\sim}{x}'_p$. Call $\underset{\sim}{x}_{p+1}$ the sought optimising value, and return to step 1 with $\underset{\sim}{x}_{p+1}$ being set equal to $\underset{\sim}{x}_p$.
4. In case $\underset{\sim}{e}$ vanishes, then it could happen that the cpmponents of $\underset{\sim}{\lambda}$ have all the same sign or some are zero, thus complying with the Kuhn-Tucker conditions (6.20, p. 282). If this sign is nonnegative, then a minimum has been reached; if nonpositive, then a maximum.

If not all components of $\underset{\sim}{\lambda}$ bear the same sign, then some of these components are deleted from it, keeping only those λ_i (i=1,2,...,q<r) with the same sign. Next, drop the r-q components of $\underset{\sim}{g}(\underset{\sim}{x})$, i.e. those corresponding to the λ's with a sign different from those of the λ's that have been kept. Then restart the procedure at step 2. The PROJG program (6.13, pp. 399-411) implements Rosen's algorithm.

The complex method, that is described next, is due to Box (6.21). Before proceeding, however, some definitions are in order:

Given a vector space V of dimension n, the (closed) polyhedron imbedded in it with the smallest number of vertices (n+1) is called a simplex. A polyhedron with a larger number of vertices is called a complex. Simplexes in two and three dimensions are the triangle and the tetrahedron, respectively. Complexes in these spaces could be the quadrilateron and the hexahedron, respectively. The complex method of Box proceeds as follows:

1. A complex of 2n vertices is defined within the feasible region. There exist some techniques to define these vertices in such a way that they quarantee that all of them fall into the said region, but they are applicable to only a few particular cases. In the absence of a criterion to choose the said vertices, it is advised that they be assigned randomly, rejecting those that fall outside of the feasible region until the complex is completed.
2. Evaluate the objective function at each of the 2n vertices and let $\underset{\sim}{x}_w$ be the vertex where this function attains its worst value.
3. Compute the centroid, $\underset{\sim}{x}_c$, of the remaining vertices, i.e.

$$\underset{\sim}{x}_c = \frac{1}{2n-1}\left(\sum_{1}^{2n} \underset{\sim}{x}_i - \underset{\sim}{x}_w\right) \tag{6.6.20}$$

4. Replace the vertex at $\underset{\sim}{x}_w$ by a new one, $\underset{\sim}{x}'_w$ in the following way

$$\underset{\sim}{x}'_w = \underset{\sim}{x}_c - \alpha(\underset{\sim}{x}_w - \underset{\sim}{x}_c) \tag{6.6.21}$$

   where α is a real positive number whose value is recommended by Box to be taken as 1.3
5. Two possibilities can arise at this stage: either $\underset{\sim}{x}'_w$ lies within the feasible region or not. If it does, a new iteration can be restarted at stage 2. If it does not, then a new value, $\underset{\sim}{x}''_w$, is defined as

$$x''_w = \frac{1}{2}(x_c + x'_w)$$

If x''_w again happens to lie outside of the feasible region, a new value is again defined as the middle point between x_c and x''_w, until the said new value falls within the feasible region. Once the rejected vertex has been regenerated, the procedure is restarted at step 2. It can happen, however, that the regenerated feasible vertex is again the worst one. This situation would slow down the convergence of the method. Box's algorithm takes account of such a case. In fact, if the regenerated feasible vertex x'_w happens to be again the worst vertex, then it is rejected and a new vertex x''_w is defined by shifting x'_w towards x_c by a certain amount β, i.e.

$$x''_w = x_c + \beta(x_w - x_c)$$

where $0 < \beta < 1$.

Box's algorithm has been implemented in some computer programs, for instance the one appearing in (6.13, pp. 368-385) and the OPTIM package (6.22,23). Two examples of linkage optimisation subject to inequality constraints are next presented, both of which were solved using the OPTIM package.

Example 6.6.1 (Synthesis of an RRRR plane linkage for function generation with optimal transmission angle), The problem of Example 6.4.2 is repeated here, except that now one constraint is introduced, namely that the input link be a crank, the output one being left free to be either a crank or a rocker. Moreover, the transmission angle is required to be the best possible one throughout a complete revolution of the input crank.

Solution:

The new objetive function is

$$z = \frac{1}{2\pi}\int_0^{2\pi} \cos^2\mu d\psi \tag{6.6.22}$$

where $\cos\mu$ is given as in (6.4.11). The decision variables are α and β (see Fig 6.4.1), as in Example 6.4.2. The objetive function then becomes

$$z = N/D \tag{6.6.23}$$

with

$$N = 2(k_1k_3+k_2)^2+k_3^4 \quad \text{and} \quad D = 2(2k_1k_2k_3+k_2^2+k_3^2+k_2^2k_3^2) \qquad (6.6.24)$$

The constraints guaranteeing that the input link is a crank are now derived. Throughout, reference will be made to Fig 6.4.1. In that figure angles ψ and ϕ will now be measured from line R_{12} R_{41}. In order for the configuration shown in that figure to be possible, triangle R_{23} R_{34} R_{41} should exist. For this triangle to exist, the following inequalities should hold

$$a_3+a_4 > \sqrt{a_1^2+a_2^2-2a_1a_2\cos\psi} \qquad \text{(i)}$$

$$|a_3-a_4| < \sqrt{a_1^2+a_2^2-2a_1a_2\cos\psi} \qquad \text{(ii)}$$

At this stage, a word of caution is in order: In writing relations (i) and (ii), a_1, a_2, a_3 and a_4 were assumed to be lengths, i.e. positive definite variables. If, however, these are computed by inverting relations (6.4.8), which in turn require k_1, k_2 and k_3 to be computed by solving the linear system (6.4.23), it could well happen that some k_i turned out to be negative, which, in turn, would yield some a_i negative. Now, a_1 can be forced to be positive, as well as a_3, for negative values of these variables would be meaningless. Thus, all lengths can be normalised with respect to a_1 by defining $a_1 \equiv 1$. Variables a_2 and a_4, however, can attain negative values, which means angles ψ and ϕ of Fig 6.4.1 should be measured not referring to the positions of lines R_{12} R_{23} and R_{41} R_{34}, respectively, but to lines R_{12} R'_{23} and R_{41} R'_{34}, where R'_{23} and R'_{34} are the symmetric points with respect to R_{12} and R_{41}, respectively, of points R_{23} and R_{34}. Thus, in inverting relations (6.4.8), care should be taken in order to obtain all a_i's positive. These relations are then inverted as:

$$a_1 = 1, \; a_2 = \frac{1}{|k_2|}, \; a_3 = \frac{\sqrt{D}}{|k_2k_3|}, \; a_4 = \frac{1}{|k_3|} \qquad \text{(iii)}$$

Squaring both sides of inequalities (i) and (ii) and rearranging terms lead, respectively, to

$$\cos\psi > \frac{(a_3+a_4)^2-a_1^2-a_2^2}{2a_1a_2} \qquad \text{(iv)}$$

and

$$\cos\psi < \frac{(a_3-a_4)^2-a_1^2-a_2^2}{2a_1a_2} \qquad \text{(v)}$$

Now, since

$$-1 \leq \cos\psi \leq +1 \qquad \text{(vi)}$$

inequality (iv) still holds if $\cos\psi$ is substituted by +1. Similarly, inequality (v) still holds if $\cos\psi$ is substituted by -1, thus obtaining, after some simplifications,

$$a_3 + a_4 > a_1+a_2 \qquad \text{(vii)}$$

and

$$|a_3-a_4| < |a_1-a_2| \qquad \text{(viii)}$$

respectively. Hence, inequalities (vii) and (viii) are the sought constraints. These are introduced into the OPTIM program, together with the constraints

$$0 \leq \alpha \leq 2\pi,\ 0 \leq \beta \leq 2\pi \qquad \text{(ix)}$$

since OPTIM requires that the decision variables be bounded from above and from below.

OPTIM also requires that one vertex of the initial (feasible) complex be supplied.

This vertex was supplied as

$\alpha = 15°$ $\beta = 45°$

which led to the following synthesis: $k_1 =- 0.8799$, $k_2 = 2.3796$, $k_3 = 0.2443$, the corresponding linkage lengths being

$$a_1=1.0,\ a_2=0.4202,\ a_3=3.8600,\ a_4=4.0926 \qquad \text{(x)}$$

the OPTIM package automatically generates the other (three) vertices of the intial complex. Since it sets the number of such vertices equal to 2n, n being the number of independent decision variables, in this case the said complex is a quadrilateron. The linkage described in (x) constitutes the "first vertex" of the complex, the remaining three appearing in Table 6.6.1 .

Table 6.6.1 Vertices of the initial complex

Vertex No.	2	3	4
α (rad)	0.4601	1.4054	1.5161
β (rad)	3.0780	1.8170	3.9932
a_1	1.0000	1.0000	1.0000
a_2	0.3895	0.1714	0.1825
a_3	1.4551	0.4970	1.4241
a_4	1.6143	1.2779	0.7065

The values of the objetive function at each vertex are shown in Table 6.6.2

TABLE 6.6.2 Values of the objective function at each of the vertices of the initial complex

Vertex No.	1	2	3	4
Objective function	0.9306	0.5917	0.4849.	0.5676

The procedure converged after 30 iterations, the results being the following:

Optimising values of the decision variables:

$\alpha = 103.30°$ $\beta = 198.83°$

Dimensions of the optimal linkage:

$a_1 = 1.0000$, $a_2 = 0.0515$, $a_3 = 0.9945$, $a_4 = 0.1154$

The minimum value attained by the objective function was z=0.1008. The procedure thus reduced z to a 10.83% of its initial value.

The foregoing linkage was analysed for a whole revolution of its input crank. Its transmission angle attained a mean value of 90°. Its cosine thus attained an r.m.s. value of $\sqrt{z}$, i.e. 0.3175.

Example 6.6.2 Synthesise an RSSR spatial linkage of the type of that appearing in Fig 5.2.1, which is here reproduced in Fig 6.6.1 for quick reference, to generate a given set $\{\phi_i, \psi_i\}_1^n$ of input-output values. Furthermore, its input link should be a crank, whereas its mechanical advantage, as large as possible.

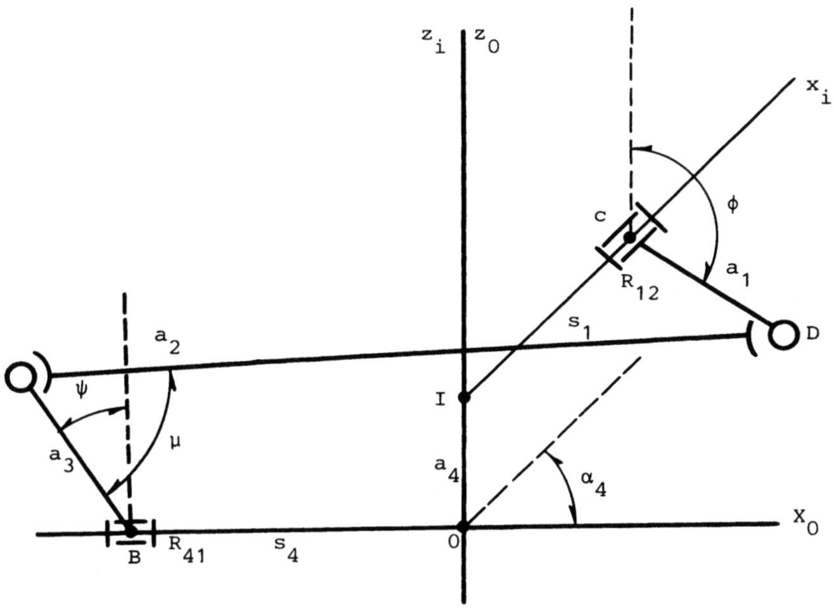

Fig 6.6.1 RSSR function generating linkage

Solution:

The input-output synthesis equation for the given linkage was written down as eq. (5.2.15). The input is now measured from a given reference line making an angle ϕ_0 with the Z_i ($||z_o$)axis, counterclockwise as viewed from axis X_i. The output angle is measured, in turn, from a given reference line making an angle ψ_0 with the Z_i axis, counterclockwise as viewed from axis X_0. Performing transformation (5.2.21), the input-output equation is obtained as eq. (5.2.20), which is here reproduced as eq. (6.6.26)

$$k_1 c\psi_i + k_2 s\psi_i + k_3 c\phi_i + k_4 s\phi_i - k_5 c\phi_i s\psi_i$$

$$-k_6 c\psi_i s\phi_i - k_7 s\phi_i s\psi_i + k_8 = c\psi_i c\phi_i,\quad i=1,2,\ldots,\ n \tag{6.6.26}$$

The given set of input-output values is that appearing in Table 6.6.2.

TABLE 6.6.2 Input-output prescribed values

ϕ(degrees)	ψ(degrees)
30	45
45	60
60	90
90	60
150	30

The objective function defined in (6.4.9), where the integral would range from 0 to 2π, is not valid here, because a value of 90°(or 270°) of the transmission angle, does not guarantee a good torque ratio. In fact, due to the 3-dimensionality of the linkage, it could happen that, even though the transmission angle attains values close to 90°, the force transmitted to the output link by the coupler link has a very large component parallel to the output axis X_o. Hence, a more direct approach to the problem is proposed, i.e. maximize a measure of the mechanical advantage. Now, this measure should be positive definite, i.e. a norm of this function. The maximal norm, max $|v|$, of the mechanical advantage is ruled out, because it could happen that no matter how low this value is within the interval of interest, if the output link happens to be a rocker, v will become infinity at its deadpoints. Something similar could happen if, say, a quadratic norm is used. In order to cope with this problem, the integral of the square of the inverse of the mechanical advantage is minimized. Thus, the objective function is now

$$z = \frac{1}{2\pi}\int_0^{2\pi} \left(\frac{1}{v}\right)^2 d\psi \qquad (6.6.27)$$

where the integrand never becomes infinity. In fact, if friction is neglected, the mechanical advantage turns out to be

$$v = \frac{d\phi}{d\psi} = \frac{\dot{\phi}}{\dot{\psi}} \qquad (6.6.27a)$$

Thus, the inverse of the mechanical advantage becomes infinity only for values of $\dot{\phi} = 0$, which happens at the deadpoints of an input link of the rocker type. Because of the constraints imposed on the problem, this possibility has been ruled out, however.

Introducing the results of Exercise 5.2.4, the conditions under which the input link is a crank are

$$\ell_{max} \leq a_2 \leq L_{min} \qquad (6.6.28)$$

where ℓ and L are the smallest and the largest values of the length of the coupler link, to guarantee mobility of the linkage for a given value of the input ϕ. Now, ℓ and L can readily be found as (See Exercise 5.2.4):

$$\ell(\phi) = \left((r-a_3)^2 + b^2\right)^{1/2} \qquad (6.6.29a)$$

$$L(\phi) = \left((r+a_3)^2 + b^2\right)^{1/2} \qquad (6.6.29b)$$

where

$$r = \sqrt{a_1^2+s_1^2+a_4^2+2a_1a_4c\phi-(s_1c\alpha_4+a_1s\alpha_4s\phi)^2}$$

and

$$b = s_4+s_1c\alpha_4+a_1s\alpha_4s\phi$$

The maximum of ℓ and the minimum of L can be obtained by zeroing the derivatives of ℓ and L with respect to ϕ. This leads to the equations (6.24)

$$(u^{1/2} \pm a_3)u^{-1/2} \frac{du}{d\phi} + 2t \frac{dt}{d\phi} = 0 \tag{6.6.30}$$

where

$$u \equiv a_1^2+s_1^2+a_4^2+2a_1a_4c\phi-(s_1c\alpha_4+a_1s\alpha_4s\phi)^2 \tag{6.6.30a}$$

and

$$t \equiv s_4+s_1c\alpha_4+a_1s\alpha_4s\phi \tag{6.6.30b}$$

Squaring both sides of eqs. (6.6.30), substituting $\sin^2\phi$ by $1-\cos^2\phi$ and simplifying the resulting expressions one obtains the single equation

$$b_0+b_1c\phi+b_2c^2\phi+b_3c^3\phi+b_4c^4\phi=s\phi(c_0+c_1c\phi+c_2c^2\phi+c_3c^3\phi) \tag{6.6.31}$$

with the following definitions:

$$b_0=a_4^2(a_3^2-a_1^2-s_1^2-a_4^2+s_1^2c^2\alpha_4+a_1^2s^2\alpha_4) \tag{6.6.32a}$$

$$b_1=s^2\alpha_4(2a_1a_3^2a_4-4a_1a_4s_1s_4c\alpha_4)-2a_1a_4^3 \tag{6.6.32b}$$

$$b_2=s^4\alpha_4\left((a_3^2+s_4^2)(a_1^2-s_1^2)\right)+s^2\alpha_4\left(a_4^2(s_1^2-s_4^2-2a_1^2)+a_3^2s_1^2-a_1^2s_4^2\right)+$$
$$+ a_4^2(a_1^2-a_3^2+a_4^2) \tag{6.6.32c}$$

$$b_3=2a_1a_4^3+s^2\alpha_4(4a_1a_4s_1s_4c\alpha_4-2a_1a_3^2a_4-2a_1a_4s_4^2) \tag{6.6.32d}$$

$$b_4=a_1^2s^2\alpha_4(a_4^2-a_3^2s^2\alpha_4-s_4^2s^2\alpha_4) \tag{6.6.32e}$$

$$c_0=-2a_1a_4^2s_1s\alpha_4c\alpha_4 \tag{6.6.32f}$$

$$c_1=2a_4s_4s^3\alpha_4(a_1^2-s_1^2)-2a_3^2a_4s_1s\alpha_4c\alpha_4-2a_4s_4s\alpha_4(a_1^2+a_4^2) \tag{6.6.32g}$$

$$c_2=2a_1s\alpha_4\left(a_4^2s_1c\alpha_4-s_1(a_3^2+s_4^2)s^2\alpha_4c\alpha_4-2a_4^2s_4\right) \tag{6.6.32h}$$

$$c_3=-2a_1^2a_4s_4s^3\alpha_4 \tag{6.6.32i}$$

Squaring both sides of eq. (6.6.31) and simplifying the resulting expression leads to

$$\sum_{i=0}^{8} A_i c^i \phi = 0 \qquad (6.6.33)$$

where

$$A_0 = b_0^2 - c_0^2 \qquad (6.6.34a)$$

$$A_1 = 2(b_0 b_1 - c_0 c_1) \qquad (6.6.34b)$$

$$A_2 = b_1^2 + 2b_0 b_2 + c_0^2 - 2c_0 c_2 - c_1^2 \qquad (6.6.34c)$$

$$A_3 = 2(b_0 b_3 + b_1 b_2 + c_0 c_1 - c_1 c_2 - c_0 c_3) \qquad (6.6.34d)$$

$$A_4 = 2b_1 b_3 + 2b_0 b_4 + b_2^2 + 2c_0 c_2 + c_1^2 - 2c_1 c_3 - c_2^2 \qquad (6.6.34e)$$

$$A_5 = 2(b_1 b_4 + b_2 b_3 + c_0 c_3 + c_1 c_2 - c_2 c_3) \qquad (6.6.34f)$$

$$A_6 = b_3^2 + 2b_2 b_4 + 2c_1 c_3 + c_2^2 - c_3^2 \qquad (6.6.34g)$$

$$A_7 = 2(b_3 b_4 + c_2 c_3) \qquad (6.6.34h)$$

$$A_8 = b_4^2 + c_3^2 \qquad (6.6.34i)$$

The real roots of eq. (6.6.33) provide the global maximum of ℓ and the global minimum of L. These values are then substituted into relation (6.6.27) to prove the feasibility of the linkage, i.e. the full rotatability of its input link.

The synthesis procedure is next outlined. Since Table 6.6.2 contains five input-output pairs, n is set equal to 5 in eq. (6.6.26). This equation, contains 8 unknowns, namely k_i (i=1,2,...,8). Hence, the designer can choose freely three of these unknowns. These are not arbitrarily chosen, however. Instead, they are chosen so as to render the objective function z, given in expression (6.6.27), a minimum. By inspection of transformation (5.2.21), one can readily notice that k_5, k_6 and k_7 contain only the angles α_4, ϕ_0 and ψ_0. Thus, these angles are henceforth chosen as the decision variables. Once these are defined, k_5, k_6 and k_7 are uniquely determined. The remaining unknowns, k_1, k_2, k_3, k_4 and k_8 are computed by solving the system of 5 linear equations arising by substituting the 5 input-output pairs of Table 6.6.2 into eq. (6.6.26). Thus, one obtains the linear system

$$\underset{\sim}{A}\underset{\sim}{x} = \underset{\sim}{b} \tag{6.6.35}$$

where

$$\underset{\sim}{x} = (k_1, k_2, k_3, k_4, k_8)^T, \quad \underset{\sim}{b} = (b_1, b_2, b_3, b_4, b_5)^T \tag{6.6.36a}$$

with

$$b_j = c\phi_j c\psi_j + k_5 c\phi_j s\psi_j + k_6 s\phi_j c\psi_j + k_7 s\phi_j s\psi_j, \quad j=1,2,\ldots,5 \tag{6.6.36b}$$

and

$$\underset{\sim}{A} = \begin{pmatrix} c\psi_1 & s\psi_1 & c\phi_1 & s\phi_1 & 1 \\ c\psi_2 & s\psi_2 & c\phi_2 & s\phi_2 & 1 \\ c\psi_3 & s\psi_3 & c\phi_3 & s\phi_3 & 1 \\ c\psi_4 & s\psi_4 & c\phi_4 & s\phi_4 & 1 \\ c\psi_5 & s\psi_5 & c\phi_5 & s\phi_5 & 1 \end{pmatrix} \tag{6.6.36c}$$

From expression (6.6.27a), the mechanical advantage is obtained as

$$v = \frac{s\psi(k_1 - k_6 s\phi - c\phi) + c\psi(k_5 c\phi + k_7 s\phi - k_2)}{s\phi(k_5 s\psi + c\psi - k_3) + c\phi(k_4 - k_6 c\psi - k_7 s\psi)} \tag{6.6.37}$$

where ψ is computed for a given value of ϕ from manipulation of eq. (6.6.26), as

$$\psi = 2\tan^{-1}(-B \pm \sqrt{B^2 - AC}) \tag{6.6.38}$$

In the latter expression, the two arising values of ψ correspond to the conjugate configurations of the linkage. In this equation,

$$A = (k_3 + a)c\phi + (k_4 + k_6)s\phi + k_8 - k_1 \tag{6.6.39a}$$

$$B = k_2 - k_5 c\phi - k_7 s\phi \tag{6.6.39b}$$

$$C = (k_3 - 1)c\phi - (k_4 - k_6)s\phi + k_8 + k_1 \tag{6.6.39c}$$

The OPTIM package, refered to in Example 6.6.1, was used also to solve the present problem. This requires that the decision variables be bounded from below and from above. In this problem, these variables are then bounded as

$$0 \leq \phi_0 \leq 2\pi, \; 0 \leq \psi_0 \leq 2\pi, \; 0 \leq \alpha_4 \leq 2\pi \tag{6.6.40}$$

This way the optimisation problem is now fully stated as a nonlinear programming problem. This reads: "Minimize z as given by eq. (6.6.27) subject to the 5 equality constraints (6.6.35) and to the inequality constraints (6.6.28) and (6.6.40)."

OPTIM requires that the user provides one vertex of the inicial feasible complex. This was generated randonly, as

$$\phi_0 = 37.7281° \; , \; \psi_0 = 307.7127° \; , \; \alpha_4 = 187.7812° \tag{6.6.41}$$

In Table 6.6.4 it appears that the linkage corresponding to the second vertex is identical to that corresponding to the first one. They are distinct linkages, however, the differences not showing up on the values of ϕ_0, ψ_0 and α_4 because of the few digits that are shown. These differences are more apparent regarding their dimensions. Moreover, it should also be pointed out that all dimensions are shown normalised with respect to a_4, for which reason this dimension appears as unity.

The values of the objective function at each vertex are shown in Table 6.6.3.

TABLE 6.6.3 Values of the objective function at each of the vertices of the initial complex

Vertex No.	1	2	3	4	5	6
Objective function z	1.6680	1.6684	0.1261	2.6544	0.5692	0.2962

The procedure converged after 29 iterations, the results being as follows:

Optimising values of the decision variables:

$\phi_0 = 99.25°$, $\psi_0 = 164.10°$, $\alpha_4 = 127.65°$

Dimensions of the optimal linkage:

$a_1 = 0.5421$, $a_2 = 5.5912$, $a_3 = 4.1482$

$a_4 = 1.0000$, $s_1 = 1.3605$, $s_4 = 1.0443$

which yielded the following synthesis:

$$k_1 = 0.6308,\ k_2 =- 1.5879,\ k_3 = 1.1225,\ k_4 = 0.2112$$
$$k_5 = 0.2645,\ k_6 = 0.2549,\ k_7 =- 1.0000,\ k_8 = 0.1010 \tag{6.6.42}$$

These values, in turn, yield the following linkage dimensions (units of lenghth):

$$a_1 = 1.1926,\ a_2 = 20.8425,\ a_3 = 1.3684,\ a_4 = 1.0,\ s_1 = 12.4786,$$
$$s_4 =- 8.3149 \tag{6.6.43}$$

The OPTIM package automatically generates the other vertices of the initial complex. It sets the number of such vertices equal to 2n which, for this case, turns to be 6. The linkage described in (6.6.41) constitutes "the first vertex" of the complex, the remaining ones appearing in Table 6.6.4.

TABLE 6.6. 4 Vertices of the initial complex

Vertex No.:	2	3	4	5	6
ϕ_0 (rad)	0.6585	1.0320	1.5161	2.6494	2.4538
ψ_0 (rad)	5.3706	3.5938	3.9932	0.5924	2.2177
α_4 (rad)	3.2774	1.8604	3.0576	3.4032	2.5130
a_1	1.1928	-6.6129	2.7135	0.6063	0.3260
a_2	20.8370	19.4400	72.3380	5.3733	11.5010
a_3	1.3683	1.4976	2.8904	0.8742	6.9070
a_4	1.0000	1.0000	1.0000	1.0000	1.0000
s_1	12.4760	18.6670	-40.071	-5.1756	-0.4219
s_4	-8.3120	-1.1086	32.102	-0.0098	8.2980

The minimum value attained by the objective function was z=0.0587. Hence, the procedure reduced z to a 3.5% of its initial value.

The complex method has already been successfully applied to the optimal synthesis of plane linkages in connection with the problem of path generation (6.25).

REFERENCES

6.1 Athans M. and P.L. Falb, Optimal Control, McGraw-Hill Book Co., N.York, 1966.

6.2 Halmos P.R., Introduction to Hilbert Space, Chelsea Publishing Co., N. York, 1957, p. 17

6.3 Bowen R. M. and C. Wang, Introduction to Vectors and Tensors, Linear and Multilinear Algebra, vol. 1, Plenum Press, N. York, 1976.

6.4 Tomáš J., "The synthesis of mechanisms as a nonlinear mathematical programming problem", Journal of Mechanisms, vol. 3, 1968, pp 30, 119-130

6.5 Suh C. H. and A.W. Mecklenburg, "Optimal design of mechanisms with the use of matrices and least squares", Mechanism and Machine Theory, vol. 8, 1973, pp 479-495

6.6 Angeles J., "Optimal synthesis of linkages using Househilder reflexions", Proceedings of the V World Congress of the IFToMM, Montreal, July 8-13, 1979, pp. 111-114

6.7 Fox R.L. and K.C. Gupta, "Optimization technology as applied to mechanism design", Journal of Engineering for Industry, Trans. ASME, Series B, vol. 95, May 1973, pp. 657-663

6.8 Hartenberg R.S. and J. Denavit, Kinematic Synthesis of Linkages, McGraw-Hill Book Co., N. York, 1964, p. 297

6.9 Alizade R.I., I.G. Novruzbekov and G.N. Sandor, "Optimization of four-bar function generating mechanisms using penalty functions with inequality and equality constraints", Mechanism and Machine Theory, vol. 10, 1975, pp. 327-336

6.10 Fiacco A.V. and G.P. McCormick, Nonlinear Programming, Sequential Unconstrained Minimization Techniques, John Wiley and Sons, Inc., N. York, 1968

6.11 Fletcher R. and C.M. Reeves, "Function minimization by conjugate gradients", Computer Journal, vol. 7, No. 4, 1964, pp. 149-154

6.12 Powell M. J. D., "An efficient method for finding the minimum of a function of several variables without calculating derivatives", Computer Journal, vol. 7, No. 4, 1964, pp. 155-162

6.13 Kuester J.L. and J.H. Mize, Optimization Techniques with Fortran, McGraw-Hill Book Co., 1973, pp. 412-463

6.14 Velasco-Levy A. and A. Montalvo, Algoritmo de Tunelización para la Optimización Global de Funciones, Comunicaciones Técnicas, IIMAS-UNAM, vol. 10, No. 204, 1979.

6.15 Suh C.-H. and C.W. Radcliffe, Kinematics and Mechanisms Design, John Wiley and Sons, Inc., N. York, 1978, pp. 221-235

6.16 Zoutendijk G., Methods of Feasible Directions, Elsevier Publishing Company, Amsterdam, 1960.

6.17 Glass H. and L. Cooper, "Sequential search: A method for solving constrained optimization problems", Journal of the ACM, vol. 12, No. 1, 1965, pp. 94-120

6.18 Rosen J.B., "The gradient projection method for nonlinear programming, Part I, Linear constraints", Journal of the SIAM, vol. 8, 1960, pp

6.19 Rosen J.B., "The gradient projection method for nonlinear programming, Part II, Non-Linear constraints", Journal of the SIAM, vol. 9, 1961, pp. 514-532.

6.20 Beveridge G.S.G. and R.S. Schechter, Optimization: Theory and Practice, McGraw-Hill Book Co., N. York, 1970, pp. 469-472

6.21 Box M. J., "A new method of constrained optimization and a comparison with other methods", Computer Journal, vol. 8, 1965, pp. 42-52

6.22 Evans L.B., Optimization Techniques for Use in Analysis of Chemical Processes, A Set of Notes, Massachussetts Institute of Technology, Cambridge (U.S.A.), 1971.

6.23 Murray M.A., E. Chicurel, L.B., Evans, F. Lara, A. Guarda and J.F. Albarrán, Aplicaciones de Computación a la Ingeniería, Limusa, Mexico City, 1975.

6.24 Rojas A., Síntesis de Mecanismos RSSR generadores de función con ventaja mecánica óptima, Master's Project, División de Estudios de Posgrado, Facultad de Ingeniería, UNAM, Mexico City, 1981.

6.25 Dukkipati R.V., Sankar S. and Osman M.O.M., "On the use of complex method of constrained optimization in linkage synthesis", Proceedings of the V World Congress on the Theory of Machines and Mechanisms, Montreal, July 8-13, 1979, pp. 382-387.

Appendix 1

ALGEBRA OF DYADICS

Let U and V be m- and n- dimensional vector spaces over the same field F. Let $U=\{\underset{\sim}{u}_1,\underset{\sim}{u}_2,\ldots,\underset{\sim}{u}_m\}$ and $V=\{\underset{\sim}{v}_1,\underset{\sim}{v}_2,\ldots,\underset{\sim}{v}_n\}$ be bases for each of these spaces. The <u>tensor product</u> of U and V, in this order, is a space W represented as $U\otimes V$, a basis of which is defined as

$$W=\{\underset{\sim}{u}_1\otimes\underset{\sim}{v}_1,\underset{\sim}{u}_1\otimes\underset{\sim}{v}_2,\ldots,\underset{\sim}{u}_1\otimes\underset{\sim}{v}_n,\underset{\sim}{u}_2\otimes\underset{\sim}{v}_1,\ldots,\underset{\sim}{u}_2\otimes\underset{\sim}{v}_n,\ldots,\underset{\sim}{u}_m\otimes\underset{\sim}{v}_1,\ldots,\underset{\sim}{u}_m\otimes\underset{\sim}{v}_n\}$$

Each of $\underset{\sim}{u}_i\otimes\underset{\sim}{v}_j$ or, in general, any expression of the form $\underset{\sim}{w}=\underset{\sim}{a}\otimes\underset{\sim}{b}$, where $\underset{\sim}{a}\in U$ and $\underset{\sim}{b}\in V$ is refered to as a <u>dyadic</u>. From the above definition, then any dyadic such as $\underset{\sim}{w}$ can be expressed as a linear combination of the dyadics of W, i.e. as

$$\begin{aligned}\underset{\sim}{w}= & w_{11}\underset{\sim}{u}_1\otimes\underset{\sim}{v}_1+\ldots+w_{1n}\underset{\sim}{u}_1\otimes\underset{\sim}{v}_n+\\ & +w_{21}\underset{\sim}{u}_2\otimes\underset{\sim}{v}_1+\ldots+w_{2n}\underset{\sim}{u}_2\otimes\underset{\sim}{v}_n+\\ & +w_{m1}\underset{\sim}{u}_m\otimes\underset{\sim}{v}_1+\ldots+w_{mn}\underset{\sim}{u}_m\otimes\underset{\sim}{v}_n\end{aligned} \tag{A1.1}$$

Just in the same way as vectors can be expressed as one-dimensional arrays of numbers (of the field over which the corresponding vector space is defined), these numbers being the vector components referred to a given basis, dyadics are also expressed as arrays (in this case, two-dimensional) of their components. Thus, the above dyadic $\underset{\sim}{w}$ is represented as

$$\begin{matrix} w_{11} & w_{12}\cdots w_{1n}\\ w_{21} & w_{22}\cdots w_{2n}\\ \vdots & \vdots \quad\quad \vdots\\ w_{m1} & w_{m2}\cdots w_{mn}\end{matrix} \tag{A1.2}$$

Hence, dyadics or <u>second-rank tensors</u>, as they are also referred to, are <u>isomorphic</u> to matrices. In particular, dyadics belonging to the space W as defined above, are isomorphic to mxn matrices, that is to say, those operations defined for matrices find their counterparts in the case of dyadics. For instance, given two dyadics, $\underset{\sim}{p}$ and $\underset{\sim}{q}$, both of W, they are equal if, and only if

$$p_{ij}=q_{ij}\ ;\ i=1,\ldots,m;\ j=1,\ldots,n$$

where subindices i, j refer to the (i,j)th component of the corresponding dyadic. The operations of addition of dyadics and multiplication by a scalar are defined in a similar fashion. Sometimes, for short, a dyadic $\underset{\sim}{a}\otimes\underset{\sim}{b}$ is represented simply as $\underset{\sim}{a}\underset{\sim}{b}$. The latter practice is followed throughout. The usual multiplication of a matrix $\underset{\sim}{A}$ by a vector $\underset{\sim}{c}$ corresponds to the dot multiplication of a dyadic $\underset{\sim}{a}\underset{\sim}{b}$ times the same vector $\underset{\sim}{c}$ where the components of $\underset{\sim}{a}\underset{\sim}{b}$ are identical to the corresponding entries of $\underset{\sim}{A}$. This product is represented as

$$\underset{\sim}{a}\underset{\sim}{b}.\underset{\sim}{c}=(\underset{\sim}{b}.\underset{\sim}{c})\underset{\sim}{a}$$

i.e. when a dyadic is dot-multiplied from the right by a vector, the result is a new vector whose value equals that of the left vector of the dyadic times the dot product of its right vector and the vector multiplying the dyadic. Similarly, the left dot multiplication of a dyadic times a vector is defined as

$$\underset{\sim}{c}.\underset{\sim}{a}\underset{\sim}{b}=(\underset{\sim}{c}.\underset{\sim}{a})\underset{\sim}{b}$$

Notice, however, that if $\underset{\sim}{a}$ and $\underset{\sim}{b}$ are vectors of dimensions m and n, respectively, then $\underset{\sim}{c}$ is n-dimensional in the first case, whereas m-dimensional in the latter.

Exercise A1.1 Letting U and V be equal to the three-dimensional Euclidean space, in the foregoing discussion, define the right- and the left- cross product of a dyadic by a vector.

Given two dyadics, $\underset{\sim}{a}\underset{\sim}{b}$ and $\underset{\sim}{c}\underset{\sim}{d}$, both of the same space W, their product is the dyadic $\underset{\sim}{E}$ defined as

$$\underset{\sim}{E}=\underset{\sim}{a}\underset{\sim}{b}.\underset{\sim}{d}\underset{\sim}{c}=(\underset{\sim}{b}.\underset{\sim}{d})\underset{\sim}{a}\underset{\sim}{c}$$

Under the above definition, if W is an mxn dimensional space, dyadic E is in an mxn dimensional space, thereby paralleling the definition of matrix multiplication. Corresponding to the transpose of a matrix, the transpose or conjugate of a dyadic $\underset{\sim}{D}=\underset{\sim}{a}\underset{\sim}{b}$ is defined as $\underset{\sim}{D}_c=\underset{\sim}{b}\underset{\sim}{a}$. Self-conjugate and antiself-conjugate dyadics are the counterparts of symmetric and antisymmetric matrices. Clearly, square matrices correspond to dyadics arising from the tensor products of one vector space times itself. The trace of such a dyadic is defined in a similar fashion to that of a square matrix. Letting this dyadic be $\underset{\sim}{a}\underset{\sim}{b}$, its trace is then

$$Tr(\underset{\sim}{a}\underset{\sim}{b})=a_i b_i=\underset{\sim}{a}.\underset{\sim}{b}$$

Similarly to the way the trace of a dyadic is defined, the vector of a 3x3 dyadic is given as

$$\underset{\sim}{v}=\underset{\sim}{a}x\underset{\sim}{b}$$

Exercise A1.2 Show that the vector of a nonzero 3x3 dyadic vanishes if the dyadic is self-conjugate

Exercise A1.3 Define a suitable inner product of dyadic that allows you to define the angle between two dyadics. Hence show that self-conjugate and antiself-conjugate dyadics are orthogonal.

Exercise A1.4 How could you define the magnitude of a dyadic?

In just the same way as dyadics were defined, n-adics can be defined as the tensor product of n vector spaces. The study of such algebraic entities falls into the realm of multilinear algebra. The topic is widely discussed in reference A1.1

REFERENCES

A1.1 Bowen R.M. and C.C. Wang, Introduction to Vectors and Tensors, vol. I, Plenum Press, N. York, 1976.

Appendix 2

DERIVATIVE OF A DETERMINANT WITH RESPECT TO A SCALAR ARGUMENT.

THEOREM A1.1 *Let $f=f(\underset{\sim}{A})$ be a scalar function of an nxn matrix argument $\underset{\sim}{A}$. Let $\underset{\sim}{A}$ in turn be a matrix function of a scalar argument t, i.e. let $\underset{\sim}{A}=\underset{\sim}{A}(t)$. The total derivative of f with respect to t, denoted by $\dot{f}(t)$, is given by*

$$\dot{f}(t)=Tr\{\frac{\partial f}{\partial \underset{\sim}{A}}(\dot{A}(t))^T\}$$

Proof:

Applying the "chain rule",

$$f'(t)\equiv\frac{df}{dt}=\frac{\partial f}{\partial a_{1i}}\frac{da_{1i}}{dt}+\frac{\partial f}{\partial a_{2i}}\frac{da_{2i}}{dt}+\ldots+\frac{\partial f}{\partial a_{1n}}\frac{da_{1n}}{dt} \tag{A2.2}$$

By means of the index notation then, the right hand side of eq. (A2.1) can be simplified as

$$f'(t)=\frac{\partial f}{\partial a_{ji}}\frac{da_{ji}}{dt} \tag{A2.3}$$

On the other hand, defining the matrices $\underset{\sim}{C}$ and $\underset{\sim}{D}$ as

$$c_{ji}=\frac{\partial f}{\partial a_{ji}},\quad d_{ij}=\frac{da_{ij}}{dt} \tag{A2.4}$$

then

$$c_{ij}d_{ji}\equiv Tr(\underset{\sim}{C}\underset{\sim}{D}^T) \tag{A2.5}$$

Substitution of (A2.4) and (A2.5) into eq. (A1.3) leads to

$$f'(t)=Tr\{(\frac{\partial f}{\partial \underset{\sim}{A}})\dot{\underset{\sim}{A}}(t)^T\},\ \text{q.e.d.} \tag{A2.6}$$

THEOREM A1.2 *Let $\underset{\sim}{A}$ be an nxn matrix. Its determinant is a scalar function of a matrix argument whose derivative with respect to $\underset{\sim}{A}$ is a matrix whose (i,j) element is given as*

$$\frac{\partial det\underset{\sim}{A}}{\partial a_{ij}}=(det\underset{\sim}{A})(\underset{\sim}{A}^{-1})_{ji}$$

Proof:

Expanding det $\underset{\sim}{A}$ in terms of the cofactors of its ith column,

$$\det\underset{\sim}{A}=a_{i1}c_{i1}+a_{i2}c_{i2}+\ldots+a_{in}c_{in} \qquad \text{(A2.7)}$$

where, according to the procedure to compute a determinant, the repeated index does not imply sum, as it usually does in the contex of the index notation. Without having to resort to the index notation, the right hand side of eq. (A2.7) can be written in the compact form (i refers to any particular row)

$$\det\underset{\sim}{A}=\sum_{j} a_{ij}c_{ij} \qquad \text{(A2.8)}$$

Differentiating the above expression with respect to a_{ij},

$$\frac{\partial\det\underset{\sim}{A}}{\partial a_{ij}} = c_{ij}$$

Letting $\underset{\sim}{C}$ be the matrix whose (i,j) entry is c_{ij} eq. (A2.9) can be written in the form

$$\frac{\partial\det\underset{\sim}{A}}{\partial\underset{\sim}{A}} = \underset{\sim}{C} \qquad \text{(A2.10)}$$

On the other hand, since the inverse of $\underset{\sim}{A}$ is given as

$$\underset{\sim}{A}^{-1}= \frac{1}{\det\underset{\sim}{A}}\ \underset{\sim}{C}^{T} \qquad \text{(A2.11)}$$

then

$$\underset{\sim}{C}=(\det\underset{\sim}{A})(\underset{\sim}{A}^{-1})^{T} \qquad \text{(A2.12)}$$

Substituting (A2.12) into eq. (A2.10), yields then

$$\frac{\partial\det\underset{\sim}{A}}{\partial\underset{\sim}{A}} = (\det A)(A^{-1})^{T},\ \text{q.e.d.} \qquad \text{(A2.13)}$$

THEOREM A2.3 *Let $\underset{\sim}{A}$ be an nxn matrix function of a scalar argument t. The derivative of $\det\underset{\sim}{A}$ with respect to t is given as*

$$\frac{d\,\det \underset{\sim}{A}}{dt} = (\det \underset{\sim}{A})\,Tr(\underset{\sim}{A}^{-1}\dot{\underset{\sim}{A}}\;(t))$$

the proof of Theorem A2.3 follows directly from Theorems A.2.1. and A.2.2, and is left as an exercise.

Exercise A2.1 Prove Theorem A2. 3

Appendix 3

COMPUTATION OF $\varepsilon_{ijk}\varepsilon_{\ell mn}$

Let $\underset{\sim}{i},\underset{\sim}{j},\underset{\sim}{k}$ (alternatively $\underset{\sim}{\ell},\underset{\sim}{m},\underset{\sim}{n}$) be values attained by an arthogonal basis $E=\{\underset{\sim}{e}_1,\underset{\sim}{e}_2,\underset{\sim}{e}_3\}$. Thus,

$$\varepsilon_{ijk}=\det\begin{pmatrix}\underset{\sim}{i}\\ \hline \underset{\sim}{j}\\ \hline \underset{\sim}{k}\end{pmatrix}, \quad \varepsilon_{\ell mn}=\det\begin{pmatrix}\underset{\sim}{\ell}\\ \hline \underset{\sim}{m}\\ \hline \underset{\sim}{n}\end{pmatrix} \tag{A3.1}$$

where the brackets contain 3x3 matrices whose rows are the components of each of the corresponding vectors. Hence

$$\varepsilon_{ijk}\varepsilon_{\ell mn}=\det\begin{pmatrix}\underset{\sim}{i}\\ \hline \underset{\sim}{j}\\ \hline \underset{\sim}{k}\end{pmatrix}\det\begin{pmatrix}\underset{\sim}{\ell}\\ \hline \underset{\sim}{m}\\ \hline \underset{\sim}{n}\end{pmatrix}$$

Since transposing a matrix does not alter the value of its determinant, the latter expression can be written as

$$\varepsilon_{ijk}\varepsilon_{\ell mn}=\det\left(\underset{\sim}{i}^T \,\vdots\, \underset{\sim}{j}^T \,\vdots\, \underset{\sim}{k}^T\right)\det\begin{pmatrix}\underset{\sim}{\ell}\\ \hline \underset{\sim}{m}\\ \hline \underset{\sim}{n}\end{pmatrix}$$

Now, introducing the fact that the product of the determinants of two matrices equals the determinant of the product (A3.1),

$$\varepsilon_{ijk}\varepsilon_{\ell mn}=\det\left\{\left(\underset{\sim}{i}^T \,\vdots\, \underset{\sim}{j}^T \,\vdots\, \underset{\sim}{k}^T\right)\begin{pmatrix}\underset{\sim}{\ell}\\ \hline \underset{\sim}{m}\\ \hline \underset{\sim}{n}\end{pmatrix}\right\}=$$

$$=\det\begin{pmatrix}\underset{\sim}{i}^T\underset{\sim}{\ell} & \underset{\sim}{j}^T\underset{\sim}{\ell} & \underset{\sim}{k}^T\underset{\sim}{\ell}\\ \underset{\sim}{i}^T\underset{\sim}{m} & \underset{\sim}{j}^T\underset{\sim}{m} & \underset{\sim}{k}^T\underset{\sim}{m}\\ \underset{\sim}{i}^T\underset{\sim}{n} & \underset{\sim}{j}^T\underset{\sim}{n} & \underset{\sim}{k}^T\underset{\sim}{n}\end{pmatrix}$$

Expanding the above determinant with respect to its first row,

$$\begin{aligned}\varepsilon_{ijk}\varepsilon_{\ell mn} = {} & \delta_{i\ell}(\delta_{jm}\delta_{kn}-\delta_{jn}\delta_{kn})+ \\ & +\delta_{j\ell}(\delta_{kn}\delta_{in}-\delta_{im}\delta_{kn})+ \\ & +\delta_{k\ell}(\delta_{im}\delta_{jn}-\delta_{jm}\delta_{in})\end{aligned}$$

which is the sought expression

REFERENCES

A3.1 Halmos P.R., Finite-Dimensional Vector Space, Springer-Verlag, N.York, 1974, p.99.

Appendix 4

SYNTHESIS OF PLANE LINKAGES FOR RIGID BODY GUIDANCE.

Referring to Fig A4.1, let X,Y be rectangular cartesian coordinate axes fixed to the rigid body that is intended to be guided through a sequence of n configurations under plane motion. In that figure, X_0,Y_0 correspond to the said axes in their reference configuration, whereas X_j,Y_j correspond to their j^{th} configuration. Angle θ_j measures the orientation of the body in the jth configuration with respect to the reference one.

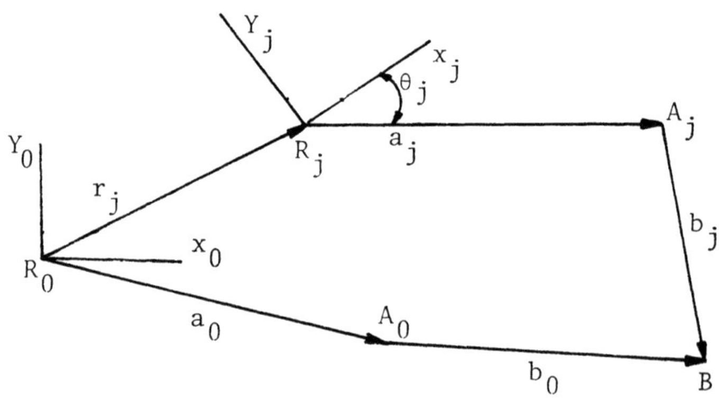

Fig A4.1 Rigid-body guidance through n configurations under plane motion

In Fig A4.1 let A and B be the circular and center points of Burmester theory (A4.1),i.e. all positions $A_0,A_1,\ldots,A_n$ of point A lie on a circle centered at B. Let $r_j(j=1,\ldots,n)$ denote the complex number corresponding to point R_j in the complex plane. Complex numbers a_0,b_0,a_j,b_j are defined corrrespondingly. Clearly, from Fig A4.1,

$$r_j+a_j+b_j=a_0+b_0 \qquad (A4.1)$$

Hence

$$b_j=a_0+b_0-r_j-a_j \qquad (A4.2)$$

But

$$a_j = e^{i\theta_j} a_0 \tag{A4.3}$$

Thus, eq. (A4.2) becomes

$$b_j = (1-e^{i\theta_j}) a_0 + b_0 - r_j \tag{A4.4}$$

Since all points A_j lie on a circle centered at B,

$$||b_j||^2 = ||b_0||^2 \tag{A4.5}$$

and substituting eq. (A4.4) into eq. (A4.5),

$$||(1-e^{i\theta_j}) a_0 + b_0 - r_j||^2 = ||b_0||^2, j=1,\ldots,n \tag{A4.6}$$

The corresponding synthesis equations for spatial mechanisms derived in section 5.4 is next reproduced for quick reference

$$||(\underset{\sim}{I} - \underset{\sim}{Q}_j) \underset{\sim}{a}_0 + \underset{\sim}{b}_0 - \underset{\sim}{r}_j||^2 = ||\underset{\sim}{b}_0||^2, \ j=1,\ldots,n \tag{A4.7}$$

Utilizing the result of Exercise 2.3.8, $\underset{\sim}{Q}_j$ can be written as

$$\underset{\sim}{Q}_j = e^{\underset{\sim}{A}_j \theta_j} \tag{A4.8}$$

Introducing eq. (A4.8) into eq. (A4.7), one finally obtains

$$||(\underset{\sim}{I} - e^{\underset{\sim}{A}_j \theta_j}) \underset{\sim}{a}_0 + \underset{\sim}{b}_0 - \underset{\sim}{r}_j||^2 = ||\underset{\sim}{b}_0||^2, \ j=1,\ldots,n \tag{A4.9}$$

The striking formal similitude between the plane and the space synthesis equations for rigid-body guidance is thus made apparent by eqs. (A4.6) and (A4.9)

REFERENCES

A4.1 Burmester R., Lehrbuch der Kinematik vol. 1, "Die ebene Bewegung", Verlag von Arthur Felix, Berlin, 1886.

Subject Index

MIX
Papier aus verantwortungsvollen Quellen
Paper from responsible sources
FSC® C105338

If you have any concerns about our products,
you can contact us on
ProductSafety@springernature.com

In case Publisher is established outside the EU,
the EU authorized representative is:
Springer Nature Customer Service Center GmbH
Europaplatz 3, 69115 Heidelberg, Germany

Printed by Libri Plureos GmbH
in Hamburg, Germany